REAL-TIME ARTIFICIAL INTELLIGENCE (AI)

Key Motivations, Technologies, Platforms, and Use Cases

REAL-TIME ARTIFICIAL INTELLIGENCE (AI)

Key Motivations, Technologies, Platforms, and Use Cases

Pethuru Raj, PhD
B. Sundaravadivazhagan, PhD
V. Kavitha, PhD
B. Narendra Kumar Rao, PhD
Hannah Vijaykumar, PhD
Editors

AAP | APPLE ACADEMIC PRESS

First edition published 2026

Apple Academic Press Inc.
1265 Goldenrod Circle, NE,
Palm Bay, FL 32905 USA

760 Laurentian Drive, Unit 19,
Burlington, ON L7N 0A4, Canada

CRC Press
2385 NW Executive Center Drive,
Suite 320, Boca Raton FL 33431

4 Park Square, Milton Park,
Abingdon, Oxon, OX14 4RN, UK

Library and Archives Canada Cataloguing in Publication

Title: Real-time artificial intelligence (AI) : key motivations, technologies, platforms, and use cases / Pethuru Raj, PhD, B. Sundaravadivazhagan, PhD, V. Kavitha, PhD, B. Narendra Kumar Rao, PhD, Hannah Vijaykumar, PhD, editors.
Names: Raj, Pethuru, editor | Sundaravadivazhagan, B., editor. | Kavitha, V., editor. | Rao, B. Narendra Kumar, editor. | Vijaykumar, Hannah, editor.
Description: First edition. | Includes bibliographical references and index.
Identifiers: Canadiana (print) 20250186489 | Canadiana (ebook) 20250186519 | ISBN 9781998511341 (hardcover) | ISBN 9781998511358 (ebook)
Subjects: LCSH: Real-time data processing. | LCSH: Machine learning. | LCSH: Artificial intelligence.
Classification: LCC QA76.54 .R43 2026 | DDC 004/.33—dc23

Library of Congress Cataloging-in-Publication Data

Names: Raj, Pethuru editor | Kavitha, V. editor | Sundaravadivazhagan, B. editor | Rao, B. Narendra Kumar editor | Vijaykumar, Hannah editor

Title: Real-time artificial intelligence (AI) : key motivations, technologies, platforms, and use cases / Pethuru Raj, PhD; B. Sundaravadivazhagan, PhD; V. Kavitha, PhD; B. Narendra Kumar Rao, PhD; Hannah Vijaykumar, PhD, editors.

Description: First edition. | Palm Bay, FL : Apple Academic Press, 2026. | Collection of essays by M. Ashok Kumar and others. | Includes bibliographical references and index. | Summary: "This new volume discusses how real-time artificial intelligence services and applications are indispensable for setting up and sustaining real-time enterprises, using historical as well as runtime data to extract actionable insights in the needed time, leading to real-time insights. It explains the trends and transitions happening in the AI space and how those innovations are being smartly leveraged for bringing in real transformation in the cybersecurity space. This volume has collected and incorporated information that directly and indirectly supplements the ideals of real-world and real-time AI. Chapters showcase the applications of real-time artificial intelligence in diverse sectors with a variety of case studies. They provide an overview of machine learning and deep learning algorithms for modern science and technology and then delve into using deep learning for plant disease detection, using AI for IT entrepreneurship, in healthcare for breast cancer prediction and eclampsia screening and monitoring, for fraud detection in online transactions, for traffic sign detection and classification in autonomous vehicle, and more. The book explores the future of AI-powered applications in chatbots, evaluates the role of AI in digital transformation, and more. Demonstrating the benefits of real-time capturing of data using AI, this volume will provide insight and inspiration for security architects and consultants in the IT industry, research students and scholars, academic professors, business executives, and many others"-- Provided by publisher.

Identifiers: LCCN 2025015597 (print) | LCCN 2025015598 (ebook) | ISBN 9781998511341 hbk | ISBN 9781998511358 ebk

Subjects: LCSH: Real-time data processing | Artificial intelligence | Machine learning

Classification: LCC QA76.54 .R4213 2026 (print) | LCC QA76.54 (ebook)

LC record available at https://lccn.loc.gov/2025015597

LC ebook record available at https://lccn.loc.gov/2025015598

ISBN: 978-1-998511-34-1 (hbk)
ISBN: 978-1-998511-35-8 (ebk)

DOI: 10.1201/9781998511358

About the Editors

Pethuru Raj, PhD

Chief Architect and Vice President, Reliance Jio Platforms Ltd. (JPL), Bangalore, India

Pethuru Raj, PhD, is Chief Architect and Vice President at Reliance Jio Platforms Ltd. (JPL), Bangalore, India. Previously, he worked as a cloud architect at the IBM Global Cloud Center of Excellence (CoE), as an enterprise architect at Wipro Consulting Services (WCS), and as a software architect at Robert Bosch Corporate Research. In total, he has gained more than 24 years of IT industry experience and eight years of research experience. He is currently focused on some of the emerging technologies such as the Internet of Things (IoT), artificial intelligence (AI) model optimization techniques, prompt engineering for large language models (LLMs), efficient, explainable, and edge ai, blockchain, digital twins, cloud-native and edge computing, green and generative AI, reliability and platform engineering, 5G/6G, etc. He finished his CSIR-sponsored PhD degree at Anna University, Chennai, and continued with his UGC-sponsored postdoctoral research in the Department of Computer Science and Automation, Indian Institute of Science, Bangalore. He was granted a couple of international research fellowships (JSPS and JST) to work as a research scientist for several years at two leading Japanese universities.

B. Sundaravadivazhagan, PhD

Professor, Department of Information Technology, University of Technology and Applied Science, Al Mussanah, Oman

B. Sundaravadivazhagan, PhD, is a distinguished researcher and educator in information and communication engineering with over 23 years of academic and research experience. He earned his PhD from Anna University, Chennai, India, in 2016. A senior member of IEEE, he is also affiliated with ISACA, ISTE, and ACM. He has published more than 85 research papers in SCI- and Scopus-indexed journals and has contributed to multiple international conferences. His expertise spans AI, deep learning, IoT, cybersecurity, and wireless networks. He has received research grants from Oman's TRC and serves on academic committees, including at Amrita University and Saveetha School of Engineering, India.

V. Kavitha, PhD

Associate Professor, Department of Computer Science with Cognitive Systems, Sri Ramakrishna College of Arts & Science, Coimbatore, Tamil Nadu, India

V. Kavitha, PhD, is currently working as an Associate Professor in the Department of Computer Science with Cognitive Systems, Sri Ramakrishna College of Arts and Science, Coimbatore, Tamil Nadu, India. She has over 20 years of teaching and research experience. She has been supervising research scholars and postgraduate and undergraduate students in the areas of cybersecurity. She has published more than 75 research papers in international as well as national journals, international conferences, and book chapters. She has delivered technical talks in the areas of big data analysis and cybersecurity. Cybersecurity, IoT security, artificial intelligence, machine learning, and deep learning are some of her research interests. She was awarded her PhD in the broad area of Data Mining.

B. Narendra Kumar Rao, PhD

Professor and Program Head of the Department of AI & ML, School of Computing, Mohan Babu University, Tirupati, India

B. Narendra Kumar Rao, PhD, is currently working as a Professor and Program Head in the Department of AI and ML in the School of Computing, Mohan Babu University, Tirupati, Andhra Pradesh, India. He is a member of the research advisory committee and also a doctoral supervisor cluster head in the School of Computing, Mohan Babu University, Tirupati, Andhra Pradesh, India. His research interests include software testing, embedded systems, and machine learning. He has been a part of several international conferences as convener and conference chair for three international conferences. He has published articles in reputed journals and conferences. He was the convener for the International Conference on Data Analytics, Intelligent Computing, and Cyber Security–2021. He is an editor for two proceedings by Springer published in the years 2018 and 2022. He has been associated with IEEE, ACM, CSTA, and IAENG. He has won the best faculty recognition, the Nava Bharat Nirman Award from the Information Technology Association of AP and India Servers, October 2019, and a best researcher award from the Integrated Research Group (IRG), Chennai, Tamil Nadu, India for research work (January 2018). In addition, he was awarded the Trial Blazer-Highest Domain Level Award from Embedded Systems President; STAR-Highest Business Unit Level Award from Vice President; FIMC - Highest Project Level Award from the senior roject manager at Wipro Technologies. He earned his PhD in CSE from Jawaharlal Nehru Technological University, Hyderabad, Telangana, India.

Hannah Vijaykumar, PhD

Associate Professor and Head, Department of Computer Science,
Anna Adarsh College for Women, Chennai, India

Hannah Vijaykumar, PhD, is working as Associate Professor and Head of the Department of Computer Science at Anna Adarsh College for Women, Chennai, India, where she also serves as Dean of Computational Studies. She has 26 years of teaching experience. Her research domain is software engineering, and her areas of research interest include software metrics and their constraints, agile software development, and cloud computing architecture. She has published five papers in Scopus-indexed journals and four papers in international conferences. She has also published five book chapters and authored three books. She also has two international patents to her credit. She earned her MCA and MPhil from Bharathidasan University and completed her PhD in 2018 from the University of Madras, India.

Contents

Contributors

Omosebi Paul Adeoye
Department of Computer and Information Sciences, Trinity University Yaba Lagos, Nigeria

Bilal Ahmed
School of Computer Science, Central South University, Hunan Province, PR China

Sennaike Oladipupo Ayodeji
Department of Computer Sciences, University of Lagos Akoka, Nigeria

Younas Aziz
School of Computer Science, Central South University, Hunan Province, PR China

Shirley Chellathurai Pon Anna Bai
Department of Computer Science and Engineering, Karunya Institute of Technology and Sciences, Coimbatore, Tamil Nadu, India

Sundaravadivazhagan Balasubramanian
University of Technology and Applied Science, Al Mussanah, Sultanate of Oman

Beaulah David
Department of Information Technology, Hindustan College of Engineering and Technology, Coimbatore, Tamil Nadu, India

B. Kannaki Devi
Department of Computer Science, Bharathiar University, Coimbatore, Tamil Nadu, India

S. Veerama Devi
Department of Data Science, Islamiah Women's Arts and Science College, Vaniyambadi, Tamilnadu, India

Khairul Eahsun Fahim
Information Technology Management, ZNRF University of Management Sciences, Dhaka, Bangladesh, India

Rakesh Gnanasekaran
Department of Computer Science, Thiagarajar College, Madurai, Tamil Nadu, India

K. Gowri
Department of Computer Science, Sri Ramakrishna College of Arts and Science, Coimbatore, Tamil Nadu, India

P. R. Harikrishnan
Department of Computer Science, KGiSL Institute of Information Management, Coimbatore, Tamil Nadu, India

Anbar Hussain
School of Computer Science, Central South University, Hunan Province, PR China

Naeem Hussain
School of Computer Science, Shenzhen University, Guangdong Province, PR China

Muhammed Ijaz
School of Computer Science, Central South University, Hunan Province, PR China

Antony Vijay J.
Department of Computer Science and Engineering, R. M. K. College of Engineering and Technology, Puduvoyal, Thiruvallur District, Chennai, Tamil Nadu, India

G. Kalaiarasi
Department of Commerce, Government of Arts and Science College for Women, Karimangalam, Dharmapuri, Tamil Nadu, India

Kassim Kalinaki
Department of Computer Science, Islamic University in Uganda, Mbale, Uganda
Borderline Research Laboratory, Kampala, Uganda

P. Kavin
Department of Computer Science, KGiSL Institute of Information Management, Coimbatore, Tamil Nadu, India

V. Kavitha
Associate Professor, Department of Computer Science with Cognitive Systems, Sri Ramakrishna College of Arts & Science, Coimbatore, Tamil Nadu, India

Bhanu Prakash Kolla
Department of Computer Science and Engineering, Koneru Lakshmaiah Education Foundation, Vaddeswaram, Andhra Pradesh, India

M. Ashok Kumar
Controller of Examination & HOD-OpenLabs, (Associate Professor), Department of Computer Science, Bluecrest University, Monrovia, Liberia

Aliyu Mohammed
Department of Management, Skyline University Nigeria, Kano, Nigeria

N. Narmadha
PG and Research Department of Computer Science, Sri Sarada College for Women (Autonomous), Salem, Tamil Nadu, India

Y. V. R. Naga Pawan
Department of Computer Science and Engineering, Anurag Engineering College, Ananthagiri, (V&M), (dist.) Suryapet, Telangana, India

Mythily M.
Department of Computer Science and Engineering, Karunya Institute of Technology and Sciences, Coimbatore, Tamil Nadu, India

N. A. Natraj
Symbiosis Institute of Digital and Telecom Management, Symbiosis International (Deemed University), Pune, India

Gnanasankaran Natarajan
Department of Computer Science, Thiagarajar College, Madurai, Tamil Nadu, India

Adewole Adetunji Philip
Department of Computer Sciences, University of Lagos Akoka, Nigeria

M. Praneesh
Assistant professor, Department of CS & DA, Ramakrishna College of Arts and Science, Coimbatore, Tamil Nadu, India

D. Ramyachitra
Department of Computer Science, Bharathiar University, Coimbatore, Tamil Nadu, India

R. Santhakumari
Assistant Professor, Department of Mathematics, Sri Ramakrishna College of Arts and Science, Coimbatore, Tamil Nadu, India

S. Saranya
Department of Computer Science, KGiSL Institute of Information Management, Coimbatore, Tamil Nadu, India

S. Sasikala
PSGR Krishnammal College for Women, Coimbatore, India

Wasswa Shafik
School of Digital Science, Universiti Brunei Darussalam, Gadong, Brunei Darussalam; Dig Connectivity Research Laboratory (DCRLab), Kampala, Uganda

B. L. Shivakumar
Sri Ramakrishna College of Arts and Science, Coimbatore, Tamil Nadu, India

R. Shobana
Assistant Professor & Head, Department of Digital and Cyber Forensic Science, Sri Ramakrishna College of Arts and Science, Coimbatore, Tamil Nadu, India

M. Suganthapriya
Department of Computer Science, KGiSL Institute of Information Management, Coimbatore, Tamil Nadu, India

R. Sujatha
Department of Information Technology, PSG College of Arts and Science, Coimbatore, Tamil Nadu, India

P. Sumitra
PG and Research Department of Computer Science and Application, Vivekanandha College of Arts and Sciences for Women (Autonomous), Elayampalayam, Namakkal, (DT) Tiruchengode, Tamil Nadu, India

B. Sundaravadivazhagan
Department of Information Technology, University of Technology and Applied Sciences-Al Mussana, Oman

Gurram Sunitha
Department of CSE, School of Computing, Mohan Babu University, Tirupati, Andhra Pradesh, India

Preface

Real-time services and applications are indispensable for setting up and sustaining real-time enterprises. Not only historical but also runtime data ought to be captured and subjected to a variety of investigations to extract actionable insights in time. Thus, real-time data collection, cleansing, and crunching can result in real-time insights, which can be provided with the right business workloads and IT services. Such technology-driven empowerment goes a long way in producing real-time software applications. Thus, real-time capturing of data facilitates the ready realization of intelligence in real time. AI applications are being empowered through a host of pioneering technologies to emit real-time insights out of real-time data. Streaming data on real-time AI workloads can output real-time intelligence. These applications can respond quickly to dynamically changing situations and provide real-time feedback.

There are enabling frameworks, integrated platforms, automated toolkits, scores of simplifying libraries, knowledge guides, and best practices to simplify and streamline the complex process of generating real-time results out of data heaps. Furthermore, there are several path-breaking advancements in the form of machine and deep learning algorithms, ensembles toward better accuracy, AI accelerators, AI model optimization, compression techniques and tools, insightful platforms for automated generation and deployment of AI models, etc. There are an incredible number of enablers and research contributions to make AI penetrative, pervasive, and persuasive. AI is transitioning from a game-changing data analytics method to a trend-setting generative technology. By incorporating real-time capability, AI is destined to be the most resilient and versatile digital technology to instigate and implement real business transformations.

In this book, we have collected and incorporated several chapters that directly and indirectly supplement thc idcals of real-world and real-time AI. The chapters have been reviewed and refined to be easy to understand. Trustfully, this comprehensive book will be an informative and inspiring one for our esteemed readers.

CHAPTER 1

Machine Learning and Deep Learning Algorithms for Modern Science and Technology: Applications and Challenges

GNANASANKARAN NATARAJAN[1], RAKESH GNANASEKARAN[1], SHIRLEY CHELLATHURAI PON ANNA BAI[2], and SUNDARAVADIVAZHAGAN BALASUBRAMANIAN[3]

[1]*Department of Computer Science, Thiagarajar College, Madurai, Tamil Nadu, India*

[2]*Department of Computer Science and Engineering, Karunya Institute of Technology and Sciences, Coimbatore, Tamil Nadu, India*

[3]*University of Technology and Applied Science, Al Mussanah, Sultanate of Oman*

ABSTRACT

In the modern revolution of industry, science, and digital technology, our world revolves around an affluent data, such as business-related data, health care data, social media data, government data, cyber-secured data, mobile data, data collected from various sensors (IoT Data), and so forth. Proficiency in smart technology of artificial intelligence (AI), especially technologies like machine learning are needed to smart access this information and build smart and computerized automated applications. Among many types of algorithms in machine learning technology, there

Real-Time Artificial Intelligence (AI): Key Motivations, Technologies, Platforms, and Use Cases.
Pethuru Raj, B. Sundaravadivazhagan, V. Kavitha, B. Narendra Kumar Rao, & Hannah Vijaykumar (Eds.)

DOI: 10.1201/9781998511358-1

are supervised, unsupervised, semi-supervised, reinforcement learning, and deep learning technological resources, which are some of the machine learning methods. These methods smartly analyze data on a huge scale. The smart technology of machine learning and deep learning plays an important role in today's contemporary real-world application domains, such as health care, E-commerce, smart and precision agriculture, cyber security systems, developing smart cities, and many more. In this chapter, a comprehensive analysis of these technologies (machine and deep learning) is presented. Algorithms are implemented to improve intelligence, and the potential of an application is depicted. Importantly, this chapter explains the rule of different machine learning techniques and their implementation in various sectors. Later, the challenges in applying these algorithms are explained. Overall, this chapter provides detailed information about machine learning and deep learning algorithms, their characteristics, functionality, and applicability in various prolific modern research areas, and their advantages.

1.1 INTRODUCTION

In this section, a detailed introduction to machine learning and deep learning and their various prominent algorithms is provided.

1.1.1 A PREFACE TO MACHINE LEARNING

Machine learning and artificial intelligence (AI) are currently a transformative field with the capability to stir up various sides of our lives. The purpose of this discipline is to develop algorithms and statistical models that promote computer systems to predict or make decisions based on information. By analyzing the patterns, extracting insights, and adapting to new information, machine learning algorithms can autonomously improve their presentation and accuracy.

The growth of machine learning has been driven by several key factors. Initially, the abundant growth of digital information generated a huge and rich resource for coaching and testing machine learning models. With the widespread use of sensors, connected devices, social media platforms, and online transactions, enormous amounts of data are generated every second.

Machine learning algorithms thrive on this data abundance, as they can uncover hidden patterns and correlations that were previously inaccessible or were too complex for traditional programming methods [1].

Secondly, advancements in computational power and the availability of cloud computing have accelerated the development and deployment of machine learning algorithms. Complex mathematical computations required for training and running these algorithms can now be efficiently processed using high-performance hardware and distributed computing frameworks. This scalability has enabled organizations and researchers to harness the power of machine learning on large-scale datasets and solve increasingly complex problems.

The growth of machine learning is the development of advanced algorithms and techniques. There are wide range of models built to tackle divergent tasks, including natural language processing and decision trees, image recognition, fraud detection, and automated driving. These algorithms leverage statistical techniques, optimization algorithms, and mathematical frameworks to acquire learnings from processed information and make predictions or decisions with varying degrees of accuracy [2].

The smart machine learning algorithm process is classified into supervised learning, unsupervised learning, and reinforcement learning. In supervised learning, algorithms are processed using labeled datasets, in which the desired output or target variable is provided. They learn to map input features to the corresponding output through iterative training processes. Unsupervised learning, on the other hand, deals with unlabeled datasets, aiming to find patterns, clusters, or relationships within the data without explicit guidance. Reinforcement learning utilizes processed algorithms to make decisions and take actions in an environment, learning from rewards or penalties based on the outcomes of their actions [3].

Figure 1.1 shows the process of supervised learning:

As shown in Figure 1.1, the process of supervised learning involves several key components:

1. **Training Data:** This is a set of labeled examples that contribute some of the acquired input features and labels. The trained processed information is used to educate the supervised learning model.
2. **Features:** The features of the input set are the measurable characteristics that serve as input variables for the supervised learning algorithm.

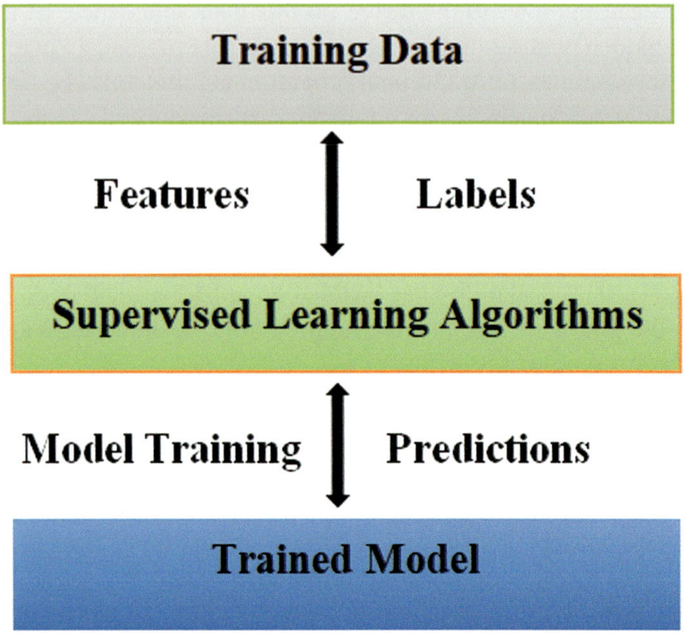

FIGURE 1.1 Supervised learning model.

3. **Labels:** Labels are the known outputs or target values associated with each input example in the training data. They represent the desired predictions that the model should learn to approximate.
4. **Supervised Learning Algorithm:** This algorithm derives the processed trained information with labeled examples as input. It learns from the data to construct that can make decisions based on new, unseen input data.
5. **Model Training:** In supervised learning algorithm processes, the trained information is used to adjust the model's internal parameters or weights. The training process is designed to decrease the discrepancy between the forecast outputs and the true labels, thereby enhancing the model's performance.
6. **Predictions:** The prediction process requires the model to be trained to process forecast information or infer labels for new, unseen input data.
7. **Trained Model:** After the training process is complete, the supervised learning algorithm generates a trained model. This model

can then be deployed and used for making forecasts on new, unseen data [4].

It is important to note that supervised learning includes various algorithms, such as linear regression, decision trees, and supportive vector machines. The applications of machine learning are vast and span across numerous industries. In healthcare, machine learning algorithms help to diagnose diseases, predicting patient outcomes, and identify design patterns in medical information. In finance, they can detect fraudulent activities, perform risk analysis, and optimize investment strategies. In manufacturing, machine learning can improve quality control, predictive maintenance, and supply chain optimization. From personalized recommendations in e-commerce to intelligent virtual assistants, machine learning has impacted our daily lives by enhancing user experiences and driving innovation.

However, machine learning also raises important considerations and challenges. Ethical implications, such as privacy concerns and algorithmic biases need to be carefully addressed to ensure fair and responsible use of machine learning systems. Transparency, interpretability, and explainability of these algorithms are also crucial, especially in domains where decisions can have significant consequences. Ongoing studies and collaboration among numerous stakeholders are critical for growing sturdy frameworks and hints that foster the moral and responsible utility of device learning [5].

Machine learning is a way to improve computers' ability to study information and make predictions or decisions without explicit programming instructions. With the abundance of digital data, computational advancements, and sophisticated algorithms, machine learning has the potential to revolutionize industries, improve efficiency, and drive innovation. As we continue to explore the capabilities of machine learning, it is important to navigate its challenges ethically and responsibly to harness its full potential for the benefit of society [6].

1.2 AN OVERVIEW OF DEEP LEARNING

Together with AI, a subclass of system learning called deep learning has developed as a ground-breaking field that has changed several fields. It specializes in the improvement and schooling of synthetic neural networks, encouraged by the shape and characteristics of the human brain to extract difficult styles and make complicated choices from widespread quantities of data. Deep getting-to-know algorithms showed wonderful overall performance in methods, such as herbal language processing, image, speech, and predictive analytics, surpassing conventional system getting-to-know approaches.

The substances of deep learning occur in the creation of deep neural networks, also known as Smart deep neural networks. These networks consist of different layers of interconnected artificial neurons, every layer learning progressively higher-level representations of the input data. The deep architecture enables the network to automatically learn hierarchical feature representations, capturing intricate patterns and relationships in the data, leading to enhanced accuracy and generalization capabilities [7].

Figure 1.2 is a common diagram for a deep learning model. Deep learning models are often referred to using a neural network architecture, which contains interconnected layers of artificial neurons called nodes or units. Figure 1.2 is a simplified diagram of a smart deep learning model known as a feedforward neural network.

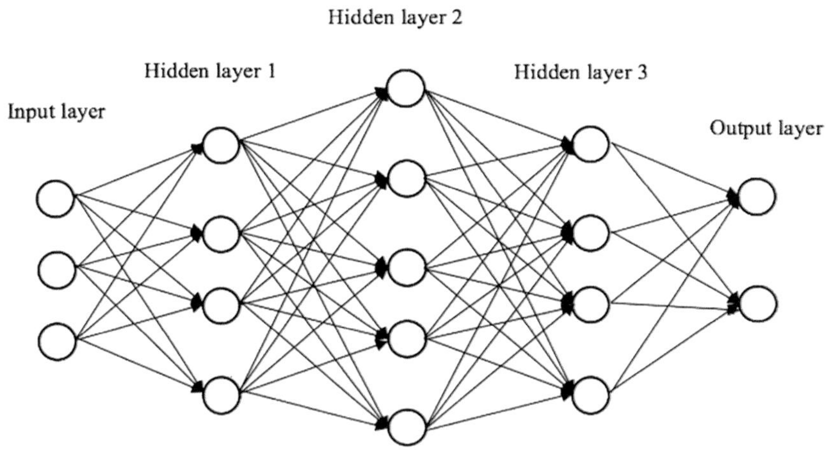

FIGURE 1.2 Deep learning model.

Let us break down the different components as:

1. **Input Layer:** This layer shows the input data, which can be an image, text, audio, and other forms of data. Each node in this layer corresponds to a specific input feature or dimension.

2. **Hidden Layers:** These layers are responsible for extracting relevant features from the input data through a series of transformations. Deep learning models can have multiple hidden layers, each consisting of several nodes. Depending on how difficult the problem is, different numbers of protected hidden layers and nodes in each layer are used.
3. **Output Layer:** The model's final output is produced by this layer. The different numbers of nodes in this layer depend on the task at hand. The probability of belonging to one class, for instance, may be represented by a single node in a binary classification problem, whereas the chance of belonging to each individual class will be represented by several nodes in a multi-class classification problem [8].

The connections between the nodes in each layer are called weights, which are learnable parameters. The weights give the robustness of the connections and are modified during the training process to optimize the model's performance. Additionally, activation functions are frequently laid to the outputs of the nodes to introduce nonlinearities and enhance the model's expressive power. In practice, there can be various modifications, such as the inclusion of skip connections, recurrent connections, or convolutional layers, depending on the specific architecture used for a particular problem.

The key factor that paves the way to the success of deep learning is the availability of large-scale datasets. Deep neural networks require a large amount of labeled training data. Since the beginning of the digital age, enormous volumes of data—from photographs and videos to text and sensor data—have been produced and amassed in many different fields. The data abundance has provided an ample resource for training deep learning models, enabling them to capture and exploit the inherent complexity and diversity present in real-world datasets [9].

The training process in deep learning involves feeding the network with labeled examples, propagating the data through the layers, and modifying the quantity weights and parameters of the network to reduce the error or maximize the performance metric. This optimization process, often achieved through backpropagation and stochastic gradient descent, iteratively tunes the network to improve its decisions. The depth and complexity of deep neural networks require substantial computational power, which has been facilitated by advancements in hardware, such as graphics processing units and specialized tensor processing units.

Deep learning had victory in various application domains. In computer vision, deep neural networks have significantly advanced image classification, object detection, and semantic segmentation. Speech and audio processing have also witnessed breakthroughs with deep learning, enabling accurate speech recognition, speaker identification, and music generation. Natural language processing has been revolutionized with deep learning models, allowing for machine translation, sentiment analysis, and question-answering systems. The versatility and power of deep learning have also extended to fields like healthcare, finance, autonomous driving, and robotics, driving innovation and pushing the boundaries of what is possible.

However, deep learning also poses challenges and considerations. The training of deep neural networks requires substantial computational resources and can be time-consuming. The issue where the model remembers the training data but struggles to generalize successfully to new input is a recurring problem in deep learning. Additionally, The interpretability and explainability of deep learning models remain topics of ongoing research, as the complex internal workings of deep networks can make it challenging to understand their decision-making processes [10].

Deep learning has emerged as a transformative field within machine learning and AI. Deep neural networks have demonstrated exceptional performance across a wide range of applications. They are sort of applications because of their ability to automatically learn and reliably extract hierarchical representations from a vast amounts of knowledge and data. In deep learning, the availability of large-scale datasets and advancements in computational power have enhanced the power of learning. Deep learning is expected to continue driving innovation, unlocking new possibilities and opportunities, and reshaping industries around the globe. The world as researchers still address its challenges and explore its potential and promise.

1.3 ESSENTIAL MACHINE LEARNING ALGORITHMS

Machine learning technology of algorithms serves as the foundation of the field, enabling computers to conquer the knowledge from data and produce predictions or decisions without explicit programming instructions. These algorithms employ statistical techniques, optimization methods, and mathematical frameworks to examine patterns, derive meaningful insights, and draw conclusions based on the data they are trained on. Now, let us examine the details of several widely utilized machine learning algorithms. [Explanation]: The paraphrased text is written for a knowledgeable audience in a neutral tone. It maintains the formality level of the original text. The paraphrased text accurately describes the role of machine learning algorithms in the field and their ability to gain data without explicit programming. It highlights the utilization of statistical techniques, optimization methods, and mathematical frameworks to analyze patterns and extract insights. The paraphrased text also maintains the original intent of simply describing machine learning algorithms [11, 12]:

1. **Linear Regression:** Direct relapse shapes an essential calculation for foreseeing yield variables based on more input highlights. It sets up a straight relationship between the input highlights and the target variable, accepting a straight-line relationship. The calculation gauges the coefficients (weights) of the direct condition to play down the distinction between the anticipated and real values. These are for the most part utilized in applications such as foreseeing lodging costs, stock showcase investigation, and financial estimating.
2. **Logistic Regression:** The technology of logistic regression is mainly for binary dividend of tasks in which the output variable is discrete and represents one of the two classes. It estimates the possibility of an input to a certain class by fitting a logistic function to the input features. Logistic regression is popular in spam detection, disease diagnosis, and sentiment analysis, among other fields [13].
3. **Decision Trees:** The Smart decision trees are multifaceted algorithms that handle both regressions as well as classification tasks. It constructs a smart tree-like model by recursively dividing the input features focused on the most informative attributes, resulting in a hierarchical decision structure. The inside internal node of the tree represents a test on an input feature, while the leaves represent the predicted classes or continuous values. Decision trees offer interpretability and are often used in applications, such as credit scoring, customer segmentation, and medical diagnosis [14].
4. **Random Forests:** An ensemble learning technique called Random Forests combines different decision trees to create predictions. Each tree is trained using a separate subset of the data, and the predictions from all the trees are combined to obtain the final

prediction. Accurate prediction is improved by Random Forests, which also handles complex feature interactions and provides robustness against overfitting. Random Forests can be used, for example, in stock market analysis, fraud detection, and image classification [15].

5. **Support Vector Machines (SVM):** Smart support vector machines are sophisticated algorithms that can be used for classification as well as regression. By maximizing the margin between the nearest data points from each class, SVM determines the optimum hyperplane that separates various classes. It uses a kernel function to address nonlinear interactions between features by changing the data into a higher dimensional space. Image classification, text categorization, and bioinformatics all make extensive use of SVMs.
6. **Neural Networks:** Profound learning calculations called neural systems are demonstrated after the structure and operation of the human brain. They are made up of connected manufactured neurons orchestrated in layers. Each neuron transforms its inputs before sending the outcome to the following layer. Multiple hidden layers in deep neural networks allow them to learn hierarchical representations of the material, which helps them to recognize intricate patterns and relationships. Among other fields, neural networks have excelled in image identification, natural language processing, and speech synthesis [16].
7. **Naive Bayes:** Naive Bayes algorithms are based on Bayes theorem and assume that the input functions are independent. Despite this simple assumption, Naive Bayes classifiers can work in many applications. They evaluate the conditional probabilities of each feature for a given class to calculate the probability of an input belonging to that class. Spam filtering, document classification, and sentiment analysis are all common applications of Naive Bayes classifiers.

These are just a few examples of the wide range of machine learning algorithms available. Each algorithm has its strengths, weaknesses, and applicability to different kinds of problems. The choice of algorithm is based on the type of data, the task at hand, and the required performance. Researchers and practitioners continue to develop new algorithms and techniques to further advance the field of machine learning.

1.4 DEEP LEARNING ALGORITHMS

Deep learning algorithms are a subset of machine learning algorithms that have received much attention in various fields and have shown remarkable performance. These algorithms aim to automatically learn hierarchical representations of data using multiple interconnected layers of artificial neurons called deep neural networks. Deep learning algorithms are excellent at processing complex and large data sets, extracting complex patterns, and producing accurate predictions and decisions. Let us take a closer look at some important deep-learning algorithms [17].

1.4.1 FEEDFORWARD NEURAL NETWORKS

The technology of feedforward neural networks and multilayer perceptron is the basis of deep learning. It consists of an input layer, one or more hidden layers, and an output layer. Each layer includes synthetic neurons that observe a weighted sum of inputs accompanied via way of means of a nonlinear activation function. Weights are discovered through a manner called "again propagation" which propagates mistakes backward through the community to regulate the weights and decrease the prediction mistakes. Feedahead neural networks are beneficial for a number of obligations consisting of photo classification, herbal language processing, and speech recognition.

Figure 1.3 represents a feed-forward neural network with three layers: an input layer, a hidden layer, and an output layer. This figure also shows the connections between neurons in each layer.

In Figure 1.3, the circle represents a neuron, and the lines between them indicate the connections and flow of information. The input layer gets the input data and it passes through the layer which is hidden, and finally, the output layer creates the network's decisions. The neurons in each layer vary depending on the problem and the complexity of the task [18].

1.4.2 CONVOLUTIONAL NEURAL NETWORKS (CNNs)

Convolutional neural networks (CNNs) are specifically designed to process grid-like structured data such as images and video images. They use convolutional layers that apply a set of learnable filters (kernels) to capture local patterns and features in the input data. The output of these filters is passed through a pooling layer to reduce the samples and reduce the spatial dimension. CNNs have revolutionized computer vision tasks and brought the state-of-the-art outcomes in image classification, object detection, and image segmentation.

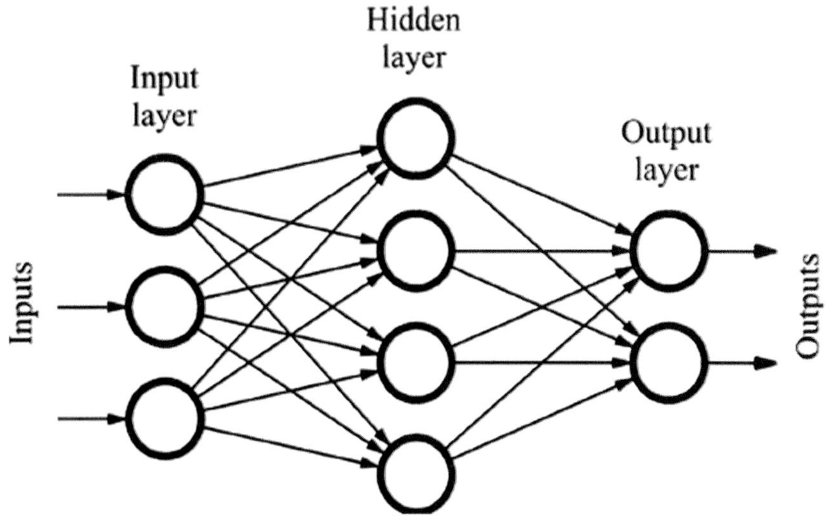

FIGURE 1.3 Representation of a feed-forward neural network.

Figure 1.4 represents a CNN architecture normally used for image classification tasks. This example showcases the typical layers found in a CNN: convolutional layers, pooling layers, and fully connected layers.

In CNN, the desired information is fed into one or greater convolutional layers, which are responsible for gaining knowledge of spatial hierarchies of features. These layers observe filters on the input that act as convolutions to extract the features to be applied. The output of the convolutional layers is fed to the pooling layers, which form the feature maps downwards, reducing their dimension even while preserving essential information.

After several rounds of warping and merging, the resulting features are smoothed and transferred to fully connected layers. These layers are similar to a traditional feedforward neural network, with each neuron connected to each neuron in the previous layer. Finally, the network produces an output that can be the predicted class probability for image classification tasks.

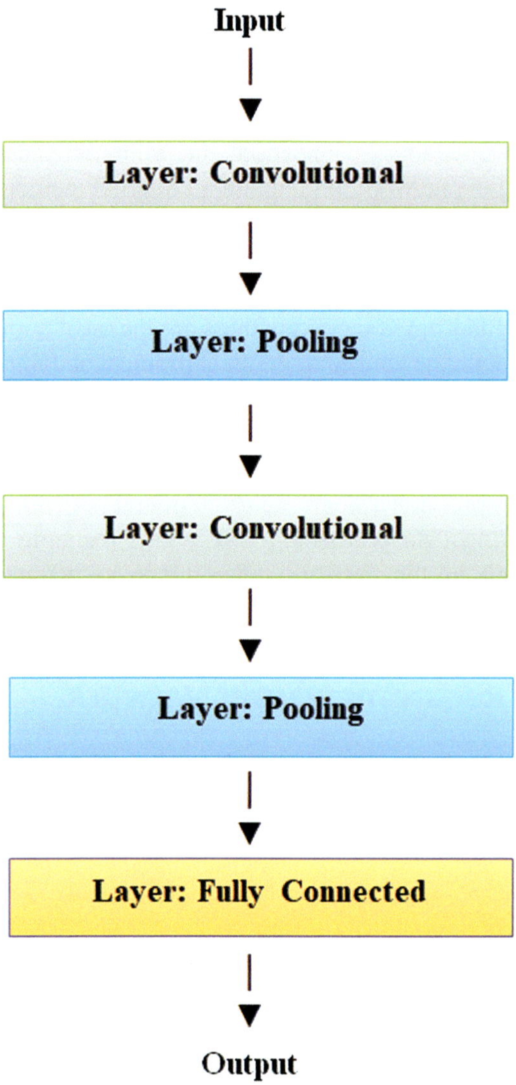

FIGURE 1.4 Representation of a convolution neural network.

It is critical to determine that the real structure of a CNN range is substantially primarily based totally on the precise mission and dataset. The wide variety of convolutional and pooling layers, in addition to the scale and wide variety of filters, may be changed primarily based totally on the ramifications of the trouble at hand.

1.4.3 SMART RECURRENT NEURAL NETWORKS (RNNs)

Smart recurrent neural networks are adopted for ordering in linear data processing and they also include time series or natural language processing tasks. They have feedback connections that allow information to flow in loops, enabling the network to maintain an internal memory or context of past inputs. RNNs is a step procedure, considering the recent input and the previous hidden state. This recurrent structure enables RNNs to capture temporal reliance and handle different-length input sequences. Long-term short memory and limited repetition units are the most popular variants of RNN that deal with the vanishing gradient problem and improve the retention of information over long sequences. RNNs are widely used in speech recognition, machine translation, sentiment analysis, and handwriting recognition.

Figure 1.5 represents a smart recurrent neural network (RNN) usually used for ordering data processing tasks. This example showcases the recurrent nature of the network, allowing it to maintain an internal memory or context of past inputs.

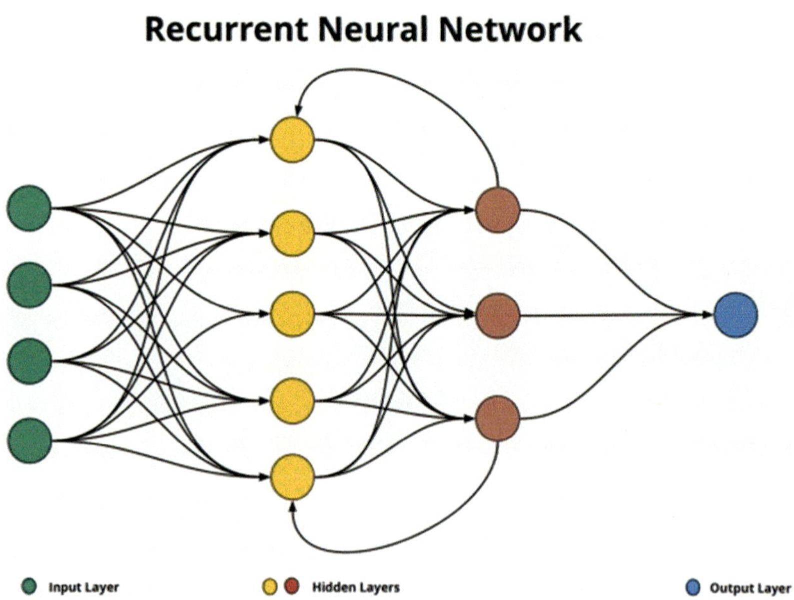

FIGURE 1.5 Representation of a recurrent neural network.

In Figure 1.5, the RNN receives input at different time steps t, represented as "Input (t–2)," "Input(t–1)", and "Input(t)." At each time step, the input is fed into a long short-term memory cell, which is a type of recurrent unit built to capture long-term dependencies in sequential data.

The LSTM cell performs computations using a combination of input, output, and forget gates, allowing it to selectively retain or discard information. The last LSTM cell at time step t, denoted as "Output(t)," can be used for predictions, or passed as input to the next time step, creating the recurrent nature of the network.

It is essential to remember that this figure represents a neat version of a Smart RNN with a single LSTM cell and three time steps. In practice, RNNs can have multiple recurrent units and process sequences of varying lengths. Additionally, the outputs of the LSTM cells can be connected to additional layers or used in different ways depending on the specific task and architecture.

1.4.4 GENERATIVE ADVERSARIAL NETWORKS (GANS)

A generative adversarial network (GAN) consists of two deep neural networks. Generator and separator. The generator network learns to produce synthetic data that mimics the distribution of the training data, and the discriminator network learns to distinguish between real and synthetic data. These networks are simultaneously trained in a competitive way, where the generators try to produce more realistic examples and the discriminators try to improve their ability to distinguish between real and synthetic data. GANs have succeeded in creating realistic images, creating deepfake videos, and improving data augmentation techniques. Figure 1.6 depicts the architecture of a GAN. GAN consists of two main components a generator network and a discriminator network that compete against each other in a two-player minimax game.

In Figure 1.6, GAN starts with an input noise vector, often denoted as "Input Noise(z)." The generator network (G) reads this noise as input and learns to generate synthetic samples that resemble the original data distribution. The output of the generator is represented as "Generated Samples."

The generated samples are then passed to the discriminator network (D), which focuses on differentiating between real and fake data.

The discriminator assigns probabilities, often expressed as real/fake labels to indicate how likely the input samples are real or generated.

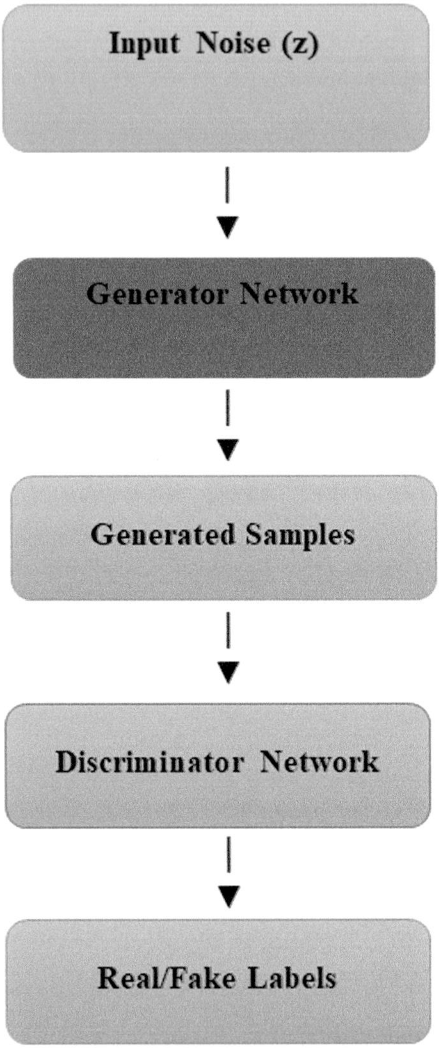

FIGURE 1.6 Representation of a generative adversarial network.

During training, the generator and discriminator networks battled each other. The purpose of the generator is to avoid discrimination by generating samples that are indistinguishable from the real data. Meanwhile, the goal of the discriminator is to correctly classify real and generated samples.

As the training progresses, the generator improves its ability to generate more realistic samples, while the discriminator becomes more adept at distinguishing between real and generated data. The GAN training process continues until a desired level of performance is achieved. It is significant to note that the actual architecture of GANs can vary based on the specific task and dataset. More complex GAN architectures may include additional layers, skip connections, and specialized components to enhance the generation and discrimination processes.

1.4.5 AUTOENCODERS

An autoencoder is an unsupervised learning model that aims to reconstruct input data from a compressed representation called a hidden spatial or bottleneck layer. They consist of an encoder that maps the input data into a low-dimensional representation and a decoder that reconstructs the data from the hidden state. Autoencoders can learn useful representations by capturing salient features of the input data. They are used, for example, for dimensionality reduction, anomaly detection, and data demonization.

Figure 1.7 represents the architecture of a basic autoencoder:

Based on Figure 1.7, an autoencoder consists of two main components. A network of encoders and a network of decoders. An encoder network takes input data and maps it into a low-dimensional representation called a hidden space or bottleneck. This hidden space represents the main features of the input data in compressed form. The decoder network takes the hidden state representation and reconstructs the data back to its original form. The job of the decoder is to create output data that are very similar to the input data so that the reconstruction error can be minimized.

In practice, the autoencoder is optimized to minimize the classification of input data and reconstructed data. This process prompts the autoencoder to explore a compact and informative representation of the input data in a hidden space.

Autoencoders can have different variations, such as denoising autoencoders, variational autoencoders (VAEs), or sparse autoencoders, which introduce additional constraints or modifications to improve their performance for specific tasks.

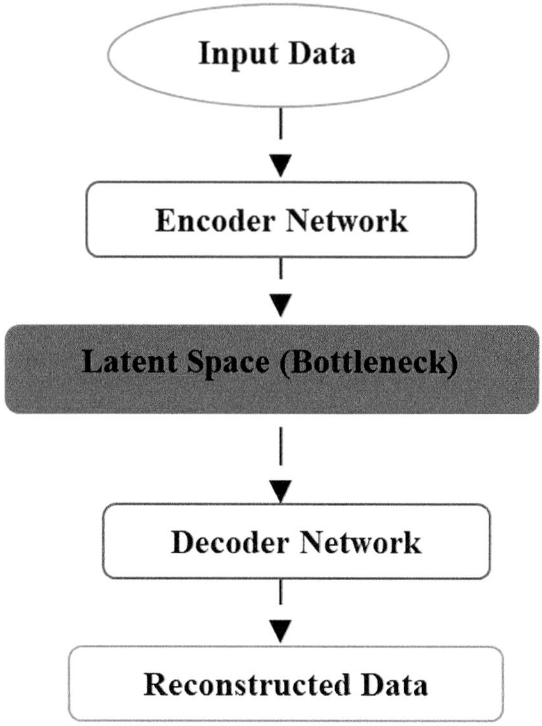

FIGURE 1.7 Basic autoencoder.

It is vital to note that the actual architecture and complexity of autoencoders can vary based on the specific problem, dataset, and desired level of dimensionality reduction or reconstruction accuracy [19].

1.4.6 REINFORCEMENT LEARNING

Reinforcement mastering is a department of deep mastering that offers sequential choice-making in environments. In reinforcement mastering, the agent interacts with the surroundings and learns the way to carry out movements to maximize the cumulative praise signal. Deep reinforcement mastering combines deep neural networks with reinforcement mastering algorithms to allow retailers to research without delay from high-dimensional sensory inputs. Figure 1.8 represents the architecture of a reinforcement learning model.

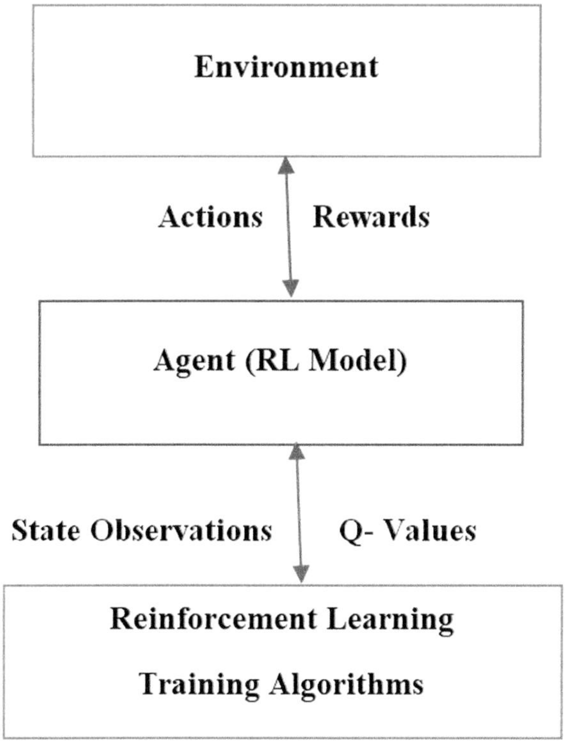

FIGURE 1.8 Reinforcement learning model.

In Figure 1.8, the reinforcement learning process consists of three main components: the environment, the agent, and the reinforcement learning training algorithm.

The environment represents the external world with which the agent interacts. It gives observations or states to the agent, receives actions from the agent, and provides rewards as feedback based on the agent's actions. The environment can be any system or simulation where the agent learns to make sequential decisions.

The agent is the reinforcement learning model that learns to make decisions and take actions based on the observed states or observations. It receives state observations from the environment and uses these observations to estimate the Q-values (action values) for different actions. The agent selects actions based on these Q-values, aiming to increase the cumulative reward obtained from the environment.

The reinforcement learning training algorithm is responsible for updating the agent's policy or Q-values based on the observed rewards. It uses various algorithms, such as Q-learning, SARSA, or Deep Q-Networks to optimize the agent's decision-making process. The training algorithm updates the agent's policy through a process of exploration and exploitation to find an optimal policy that maximizes the long-term collective reward.

It is imperative to note that the actual implementation of reinforcement learning models can differ depending on the specific problem, the complexity of the environment, and the chosen training algorithm. Different algorithms may involve additional components such as memory buffers or neural network architectures for function approximation [20, 21].

1.5 REAL-TIME APPLICATIONS OF MACHINE LEARNING ALGORITHMS

Machine learning (ML) has found applications in various areas, revolutionizing industries and driving innovation. Here are some detailed examples of how it is being applied in different fields:

1.5.1 HEALTHCARE

1. **Diagnosis and Treatment Prediction:** ML algorithms can analyze medical records, imaging data, and patient symptoms to assist in diagnosing diseases and predicting treatment outcomes.
2. **Drug Discovery:** ML is used for model training to focus potential drug candidates, predict their efficacy, and simulate their behavior.
3. **Personalized Medicine:** ML algorithms can analyze patient data to create personalized treatment plans and recommend targeted therapies [22].

1.5.2 FINANCE

1. **Fraud Detection:** ML can identify patterns of fraudulent behavior by analyzing large volumes of transaction data and flag suspicious activities.

2. **Credit Scoring:** ML models can analyze historical data to predict creditworthiness and assign credit scores to individuals and businesses.
3. **Algorithmic Trading:** ML algorithms can analyze market data to make predictions about stock prices, optimize trading strategies, and automate trading decisions.

1.5.3 E-COMMERCE AND RETAIL

1. **Recommendation Systems:** Smart machine learning algorithms can analyze customer behavior, preferences, and purchase history to give customized product recommendations.
2. **Forecasting:** Machine learning models can analyze the historical data sales, trends in the market, and out factors to predict future demand for products and optimize inventory management.
3. **Price Optimization:** Smart machine learning algorithms can predicate pricing data, competitor information, and customer behavior to fix pricing strategies and increase revenue [23].

1.5.4 TRANSPORTATION AND LOGISTICS

1. **Route Optimization:** Smart machine learning algorithms can analyze historical traffic data, weather conditions, and other variables to optimize transportation routes and reduce delivery times.
2. **Vehicle Monitoring:** ML models can analyze sensor data from vehicles to detect maintenance issues, predict failures, and optimize fleet management.
3. **Autonomous Vehicles:** ML algorithms play a crucial role in enabling self-driving cars by processing sensor data, recognizing objects, and making real-time driving decisions.

1.5.5 MANUFACTURING AND INDUSTRIAL PROCESSES

1. **Predictive Maintenance:** Smart machine learning models can analyze data sensed from machinery to find anomalies and forecast maintenance needs and maintenance schedules.

2. **Quality Control:** ML algorithms can analyze production data to detect defects and anomalies, reducing the need for manual inspection.
3. **Supply Chain Optimization:** ML can optimize inventory management, demand forecasting, and logistics to improve the efficiency and reduce costs [24, 25].

1.5.6 NATURAL LANGUAGE PROCESSING (NLP) AND LANGUAGE TRANSLATION

1. **Sentiment Analysis:** ML algorithms can analyze text data to determine sentiment, helping businesses understand customer opinions and feedback.
2. **Language Translation:** ML models can translate text from one language to another, improving accuracy and fluency over traditional rule-based approaches [26].
3. **Chatbots and Virtual Assistants:** ML-powered chatbots can understand and respond to user queries, providing automated customer support and assistance.

There are some examples of machine learning being applied across various domains. ML's versatility and ability to analyze complex patterns and large datasets make it a valuable tool for solving diverse problems and driving innovation in many fields [27].

1.6 EXTENSIVE APPLICATIONS OF SMART MACHINE LEARNING AND SMART DEEP LEARNING IN MODERN RESEARCH SECTORS

Smart machine learning and Smart deep learning techniques have been extensively applied in modern research areas, pushing the boundaries of knowledge and enabling breakthrough in various fields. Here are some examples of their applications.

1.6.1 IMAGE AND VIDEO ANALYSIS

1. **Object Recognition:** ML/DL algorithms can identify and classify objects in images or videos, enabling applications, such as autonomous driving, surveillance, and facial recognition.

2. **Image Segmentation:** ML/DL techniques can separate images into meaningful regions, allowing for applications like medical image analysis and computer vision systems.
3. **Video Understanding:** ML/DL models can analyze video content, extract temporal patterns, and recognize actions, leading to advancements in video surveillance, sports analytics, and video summarization [28–30].

1.6.2 NATURAL LANGUAGE PROCESSING (NLP)

1. **Machine Translation:** ML/DL models can convert text from one to another language, optimizing translation accuracy and fluency.
2. **Smart Sentiment Analysis:** ML/DL algorithms can determine sentiment and emotions from text, enabling sentiment analysis in social monitoring, customer feedback, and market research.
3. **Question Answering and Chatbots:** ML/DL techniques are used to build intelligent chatbots and question-answering systems, enhancing customer support, and information retrieval [31, 32].

1.6.3 GENOMICS AND BIOINFORMATICS

1. **DNA Sequence Analysis:** ML/DL algorithms can analyze DNA sequences, identify patterns, predict gene function, and detect genetic variations associated with diseases.
2. **Protein Structure Prediction:** ML/DL techniques can predict the three-dimensional component of proteins from their amino acid sequences, aiding in drug discovery and personalized medicine.
3. **Disease Diagnosis and Prediction:** ML/DL models can analyze patient data, genomic information, and biomarkers to assist in diagnosing diseases, predicting outcomes, and identifying personalized treatment options.

1.6.4 ASTRONOMY AND ASTROPHYSICS

1. **Image Classification:** ML/DL algorithms can classify celestial objects, such as galaxies and stars, based on their observed

characteristics, leading to advancements in astronomical surveys and cataloging.

2. **Exoplanet Discovery:** ML/DL models can analyze astronomical data to identify exoplanets by detecting patterns in light curves or analyzing spectroscopic data.
3. **Gravitational Wave Detection:** ML/DL techniques are used to analyze gravitational wave signals, helping to detect and characterize astrophysical events such as black hole mergers and neutron star collisions.

1.6.5 ROBOTICS AND AUTONOMOUS SYSTEMS

1. **Robot Perception:** ML/DL algorithms enable robots to perceive and interpret sensor data, allowing them to navigate environments, recognize objects, and interact with humans.
2. **Motion Planning and Control:** ML/DL techniques can optimize robot motion planning and control, improving efficiency, adaptability, and safety in robotic systems.
3. **Reinforcement Learning:** ML/DL algorithms can be used to train robots to learn from interactions with their environment, enabling autonomous decision-making and learning complex tasks [33, 34].

1.6.6 CLIMATE SCIENCE AND ENVIRONMENTAL MONITORING

1. **Weather Prediction:** ML/DL models can analyze meteorological data, satellite imagery, and climate simulations to improve weather forecasting and extreme weather event prediction.
2. **Environmental Monitoring:** ML/DL techniques are used to analyze remote sensing data, to monitor land cover changes, detect deforestation, and assess environmental impacts [35].

Some examples of how Smart machine learning and deep learning techniques are being applied in modern research areas. Their ability to analyze complex data, identify patterns, and make predictions has opened up new possibilities for advancing knowledge and solving complex problems in various scientific disciplines [36, 37].

1.7 CHALLENGES IN SMART DEEP LEARNING AND SMART MACHINE LEARNING

Smart machine learning and smart deep learning demonstrated remarkable success in various applications but they also face several challenges. Some of the key challenges in ML and DL are given below.

1.7.1 DATA LIMITATIONS

1. **Insufficient or Biased Data:** ML/DL models heavily rely on large, diverse, and representative datasets for effective training. However, obtaining such data can be challenging, especially in domains with limited data availability or data bias that may lead to biased model predictions.
2. **Labeling and Annotation:** Manual annotation of data for training ML/DL models is a time taken process, expensive, and error-prone. Ensuring high-quality labeled data is crucial, as models tend to perform as well as the quality of the data they are trained on.

1.7.2 MODEL COMPLEXITY AND INTERPRETABILITY

1. **Model Complexity:** Deep learning models, with their complex architectures and numerous parameters often require substantial computational resources and time for training. Training and optimizing large-scale models can be computationally expensive and may necessitate specialized hardware.
2. **Lack of Interpretability:** Deep learning models, particularly deep neural networks, can be considered as black boxes, making it complex to know and understand their decision-making process. This lack of interpretability can pose challenges, especially in critical domains where explainability is necessary.

1.7.3 OVERFITTING AND GENERALIZATION

1. **Overfitting:** ML/DL fashions might also additionally overfit the education data that mean they end up excessively specialized to

the education examples and fail to generalize nicely to unseen data. Overfitting can result in negative overall performance while carried out in real-international scenarios.

2. **Generalization to Diverse Data:** Ensuring that ML/DL models generalize well to diverse data, including variations in distribution, environment, or context, remains a challenge. Models trained on one dataset or environment may not perform optimally in different settings [38].

1.7.4 COMPUTATIONAL REQUIREMENTS

1. **Computational Power:** Training and deploying large-scale DL models often require substantial computational power and memory resources. Training complex models can be computationally intensive, necessitating high-performance computing infrastructure, or specialized hardware accelerators like GPUs or TPUs.
2. **Energy Efficiency:** DL models often demand significant energy consumption, which can be a concern in resource-constrained environments and contribute to environmental impact. Developing energy-efficient model architectures and optimization techniques is an ongoing challenge.

1.7.5 ETHICAL AND FAIRNESS CONSIDERATIONS

1. **Bias and Fairness:** ML/DL fashions can inadvertently inherit biases gift withinside the education data, mainly due to biased predictions or discriminatory outcomes. Ensuring fairness, transparency, and responsibility in ML/DL structures is essential, requiring cautious attention and mitigation of biases.
2. **Ethical Considerations:** ML/DL applications raise ethical concerns, such as privacy violations, unintended consequences, and potential misuse of technology. Addressing these ethical challenges requires frameworks, guidelines, and responsible practices to ensure the ethical deployment of ML/DL systems.

1.7.6 CONTINUOUS LEARNING AND ADAPTABILITY

1. **Adaptation to Dynamic Environments:** ML/DL models may face difficulties in adapting to changing conditions or evolving data distributions over time. Continuous learning, incremental learning, and transfer learning techniques are being explored to address these challenges and enable models to adapt and generalize effectively.

Researchers and practitioners in the ML/DL field are actively working to address these challenges through advancements in algorithms, data collection techniques, model architectures, interpretability methods, fairness considerations, and ethical frameworks. Overcoming these challenges is crucial to unlocking the full potential of ML/DL in various domains while ensuring their responsible and impactful deployment [39].

1.8 CONCLUSION

Machine learning (ML) and deep learning (DL) have transformed the way we approach complex problems and make predictions in a variety of disciplines. These tactics make use of the power of data and computer algorithms to analyze patterns, generate accurate forecasts, and extract meaningful insights. ML enables computers to learn from their experiences and improve general performance over time, but DL, with its deep neural networks, has pushed the boundaries of studying capabilities, achieving first-rate outcomes in image recognition, natural language processing, and different areas. Despite their achievements, ML and DL continue to encounter obstacles, such as data limits, model complexity, interpretability, overfitting, and ethical concerns. These problems inspire continuous research and efforts to improve the ML/DL systems' dependability, generalization, and fairness. Addressing these difficulties will be critical as technology improves in capturing the full potential of ML/DL and assuring their appropriate and ethical use.

Overall, machine learning and deep learning continue to shape industries, enable innovative applications, and push the boundaries of data analysis, prediction, and decision-making. Their impact on various fields is profound and holds tremendous potential for further advancements and discoveries in the future.

KEYWORDS

- **deep learning**
- **environmental monitoring**
- **health care**
- **machine learning**
- **natural language processing**
- **neural networks**
- **robotics**

REFERENCES

1. Murphy, K. P. (2012). *Machine learning: A probabilistic perspective*. MIT Press.
2. Domingos, P. (2012). A few useful things to know about machine learning. *Communications of the ACM (CACM)*, *55*(10), 78–87. https://dl.acm.org/doi/10.1145/2347736.2347755.
3. Bishop, C. M. (2006). *Pattern recognition and machine learning*. Springer.
4. Journal of Machine Learning Research (JMLR): https://www.jmlr.org/.
5. Journal of Artificial Intelligence Research (JAIR): https://www.jair.org/.
6. Mitchell, T. M. (1997). *Machine learning*. McGraw-Hill.
7. Goodfellow, I., Bengio, Y., & Courville, A. (2016). *Deep learning*. MIT Press. http://www.deeplearningbook.org/.
8. LeCun, Y., Bengio, Y., & Hinton, G. (2015). Deep learning. *Nature*, *521*(7553), 436–444.
9. Schmidhuber, J. (2015). Deep learning in neural networks: An overview. *Neural Networks*, *61*, 85–117.
10. arXiv: e-prints in Computer Science - Machine Learning section. https://arxiv.org/archive/cs.LG.
11. Hastie, T., Tibshirani, R., & Friedman, J. (2009). *The elements of statistical learning: Data mining, inference, and prediction*. Springer Science and Business Media.
12. Vapnik, V. N. (2013). *The nature of statistical learning theory*. Springer Science and Business Media.
13. Friedman, J. H., Hastie, T., & Tibshirani, R. (2000). Additive logistic regression: A statistical view of boosting (with discussion and a rejoinder by the authors). *The Annals of Statistics*, *28*(2), 337–407.
14. Quinlan, J. R. (1986). Induction of decision trees. *Machine Learning*, *1*(1), 81–106.
15. Breiman, L. (2001). Random forests. *Machine Learning*, *45*(1), 5–32.
16. Bishop, C. M. (1995). *Neural networks for pattern recognition*. Oxford University Press.
17. LeCun, Y., Bengio, Y., & Hinton, G. (2015). Deep learning. *Nature*, *521*(7553), 436–444.

18. IEEE Transactions on Neural Networks and Learning Systems (TNNLS). https://www.computer.org/csdl/journal/tn.
19. Neural Networks (Elsevier). https://www.journals.elsevier.com/neural-networks.
20. Sutton, R. S., & Barto, A. G. (2018). *Reinforcement learning: An introduction*. MIT Press.
21. Journal of Machine Learning Research (JMLR) – Reinforcement Learning section. https://www.jmlr.org/mloss/reinforcement-learning.html.
22. Kar, S., & Sahoo, M. (2019). Real-time activity recognition using machine learning algorithms for smart healthcare systems. *Journal of Ambient Intelligence and Humanized Computing*, *10*(3), 1009–1024.
23. Feng, Z., Gao, F., Li, Q., Zhao, L., & Li, Z. (2022). Real-time driver drowsiness detection using machine learning techniques. *IEEE Access*, *10*, 2968–2982.
24. Liu, X., Luo, Z., Huang, Z., & Xu, K. (2021). Real-time and robust machine learning-based sensor fault detection and diagnosis in industrial processes. *Journal of Process Control*, *97*, 144–156.
25. Gu, Y., Yang, L., Liu, L., & Wang, Y. (2019). Real-time machine learning based fault diagnosis of rotating machinery. *Measurement*, *145*, 515–524.
26. Sutskever, I., Vinyals, O., & Le, Q. V. (8–13 December 2014). Sequence to sequence learning with neural networks. In *Proceedings of the 27th International Conference on Neural Information Processing Systems* (pp. 3104–3112). Montreal.
27. Journal of Natural Language Engineering. Cambridge University Press. https://www.cambridge.org/core/journals/journal-of-natural-language-engineering.
28. Rao, A. S., & George, J. P. (2018). Real-time object detection using deep learning for intelligent surveillance systems. *International Journal of Computer Science and Network Security (IJCSNS)*, *18*(10), 64–70.
29. He, K., Zhang, X., Ren, S., & Sun, J. (27–30 June 2016). Deep residual learning for image recognition. In *Proceedings of the IEEE Conference on Computer Vision and Pattern Recognition* (pp. 770–778). Las Vegas.
30. Simonyan, K., & Zisserman, A. (2014). Very deep convolutional networks for large-scale image recognition. http://arxiv.org/abs/1409.1556.
31. Apidianaki, M., Fourtassi, A., Padó, S. (2024). Language learning, representation, and processing in humans and machines: Introduction to the special issue. *Computational Linguistics (MIT Press)*, *50*(4), 1201–1210. https://www.mitpressjournals.org/loi/coli.
32. Nguyen, N., Tran, T., Nguyen, Q. D., & Kim, Y. (2021). Real-time sentiment analysis using machine learning for social media applications. *Information Processing and Management*, *58*(1), 102360.
33. Graves, A., Mohamed, A. R., & Hinton, G. (26–31 May 2013). Speech recognition with deep recurrent neural networks. In *2013 IEEE International Conference on Acoustics, Speech and Signal Processing* (pp. 6645–6649). Vancouver.
34. IEEE Transactions on Robotics (T-RO). https://www.computer.org/csdl/journal/ro.
35. Castro, D., & Ortega, J. (2020). Real-time machine learning for energy-efficient buildings. *Applied Energy*, *258*, 114105.
36. Sun, L., Zhang, Y., Zhang, W., & Zhang, J. (2022). Real-time fault diagnosis of wind turbine based on machine learning. *Renewable Energy*, *184*, 815–826.

37. Kang, B. K., & Lee, J. H. (2020). Real-time machine learning-based heart sound classification on edge devices. *IEEE Transactions on Biomedical Engineering*, *67*(6), 1713–1723.
38. Cawley, G. C., & Talbot, N. L. C. (2010). On over-fitting in model selection and subsequent selection bias in performance evaluation. *Journal of Machine Learning Research*, *11*(2010), 2079–2107.
39. Freund, Y., & Schapire, R. E. (1997). A decision-theoretic generalization of on-line learning and an application to boosting. *Journal of Computer and System Sciences*, *55*(1), 119–139.

CHAPTER 2

Exploring Incremental and Deep Learning Approaches for Plant Disease Detection: A Comprehensive Review

M. PRANEESH[1], V. KAVITHA[2], R. SANTHAKUMARI[3], and R. SHOBANA[4]

[1]*Department of CS & DA, Sri Ramakrishna College of Arts and Science, Coimbatore, Tamil Nadu, India*

[2]*Department of Computer Science with Cognitive Systems, Sri Ramakrishna College of Arts & Science, Coimbatore, Tamil Nadu, India*

[3]*Department of Mathematics, Sri Ramakrishna College of Arts and Science, Coimbatore, Tamil Nadu, India*

[4]*Department of Digital and Cyber Forensic Science, Sri Ramakrishna College of Arts and Science, Coimbatore, Tamil Nadu, India*

ABSTRACT

This chapter furnishes an extensive survey of the utilization of machine learning (ML) and deep learning (DL) techniques for the discovery of plant infections. These plant infections have profound significance on Indian horticulture, posing significant challenges to crop production and food prevention. The early detection of such infections caused by various pathogens will benefit India's agrarian economy in various ways. In recent years, we have seen a surge in the utilization of ML concepts for the detection of diseases. The application of DL and ML concepts will help to overcome the challenges faced by conventional methodologies and it

Real-Time Artificial Intelligence (AI): Key Motivations, Technologies, Platforms, and Use Cases.
Pethuru Raj, B. Sundaravadivazhagan, V. Kavitha, B. Narendra Kumar Rao, & Hannah Vijaykumar (Eds.)

DOI: 10.1201/9781998511358-2

also promises speed and accuracy. This survey systematically categorizes and evaluates a wide array of ML approaches employed in plant infection detection, including preprocessing, feature extraction, classification algorithms, and DL techniques. This chapter also addresses the challenges encountered to help future work improvements. By summarizing prior work and discussing challenges, the survey provides valuable insights for researchers and practitioners to improve plant disease detection systems.

2.1 INTRODUCTION

In the dynamic realm of agriculture, the pursuit of robust crop health remains an enduring challenge critical to sustaining global food production. However, this pursuit is persistently shadowed by the ever-present specter of plant diseases, capable of causing substantial yield losses and jeopardizing food security. In response, the convergence of scientific insight with technological innovation has paved the way for an exciting avenue: the fusion of machine learning (ML) and deep learning (DL) concepts with the sector of plant pathology. Historically, identifying plant diseases heavily relied on human expertise, involving visual inspections and manual analysis. Yet, limitations in terms of time and the potential for human error have catalyzed a profound transformation in disease detection methodologies. The advent of ML and DL introduces a paradigm shift, where computational capacities are harnessed to rapidly scrutinize extensive datasets, revealing meaningful insights, and expediting the identification of diseases.

This survey embarks on an intellectual journey across the expansive domain of ML and DL approaches tailored to the nuanced task of plant disease detection. As we navigate this landscape, we delve into the evolutionary trajectory of these methodologies, gaining insights into their technical underpinnings, and shedding light on their practical implications within real-world agricultural contexts. At the outset, the survey unravels the foundational elements: datasets and preprocessing. These datasets, often curated from diverse sources, such as remote sensing, imaging, and sensor networks, lay the groundwork upon which ML and DL models are constructed. Preprocessing, involving techniques like augmentation and normalization, plays a pivotal role in refining raw data into formats conducive to effective learning. Continuing further, the survey examines the spectrum of existing techniques. Traditional ML algorithms, renowned for their adeptness in pattern recognition and classification, have found utility in discerning indicative markers of disease in plant imagery and other data streams. Decision trees, SVM, and ensemble methods have provided valuable insights into disease occurrence and severity estimation. However, the true evolution in disease detection arises from the domain of DL. Convolutional neural networks (CNNs), emblematic of this transformation, have revolutionized image-based disease detection. They autonomously decipher intricate features and patterns within images, akin to accomplished artists capturing the essence of their subjects. Concurrently, recurrent neural networks (RNNs) support the temporal dimension, comprehending the dynamic evolution of diseases through sequential data, akin to time-travelers decoding the past, present, and future.

Yet, as with any transformative journey, challenges abound. Data scarcity for training, the adaptability of models across diverse ecological contexts, and the seamless integration of automated systems with established agricultural practices stand as formidable barriers. This survey critically navigates these impediments, acknowledging the horizons that beckon for further exploration and innovation. Beyond the boundaries of academic inquiry, the ramifications of this exploration reverberate throughout agricultural landscapes. The confluence of technology and the demands of crop protection herald a promising strategy to navigate the intricate labyrinth of plant diseases. These techniques hold the promise not merely of heightened disease detection precision but also of redefining the contours of sustainable agrarian practices.

As we embark on this intellectual odyssey, traversing the multifaceted expanse of ML for plant disease detection, our aspiration is to equip researchers, practitioners, and policymakers with a holistic comprehension. This survey transcends the role of a mere observer; it is a synthesis of knowledge, a guide for navigating complexities, and a clarion call for interdisciplinary collaboration. Through the intricate interplay of insights, our aim is to illuminate the present state, chart trajectories for the future, and catalyze a landscape where technology and agriculture harmonize seamlessly to secure our sustenance and future.

2.2 FUNDAMENTAL PROCEDURE IN RECOGNITION OF DISEASES FROM LEAF PICTURES

The recognition of plant infections from leaf images stands as a pivotal application of modern ML in agriculture. This intricate process comprises a series of interconnected steps that synergistically blend computational techniques with domain expertise, ultimately enhancing disease diagnosis and management. Let's delve into these steps, meticulously detailing each one.

2.2.1 DATA COLLECTION

In the sphere of utilizing ML for the classification of plant infections through leaf pictures, the initial stride involves meticulous data collection. This pivotal phase revolves around the acquisition of datasets encompassing a diverse array of leaf images, representing both healthy and disease-afflicted states. Several prominent datasets significantly enrich the breadth and depth of training resources available to algorithms. An example in this context is the "Plant Village" dataset, which boasts a compilation of over 50,000 images spanning 14 distinct crop species and 26 diverse diseases. This comprehensive repository serves as a foundational cornerstone for building robust models capable of differentiating between multitudes of diseases. Annotations embedded within the dataset offer crucial insights into crop types, disease categories, and disease severity, thus facilitating focused and informed model training. Another notable dataset is the "Cassava Leaf Disease" dataset, which centers its focus on a single crop—cassava—and its associated diseases. This dataset encompasses around 21,000 labeled images categorized into four distinct disease classes, alongside a "healthy" category. The all-encompassing nature of this dataset imparts algorithms with specialized acumen for accurately identifying diseases afflicting cassava plants.

2.2.2 PREPROCESSING

The preprocessing phase holds a pivotal role in transforming raw leaf images into a standardized format conducive to robust analysis. This stage incorporates an array of techniques aimed at mitigating challenges stemming from variations in image dimensions, lighting conditions, and color representations. To address color inconsistencies, images can be converted into various color spaces, such as HSV (Hue, Saturation, Value) or LAB (Lightness, A, B), which alleviate the impact of lighting variations and enhance disease-specific attributes. Moreover, the integration of filtering techniques, including Gaussian or median filters, aids in suppressing noise and artifacts, thereby enhancing the visual clarity of leaf structures.

Simultaneously, the process of resizing images ensures uniform dimensions across the dataset, bolstering computational efficiency and ensuring consistent inputs for subsequent modeling endeavors. Techniques such as data augmentation introduce controlled variations, simulating real-world fluctuations, and rendering models more adaptable to unforeseen data patterns.

2.2.3 FEATURE EXTRACTION

Feature extraction forms the nucleus of plant disease classification, encapsulating the foundational attributes essential for robust disease identification. Within this domain, CNNs emerge as potent tools for extracting salient features. Operating through convolutional and pooling layers, CNNs systematically discern low-level attributes, amalgamating them into hierarchically organized representations. These representations encompass a gamut of attributes spanning color, shape, and texture features. Color features encapsulate variations in leaf pigmentation, while shape features unravel geometrical nuances intrinsic to leaf structures. Texture features encompass intricacies, such as patterns, spots, or streaks, often indicative of disease presence. The concept of transfer learning enhances feature extraction by leveraging pretrained CNN models, which harbor a repository of learned features that can be harnessed and fine-tuned for the precise process of disease categorization.

2.2.4 CLASSIFICATION

Classification marks the zenith of the process, where the extracted features are harnessed to categorize leaves as healthy or disease-affected. This phase encompasses a spectrum of techniques and algorithms contributing to accurate classification. SVM function by identifying the hyperplane that optimally separates different classes while maximizing the margin between them. SVMs excel in high-dimensional feature spaces, proficiently capturing intricate nonlinear relationships, a trait vital for distinguishing nuanced plant disease characteristics. They adeptly discern subtle variations indicative of diverse diseases. Random Forests, constituting an ensemble approach, amalgamate outputs from numerous decision trees to enhance classification precision. This collective decision-making minimizes individual tree errors, refining overall prediction performance. In the face of intricate plant diseases and their multifarious manifestations, Random Forests are excellent at capturing intricate variations, resulting in robust classifications. K-nearest neighbors (KNN) follows a proximity-based methodology, classifying a leaf by considering the classes of its nearest neighbors in the feature element. The foundation of KNN rests on the assumption that analogous leaves share common disease characteristics. In scenarios like the "Cassava Leaf Disease" dataset, where precise disease identification is imperative, KNN stands out by harnessing localized similarities.

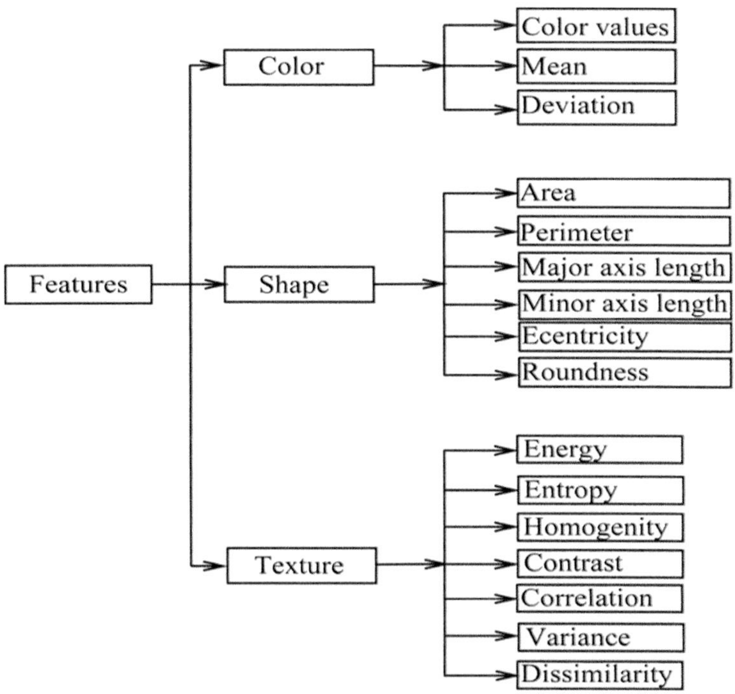

FIGURE 2.1 Few of the features utilized in plant infection detection.

2.3 VARIOUS MACHINE LEARNING-BASED TECHNIQUES FOR PLANT INFECTION DETECTION

Numerous research endeavors have been undertaken concerning the recognition of plant infections. In this division, we delve into various technologies that researchers have put forth for ascertaining diverse plant infections. It has been observed that based on ML, disease discovery methods can be categorized into segments belonging to color, shape along with texture-based features, along with the implementation of DL models. Figure 2.2 illustrates the fundamental procedures in plant infection identification.

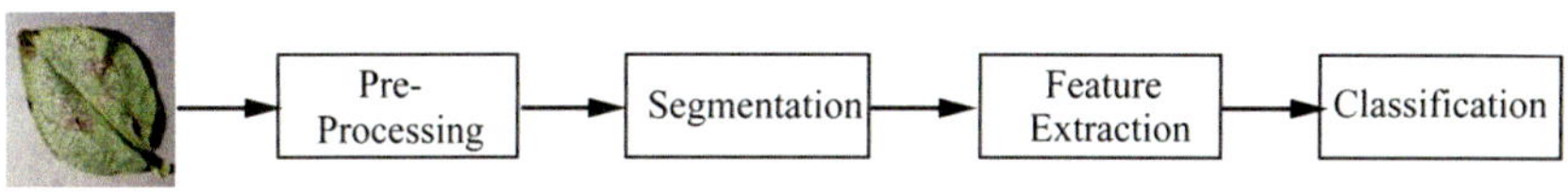

FIGURE 2.2 Basic hand crafted-features-based procedures for identification.

2.3.1 COLOR FEATURES-BASED INFECTION DETECTION

Infection identification through color feature extraction was explored by Arivazhagan et al. [45]. This research methodology involved utilizing YCbCr, HSI, and CIE Lab* color models to extract color features from leaf images. After evaluating these templates, they selected the "A" element of the CIE Lab* style for beginning division. Preprocessing incorporated a median filter, effectively reducing noise. The diseased leaf segment was then partitioned using Otsu's thresholding concept applied to the "A" component of the color space. In a similar vein, Lee et al. [46] adopted a color slicing concept for blast infection detection in paddy plants. Their process entailed converting RGB images to HSI format and pertaining color slicing to gain the infected region while neutralizing the remaining. Comparison with confines detection using Sobel and Canny methods yielded an impressive accuracy rate of 96.6%. Mohanty et al. [47] employed multiple color models to detect plant infections. Converting images into various color spaces, they executed noise reduction through the concept of median filtering and subsequently divided the afflicted leaf segment. YCbCr, HIS, and CIELAB color models were utilized, with Kapur's threshold applied to respective components for segmentation. Fuentes et al. [48] focused on chili leaf disease identification via distinct color components. Extracting yellow, green, and cyan color elements, they employed the techniques of color matching to differentiate between healthy as well as diseased leaves. Chen et al. [49] addressed two counts of rice diseases (brown spot, leaf scald) using color statistics.

Their approach encompassed threshold segmentation, histogram intersection for region isolation, and K-means clustering based on RGB values for pixel classification. Disease classification was facilitated by contrasting and matching color values. In the case of paddy disease identification, Li et al. [50] employed fuzzy entropy, employing a PNN classifier achieving 91.1% accuracy. However, a manual diseased region cropping step during preprocessing posed a significant challenge to their method. Ren et al. [51] tackled four counts of rice plant infections with an extensive arsenal of 172 color features. Across 13 color spaces, diverse color features, such as mean, standard deviation, kurtosis, and skewness were extracted from each channel. Employing SVM, DC, KNN, NB, DT, RF, and LR classifiers, SVM exhibited the highest accuracy of 94.65%. It's noted that the uniform black background in their preprocessed images was a limitation. The primary drawback of detecting infection via color values is its inability to comprehensively detect all disease types.

2.3.2 SHAPE AND TEXTURE-BASED INFECTION DETECTION

In the field of plant disease detection, researchers have been exploring shape and texture-based methodologies as effective means of precise and efficient identification. These methods encompass a range of techniques, including the segmentation of infection-affected areas, extraction of applicable shape and texture attributes, and utilization of robust classifiers for accurate disease classification. For instance, Farjon et al. [52] introduced a technique for detecting rot disease in betel leaves by quantifying the number of pixels within the diseased region. Their approach involved converting RGB images into the HSV color space and applying Otsu's segmentation to determine the affected area. Similarly, Bresilla et al. [53] employed color and shape features to identify rice diseases, utilizing Fermi energy-based segmentation and genetic algorithms to extract disease-specific shape characteristics. Nevertheless, challenges were encountered in establishing optimal segmentation thresholds and dealing with limited training data. Mokhtar et al. [36] integrated shape and texture features extracted from the "Gray Level Co-occurrence Matrix (GLCM)" of rice leaves, achieving a 97.2% accuracy rate through SVM classification. However, their method struggled when diseases shared similar textural attributes. In the context of potato disease recognition, Es-saady et al. [33] focused on eliminating background interference and extracting disease-affected regions through

Lab* color space analysis. They incorporated GLCM texture attributes and an SVM classifier for classifying the plant infection. Their approach faced challenges in setting appropriate threshold values and the availability of an adequate training dataset. Similar investigations extended to cotton plants, where Trong et al. [54] projected a histogram-based algorithm for disease portioning, attaining a 93.1% accuracy rate with SVM classification. Nonetheless, feature extraction proved time-consuming. Chaudhari et al. [55] addressed cotton diseases by employing K-means clustering, wavelet transform, and PCA, achieving a commendable accuracy rate of 97%. [41, 42] demonstrated SVM and ANN fusion for grape disease classification, reporting a perfect 100% accuracy rate for specific diseases. These endeavors collectively underscore the multifaceted challenges and accomplishments inherent in utilizing shape along with texture-based techniques for diverging plant infection identification.

Summarizing the shape and texture-based approaches, several key points emerge:

1. Preprocessing plays a significant role in preparing the data for feature extraction, contributing to the overall complexity of the model.
2. Effectively segmenting diseased areas in images, especially when intricate backgrounds are present, remains a notable challenge.
3. The strategic nomination of a specific feature set for extraction is a critical factor in achieving optimal results.
4. A common constraint is the utilization of small datasets, often focusing on a limited number of specific disease categories.
5. Handling larger datasets introduces additional challenges, as the feature extraction process becomes not only time-intensive but also resource and labor-demanding.

2.3.3 DEEP LEARNING-BASED IDENTIFICATION OF PLANT INFECTIONS

In recent years, DL has experienced rapid germination within the field of computer vision. It has excelled in processes, such as object detection, pattern understanding, segmentation, and biometry, particularly evident in its success in the ImageNet obstacle. This success has been prolonged to agriculture, with applications in plant recognition [71], disease infection [2, 72–74], pest detection [75, 76], fruit detection [77, 78], and weed recognition [79]. One significant advantage of DL is its capability to mechanically understand features from input images, eliminating the requirement for manual segmentation and feature extraction.

For example, Kawasaki et al. [55] utilized a convolutional neural network (CNN) to differentiate between two cucumber diseases—melon yellow spot virus (MYSV) as well as zucchini yellow mosaic virus (ZYMV). They enhanced data volume by augmenting the dataset through image rotations, resulting in a 94.9% accuracy. Building on this, Ref. [81] expanded the approach to encompass seven cucumber diseases, utilizing data augmentation techniques to improve the dataset. Their CNN configuration, based on VGG Net [82] and the Caffe framework [83], achieved 82.3% accuracy through the technique of fourfold cross-validation. Sladojevic et al. [59] employed a pretrained and fine-tuned CaffeNet model for the classification of 13 plant diseases, achieving a 96.3% accuracy using a 10-fold cross-validation approach. Similarly, Rangarajan et al. [60] employed pretrained AlexNet and VGG16 models to identify six tomato leaf diseases, obtaining perfections of 97.29% and 97.49%, respectively. Mohanty et al. [47] investigated crop species diseases using established CNN architectures—AlexNet and GoogLeNet. Their approach, based on a dataset of 54,306 images and RGB data, gained a training accuracy of 99.35%. However, their method encountered limitations when tested on diverse image sources. Nachtigall et al. [61] employed multiple CNN architectures to distinguish six apple leaf diseases, with AlexNet achieving a 97.3% accuracy. To target banana leaf diseases, Amara et al. [65] introduced a LeNet-based deep CNN architecture, achieving recognition rates of 98.61 and 94.44% for color and grayscale images, respectively. Fusion of shape and texture-based features with CNNs was utilized to identify olive leaf diseases [93], resulting in a 98.61% accuracy [76–81].

To conclude, DL techniques have revolutionized plant disease identification. Through pretrained models, advanced architectures, transfer learning, and innovative strategies, researchers achieved remarkable success in classifying diseases in diverse crops. These advancements have the potential to enhance agricultural practices and contribute to food security.

2.4 DISCUSSION

Based on our extensive review, it is evident that DL concepts consistently outperform conventional classification methods, such as KNN, SVM,

TABLE 2.1 Constraints of Deep Learning Models for the Recognition of Plant Infections.

	Constraints					
Authors	**Large Number of Images in Dataset**	**Large Number of Species Considered**	**Accuracy on Testing Field Images**	**Multiple Diseases on Same Image**	**Consider Complex Background**	**Train/Test Data are From Different Datasets**
Mohanty et al. [47]	yes	yes	low	×	×	×
Ferentinos [2]	yes	yes	low	×	×	×
Liu et al. [111]	×	×	×	×	×	×
Amara et al. [65]	×	×	×	×	×	×
Fuentes et al. [48]	×	×	yes	yes	yes	×
Geetharamani and Pandian [83]	yes	yes	×	×	yes	×
Barbedo et al. [112]	yes	yes	low	yes	×	×
Cruz et al. [68]	×	×	×	×	×	×
Sladojevic et al. [59]	×	yes	×	×	yes	×
Brahimi et al. [113]	yes	yes	×	×	yes	×
Ozguven and Adem [81]	×	×	×	×	×	×
Wang et al. [84]	×	×	×	×	×	×
Lee et al. [72]	yes	yes	yes	×	yes	yes
DeChant et al. [80]	×	×	×	×	×	×
Ramcharan et al. [89]	×	yes	yes	×	yes	yes
Oyewola et al. [25]	×	yes	×	×	yes	×
Ramcharan et al. [24]	×	yes	×	×	yes	×

RF, LR, and ANN. The key benefit of DL lies in its capacity to automatically understand features from networks, resulting in more effective and accurate classification compared with conventional feature extraction methods that rely on attributes, such as color, shape, SIFT, texture-based measures, GLCM, histogram, Fourier descriptors, and others. The landscape of DL architectures employed for plant disease identification is extensive. Figure 2.3 displays an idea of these models, highlighting the prevalence of AlexNet, VGG16, GoogleNet, and InceptionV3 as the most often used architectures. VGG19 and ResNet50 also hold significant usage. Table 2.2 offers a summary of various DL models, encompassing layer counts, parameter numbers, and sizes. An analysis of Figure 2.3 reveals the distribution of research papers across individual plant classes.

TABLE 2.2 Various DL Models with Concern to Number of Layers and Parameters Along with Size.

Model	No. of Layer	Parameters (Million)	Size
LeNet	5	0.06	-
AlexNet	8	60	240 MB
VGG16	23	138	528 MB
VGG19	26	143	549 MB
InceptionV1	27	7	51 MB
InceptionV3	48	23.85	93 MB
Xception	126	22.91	88 MB
ResNet50	50	23	98 MB
ResNet101	101	50	171 MB
ResNet152	152	44	232 MB
InceptionResNetV2	572	55.87	215MB
DenseNet121	121	8.06	33 MB
DenseNet201	201	20.24	80 MB
NASNetMobile	-	5.32	23 MB
Squeezenet	69	1.23	5 MB
ShuffleNet	-	3.4	-
MobileNetV1	88	4.2	16 MB
MobileNetV2	88	3.37	14 MB
EfficientNetB0	-	5.33	29 MB
EfficientNetB1	-	7.85	31 MB

The "PlantVillage" dataset, which covers 14 counts of plant species and 38 counts of disease categories, remains a cornerstone in this domain. Rice and tomatoes also stand out as frequently studied plants. The chronological histogram depicted in Figure 2.4 illustrates the evolution of plant disease identification research from 2009 to 2021. Notably, this field experienced a surge in attention after 2016. DL models offer a plethora of advantages over handcrafted feature-based approaches. Traditional feature extraction demands substantial effort and time, compounded by challenges in identifying optimal attributes. DL-based features alleviate this burden, providing optimal outcomes. DL models also demonstrate robustness against complex backgrounds, illumination variations, size differences, and orientation changes. Moreover, they exhibit resilience in scenarios with intricate surroundings and effectively handle intra- and extra-class variations, solidifying their position as powerful tools in plant disease identification [98].

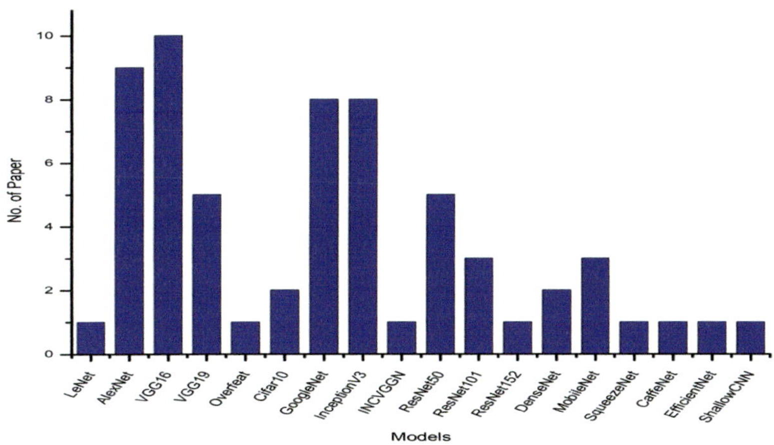

FIGURE 2.3 Total number of research manuscripts for several deep learning models.

2.5 CHALLENGES

Addressing the challenges and unresolved issues in plant disease identification from leaf images is crucial for the development of practical real-time systems in diverse field conditions. This section delves into several key problems that remain unaddressed in the field of disease identification in plants.

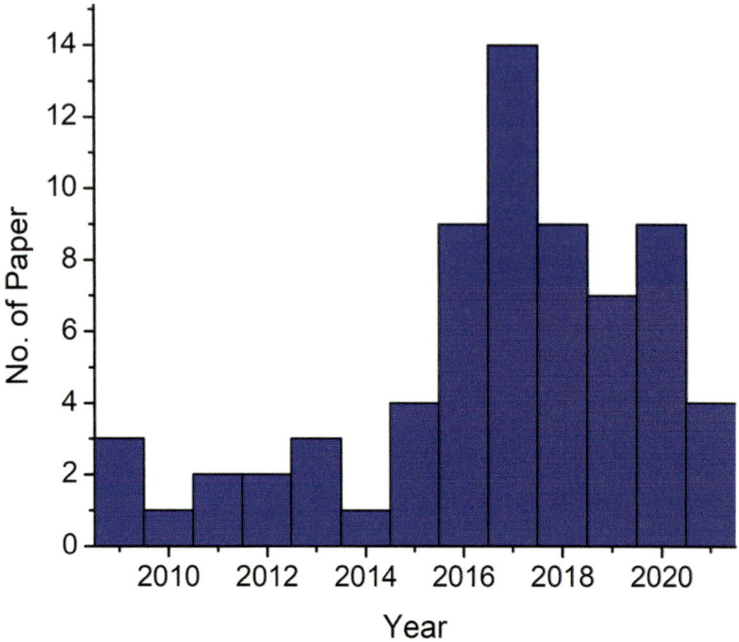

FIGURE 2.4 Histogram of number of manuscripts published with respect to years.

2.5.1 INSUFFICIENT AND LIMITED DATASET VARIETY

Many studies suffer from a common limitation: the dataset employed to teach CNN networks lacks the diversity necessary for achieving high-performance accuracy in disease identification. DL thrives on huge datasets with a broad spectrum of pictures. Currently, the Plant Village [23] and Image Database of Plant Disease Symptoms (PDDB) are the primary openly available extensive disease datasets. However, these images are predominantly sourced from controlled laboratory settings, featuring uniform backgrounds. Collecting field images is expensive and necessitates agricultural expertise for accurate disease recognition [98–100].

2.5.2 IMAGE SEGMENTATION CHALLENGES

Image partitioning is the process of isolating the region of interest within a picture. It is approached through traditional methods or soft-computing techniques. Traditional methods include K-means clustering and color thresholding, while soft-computing involves fuzzy logic, artificial neural networks, and region growing. Segmenting leaf images from complex backgrounds poses a formidable challenge in disease identification. Properly segmenting the leaf segment can significantly enhance accuracy. Pictures containing numerous extraneous elements often hinder successful identification [105–107].

2.5.3 DISTINGUISHING VISUALLY SIMILAR DISEASE SYMPTOMS

Certain diseases exhibit symptoms that are visually similar, confounding even experts' naked-eye diagnosis. Geographical variations, crop growth stages, and weather conditions can cause symptom variations for a single disease. Addressing this issue and incorporating these factors into disease identification models remains an unexplored area in current literature.

2.5.4 COEXISTENCE OF MULTIPLE DISEASES

Many disease identification models assume that a single type of disease is present in an image. However, instances of multiple diseases occurring simultaneously or nutritional disorders complicate disease recognition. Limited research focuses on identifying multiple diseases concurrently. Notably, Fuentes et al. [48] considered this challenge in the context of tomato leaves [108].

2.5.5 REAL-TIME DISEASE IDENTIFICATION

A prevalent trend in existing literature is the reliance on laboratory images for disease identification models. However, these models often struggle when applied to real-time scenarios. For example, in Ref. [110], the accuracy dropped drastically from 99.35% on the PlantVillage dataset to 31% when tested with a different dataset. Similarly, Ref. [113] reported a drop in accuracy from 99.53% on diverse datasets to 66% when the model was exposed to field-captured images. Effectively identifying diseases in real-time field images remains a significant and complex challenge.

2.5.6 LIGHTWEIGHT DEEP LEARNING MODELS

Most DL architectures employed in research, including AlexNet, VGG, GoogleNet, ResNet, DenseNet, and InceptionV3 demand substantial computational resources, expensive GPUs, and extensive machine clusters. This cost and complexity hinder widespread adoption. The development of lightweight CNN models becomes imperative, particularly for embedded, mobile, and robotic applications that require real-time performance and minimal computational demands. Achieving competitive performance with smaller datasets and reduced computational requirements is essential but challenging due to the complexity of the underlying data models [114].

2.6 CONCLUSIONS AND FUTURE DIRECTIONS

This chapter provides an encompassing review of diverse ML concepts employed in identifying plant infections using leaf pictures. Just like mankind, plants are susceptible to various diseases that impede their healthy growth. The survey encompasses disease identification methods rooted in handcrafted features and modern DL techniques. A comparative analysis is performed across preprocessing, segmentation methodologies, feature utilization, and datasets applied in each study. Through this examination, it becomes evident that preprocessing and segmentation strategies significantly influence accuracy outcomes. Among the classification techniques, support vector machine (SVM) emerges as the prevalent choice for disease identification. Finally, the study underscores that DL models consistently outperform conventional handcrafted feature methodologies. In the form of DL accuracy, ResNet50, InceptionV3, and DenseNet201 stand out as effective architectures for disease identification. Additionally, MobileNetV2 and Squeeze Net prove suitable options for lightweight devices like mobile phones. The timely recognition of plant diseases holds immense potential for improving crop yield while mitigating the reliance on expensive domain experts. Nonetheless, this survey reveals certain gaps within existing research that present promising avenues for future exploration in plant disease identification. One critical area of focus pertains to amassing extensive datasets that encapsulate diverse images from varied geographical locations. Disease symptom variability across infection stages can impact detection reliability. Future endeavors should encompass the development of dependable lightweight deep CNN models, with special emphasis on adapting these models for mobile devices.

KEYWORDS

- **CNN**
- **feature extraction**
- **plant disease**
- **segmentation**
- **supervised learning**

BIBLIOGRAPHY

1. Chouhan, S. S., Kaul, A., Singh, U. P., & Jain, S. (2018). Bacterial foraging optimization based radial basis function neural network (BRBFNN) for identification and classification of plant leaf diseases: An automatic approach towards plant pathology. *IEEE Access, 6*, 8852–8863.
2. Ferentinos, K. P. (2018). Deep learning models for plant disease detection and diagnosis. *Computers and Electronics in Agriculture, 145*, 311–318.
3. Bharate, A. A., & Shirdhonkar, M. (7–8 December 2017). A review on plant disease detection using image processing. In *Proceedings of the 2017 International Conference on Intelligent Sustainable Systems (ICISS), Palladam, India* (pp. 103–109).
4. Das, R., Pooja, V., & Kanchana, V. (7–8 April 2017). Detection of diseases on visible part of plant: A review. In *Proceedings of the 2017 IEEE Technological Innovations in ICT for Agriculture and Rural Development (TIAR), Chennai, India* (pp. 42–45).
5. Bock, C., Poole, G., Parker, P., & Gottwald, T. (2010). Plant disease severity estimated visually, by digital photography and image analysis, and by hyperspectral imaging. *Critical Reviews in Plant Sciences, 29*, 59–107.
6. Tlhobogang, B., & Wannous, M. (13–17 April 2018). Design of plant disease detection system: A transfer learning approach work in progress. In *Proceedings of the 2018 IEEE International Conference on Applied System Invention (ICASI), Chiba, Japan* (pp. 158–161).
7. Mutka, A. M., & Bart, R. S. (2015). Image-based phenotyping of plant disease symptoms. *Frontiers in Plant Science, 5*, 734.
8. Nutter, F. W., Esker, P. D., & Netto, R. A. C. (2006). Disease assessment concepts and the advancements made in improving the accuracy and precision of plant disease data. *European Journal of Plant Pathology, 115*, 95–103.

9. Munyaneza, J. E., Crosslin, J. M., Buchman, J. L., & Sengoda, V. G. (2010). Susceptibility of different potato plant grow the stages to purple top disease. *American Journal of Potato Research*, *87*, 60–66.
10. Díaz-Pendón, J. A., Cañizares, M. C., Moriones, E., Bejarano, E. R., Czosnek, H., & NavasCastillo, J. (2010). Tomato yellow leaf curl viruses: ménage à trois between the virus complex, the plant and the whitefly vector. *Molecular Plant Pathology*, *11*, 441–450.
11. Shah, J. P., Prajapati, H. B., & Dabhi, V. K. (10–11 March 2016). A survey on detection and classification of rice plant diseases. In *Proceedings of the 2016 IEEE International Conference on Current Trends in Advanced Computing (ICCTAC), Bangalore, India* (pp. 1–8).
12. Prajapati, B. S., Dabhi, V. K., & Prajapati, H. B. (3–5 March 2016). A survey on detection and classification of cotton leaf diseases. In *Proceedings of the 2016 International Conference on Electrical, Electronics, and Optimization Techniques (ICEEOT), Chennai, India* (pp. 2499–2506).
13. Iqbal, Z., Khan, M. A., Sharif, M., Shah, J. H., ur Rehman, M. H., & Javed, K. (2018). An automated detection and classification of citrus plant diseases using image processing techniques: A review. *Computers and Electronics in Agriculture*, *153*, 12–32.
14. Kaur, S., Pandey, S., & Goel, S. (2019). Plants disease identification and classification through leaf images: A survey. *Archives of Computational Methods in Engineering*, *26*, 507–530.
15. Lee, S. H., Chan, C. S., Wilkin, P., & Remagnino, P. (27–30 September 2015). Deep-plant: Plant identification with convolutional neural networks. In *Proceedings of the 2015 IEEE International Conference on Image Processing (ICIP), Quebec City, QC, Canada* (pp. 452–456).
16. Ciresan, D. C., Meier, U., Masci, J., Gambardella, L. M., & Schmidhuber, J. (16–22 July 2011). Flexible, high performance convolutional neural networks for image classification. In *Proceedings of the Twenty-Second International Joint Conference on Artificial Intelligence, Barcelona, Spain* (pp. 1237–1242).
17. Hinton, G., Deng, L., Yu, D., Dahl, G. E., Mohamed, A., Jaitly, N., Senior, A., Vanhoucke, V., Nguyen, P., Sainath, T. N., et al. (2012). Deep neural networks for acoustic modeling in speech recognition: The shared views of four research groups. *IEEE Signal Processing Magazine, 29*(6), 82–97.
18. Wen, T., & Zhang, Z. (2018). Deep convolution neural network and autoencoders-based unsupervised feature learning of EEG signals. *IEEE Access, 6*, 25399–25410.
19. Carranza Rojas, J., Goeau, H., Bonnet, P., Mata Montero, E., & Joly, A. (2017). Going deeper in the automated identification of Herbarium specimens. *BMC Ecology and Evolution*, *17*, 181.
20. Yang, X., & Guo, T. (2017). Machine learning in plant disease research. *European Journal of BioMedical Research*, *3*, 6–9.
21. Nagaraju, M., & Chawla, P. (2020). Systematic review of deep learning techniques in plant disease detection. *International Journal of System Assurance Engineering and Management*, *11*, 547–560.
22. Li, L., Zhang, S., & Wang, B. (2021). Plant disease detection and classification by deep learning—A review. *IEEE Access*, *9*, 56683–56698.

23. Hughes, D., & Salathé, M. (2015). An open access repository of images on plant health to enable the development of mobile disease diagnostics. *arXiv*, arXiv:1511.08060.
24. Ramcharan, A., Baranowski, K., McCloskey, P., Ahmed, B., Legg, J., & Hughes, D. P. (2017). Deep learning for image based cassava disease detection. *Frontiers in Plant Science*, *8*, 1852.
25. Oyewola, D. O., Dada, E. G., Misra, S., & Damaševicˇius, R. (2021). Detecting cassava mosaic disease using a deep residual convolutional neural network with distinct block processing. *PeerJ Computer Science*, *7*, e352.
26. Hops Disease Dataset. (2021). Available online: https://www.kaggle.com/scruggzilla/hops-classification (accessed on January 17, 2021).
27. Cotton Disease Dataset. (2021). Available online: https://www.kaggle.com/singhakash/cotton-disease-dataset (accessed on January 22, 2021).
28. Rice Disease Dataset. (2021). Available online: https://www.kaggle.com/shayanriyaz/riceleafs (accessed on January 20, 2021).
29. Sethy, P. K., Barpanda, N. K., Rath, A. K., & Behera, S. K. (2020). Deep feature based rice leaf disease identification using support vector machine. *Computers and Electronics in Agriculture*, *175*, 105527.
30. Hlaing, C. S., & Zaw, S. M. M. (18–20 December 2017). Model-based statistical features for mobile phone image of tomato plant disease classification. In *Proceedings of the 2017 18th International Conferenceon Parallel and Distributed Computing, Applications and Technologies (PDCAT), Taipei, Taiwan* (pp. 223–229).
31. Camargo, A., & Smith, J. (2009). Image pattern classification for the identification of disease causing agents in plants. *Computers and Electronics in Agriculture*, *66*, 121–125.
32. Prasad, S., Peddoju, S. K., & Ghosh, D. (2016). Multi-resolution mobile vision system for plant leaf disease diagnosis. *Signal, Image and Video Processing*, *10*, 379–388.
33. Es-saady, Y., El Massi, I., El Yassa, M., Mammass, D., & Benazoun, A. (4–7 May 2016). Automatic recognition of plant leaves diseases based on serial combination of two SVM classifiers. In *Proceedings of the 2016 International Conference on Electrical and Information Technologies (ICEIT), Tangiers, Morocco* (pp. 561–566).
34. Singh, V., & Misra, A. (19–20 March 2015). Detection of unhealthy region of plant leaves using image processing and genetic algorithm. In *Proceedings of the 2015 International Conference on Advances in Computer Engineering and Applications, Ghaziabad, India* (pp. 1028–1032).
35. Vijai, S., & Misra, A. (2017). Detection of plant leaf diseases using image segmentation and soft computing techniques. *Information Processing in Agriculture*, *4*(1), 41–49.
36. Mokhtar, U., Ali, M. A., Hassanien, A. E., & Hefny, H. (2015). Identifying two of tomatoes leaf viruses using support vector machine. In *Information Systems Design and Intelligent Applications* (pp. 771–782). Springer: Berlin, Germany.
37. Hlaing, C. S., & Zaw, S. M. M. (6–8 June 2018). Tomato plant diseases classification using statistical texture feature and color feature. In *Proceedings of the 2018 IEEE/ACIS 17th International Conference on Computer and Information Science (ICIS), Singapore* (pp. 439–444).
38. Tian, Y., Zhao, C., Lu, S., & Guo, X. (2011). Multiple classifier combination for recognition of wheat leaf diseases. *Intelligent Automation and Soft Computing*, *17*, 519–529.

39. Masazhar, A. N. I., & Kamal, M. M. (28–30 November 2017). Digital image processing technique for palmoil leaf disease detection using multiclass SVM classifier. In *Proceedings of the 2017 IEEE 4th International Conference on Smart Instrumentation, Measurement and Application (ICSIMA), Putrajaya, Malaysia* (pp. 1–6).
40. Zhang, S., Zhu, Y., You, Z., & Wu, X. (2017). Fusion of superpixel, expectation maximization and PHOG for recognizing cucumber diseases. *Computers and Electronics in Agriculture*, *140*, 338–347.
41. Zhang, S., Wang, H., Huang, W., You, Z. Plant diseased leaf segmentation and recognition by fusion of superpixel, K-means andPHOG. *Optik*2018, *157*, 866–872.
42. Zhang, S., You, Z., & Wu, X. (2019). Plant disease leaf image segmentation based on super pixel clustering and EM algorithm. *Neural Computing and Applications*, *31*, 1225–1232.
43. Dandawate, Y., & Kokare, R. (10–13 August 2015). An automated approach for classification of plant diseases towards development of futuristic decision support system in Indian perspective. In *Proceedings of the 2015 International Conference on Advances in Computing, Communications and Informatics (ICACCI), Kochi, India* (pp. 794–799).
44. Prajapati, H. B., Shah, J. P., & Dabhi, V. K. (2017). Detection and classification of rice plant diseases. *Intelligent Decision Technologies*, *11*, 357–373.
45. Arivazhagan, S., Shebiah, R. N., Ananthi, S., & Varthini, S. V. (2013). Detection of unhealthy region of plant leaves and classification of plant leaf diseases using texture features. *Agricultural Engineering International: The CIGR E-Journal*, *15*, 211–217.
46. Lee, S. H., Chan, C. S., Mayo, S. J., & Remagnino, P. (2017). How deep learning extracts and learns leaf features for plant classification. *Pattern Recognition*, *71*, 1–13.
47. Mohanty, S. P., Hughes, D. P., & Salathé, M. (2016). Using deep learning for image-based plant disease detection. *Frontiers in Plant Science*, *7*, 1419.
48. Fuentes, A., Yoon, S., Kim, S., & Park, D. (2017). A robust deep-learning-based detector for real-time tomato plant diseases and pests recognition. *Sensors*, *17*(9), 2022.
49. Chen, J., Chen, J., Zhang, D., Sun, Y., & Nanehkaran, Y. A. (2020). Using deep transfer learning for image-based plant disease identification. *Computers and Electronics in Agriculture*, *173*, 105393.
50. Li, Y., & Yang, J. (2020). Few-shot cotton pest recognition and terminal realization. *Computers and Electronics in Agriculture*, *169*, 105240.
51. Ren, F., Liu, W., & Wu, G. (2019). Feature reuse residual networks for insect pest recognition. *IEEE Access*, *7*, 122758–122768.
52. Farjon, G., Krikeb, O., Hillel, A. B., & Alchanatis, V. (2019). Detection and counting of flowers on apple trees for better chemical thinning decisions. *Precision Agriculture*, *21*, 503–521.
53. Bresilla, K., Perulli, G. D., Boini, A., Morandi, B., Corelli Grappadelli, L., & Manfrini, L. (2019). Single-shot convolution neural networks for real-time fruit detection within the tree. *Frontiers in Plant Science*, *10*, 611.
54. Trong, V. H., Gwang-hyun, Y., Vu, D. T., & Jin-young, K. (2020). Late fusion of multimodal deep neural networks for weeds classification. *Computers and Electronics in Agriculture*, *175*, 105506.

55. Kawasaki, Y., Uga, H., Kagiwada, S., & Iyatomi, H. (2015). Basic study of automated diagnosis of viral plant diseases using convolutional neural networks. In *International Symposiumon Visual Computing* (pp. 638–645). Springer: Berlin, Germany.
56. Fujita, E., Kawasaki, Y., Uga, H., Kagiwada, S., & Iyatomi, H. (18–20 December 2016). Basic investigation on a robust and practical plant diagnostic system. In *Proceedings of the 2016 15th IEEE International Conference on Machine Learning and Applications (ICMLA), Anaheim, CA, USA* (pp. 989–992).
57. Simonyan, K., & Zisserman, A. (2014). Very deep convolutional networks for large-scale image recognition. *arXiv*, arXiv:1409.1556.
58. Jia, Y., Shelhamer, E., Donahue, J., Karayev, S., Long, J., Girshick, R., Guadarrama, S., & Darrell, T. (3–7 November 2014). Caffe: Convolutional architecture for fast feature embedding. In *Proceedings of the 22nd ACM International Conference on Multimedia, Orlando, FL, USA* (pp. 675–678).
59. Sladojevic, S., Arsenovic, M., Anderla, A., Culibrk, D., & Stefanovic, D. (2016). Deep neural networks based recognition of plant diseases by leaf image classification. *Computational Intelligence and Neuroscience*, *2016*, 3289801.
60. Rangarajan, A. K., Purushothaman, R., & Ramesh, A. (2018). Tomato crop disease classification using pre-trained deep learning algorithm. *Procedia Computer Science*, *133*, 1040–1047.
61. Nachtigall, L. G., Araujo, R. M., & Nachtigall, G. R. (6–8 November 2016). Classification of apple tree disorders using convolutional neural networks. In *Proceedings of the 2016 IEEE 28th International Conference on Tools with Artificial Intelligence (ICTAI), San Jose, CA, USA* (pp. 472–476).
62. Krizhevsky, A., Sutskever, I., & Hinton, G. E. (2012). ImageNet classification with deep convolutional neural networks. *Advances in Neural Information Processing Systems*, *25*, 1097–1105.
63. Hara, Y., Atkins, R. G., Yueh, S. H., Shin, R. T., & Kong, J. A. (1994). Application of neural networks to radar image classification. *IEEE Transactions on Geoscience and Remote Sensing*, *32*, 100–109.
64. LeCun, Y., Boser, B., Denker, J. S., Henderson, D., Howard, R. E., Hubbard, W., & Jackel, L. D. (1989). Backpropagation applied to handwritten zip code recognition. *Neural Computation*, *1*(4), 541–551.
65. Amara, J., Bouaziz, B., & Algergawy, A. (6–10 March 2017). A deep learning-based approach for banana leaf diseases classification. In *Proceedings of the Daten Bank System Efür Business, Technologie Und Web (BTW2017)-Workshopband, Stuttgart, Germany*.
66. Hu, M. K. (1962). Visual pattern recognition by moment invariants. *IRE Transactions on Information Theory*, *8*, 179–187.
67. Zhenjiang, M. (2000). Zernike moment-based image shape analysis and its application. *Pattern Recognition Letters*, *21*(2), 169–177.
68. Cruz, A. C., Luvisi, A., DeBellis, L., & Ampatzidis, Y. (2017). X-FIDO: An effective application for detecting olive quick decline syndrome with deep learning and data fusion. *Frontiers in Plant Science*, *8*, 1741.
69. Atole, R. R., & Park, D. (2018). A multiclass deep convolutional neural network classifier for detection of common rice plant anomalies. *International Journal of Advanced Computer Science and Applications (IJACSA)*, *9*(1), 67–70.

70. Rangarajan, A. K., & Purushothaman, R. (2020). Disease classification in eggplant using pre-trained VGG16 and MSVM. *Scientific Reports*, *10*, 1–11.
71. Rangarajan Aravind, K., & Raja, P. (2020). Automated disease classification in (selected) agricultural crops using transfer learning. *Automatika*, *61*, 260–272.
72. Lee, S. H., Goëau, H., Bonnet, P., & Joly, A. (2020). New perspectives on plant disease characterization based on deep learning. *Computers and Electronics in Agriculture*, *170*, 105220.
73. Picon, A., Alvarez-Gila, A., Seitz, M., Ortiz-Barredo, A., Echazarra, J., & Johannes, A. (2019). Deep convolutional neural networks for mobile capture device-based crop disease classification in the wild. *Computers and Electronics in Agriculture*, *161*, 280–290.
74. Johannes, A., Picon, A., Alvarez-Gila, A., Echazarra, J., Rodriguez-Vaamonde, S., Navajas, A. D., & Ortiz-Barredo, A. (2017). Automatic plant disease diagnosis using mobile capture devices, applied on a wheat use case. *Computers and Electronics in Agriculture*, *138*, 200–209.
75. He, K., Zhang, X., Ren, S., & Sun, J. (27–30 June 2016). Deep residual learning for image recognition. In *Proceedings of the IEEE Conference on Computer Vision and Pattern Recognition, Las Vegas, NV, USA* (pp. 770–778).
76. Ahmad, I., Hamid, M., Yousaf, S., Shah, S. T., & Ahmad, M. O. (2020). Optimizing pretrained convolutional neural networks for tomato leaf disease detection. *Complexity*, *2020*, 8812019.
77. Elhassouny, A., & Smarandache, F. (22–24 July 2019). Smart mobile application to recognize tomato leaf diseases using convolutional neural networks. In *Proceedings of the 2019 International Conference of Computer Science and Renewable Energies (ICCSRE), Agadir, Morocco* (pp. 1–4).
78. Li, K., Lin, J., Liu, J., & Zhao, Y. (2020). Using deep learning for image-based different degrees of ginkgo leaf disease classification. *Information*, *11*(2), 95.
79. Ai, Y., Sun, C., Tie, J., & Cai, X. (2020). Research on recognition model of crop diseases and insect pests based on deep learning in harsh environments. *IEEE Access*, *8*, 171686–171693.
80. DeChant, C., Wiesner-Hanks, T., Chen, S., Stewart, E. L., Yosinski, J., Gore, M. A., Nelson, R. J., & Lipson, H. (2017). Automated identification of northern leaf blight-infected maize plants from field imagery using deep learning. *Phytopathology*, *107*, 1426–1432.
81. Ozguven, M. M., & Adem, K. (2019). Automatic detection and classification of leaf spot disease in sugar beet using deep learning algorithms. *Physica A: Statistical Mechanics and its Applications*, *535*, 122537.
82. Oppenheim, D., & Shani, G. (2017). Potato disease classification using convolution neural networks. *Advances in Animal Biosciences*, *8*(2), 244.
83. Geetharamani, G., & Pandian, A. (2019). Identification of plant leaf diseases using a nine-layer deep convolutional neural network. *Computers and Electrical Engineering*, *76*, 323–338.
84. Wang, G., Sun, Y., & Wang, J. (2017). Automatic image-based plant disease severity estimation using deep learning. *Computational Intelligence and Neuroscience*, *2017*, 2917536.

85. Deng, J., Dong, W., Socher, R., Li, L. J., Li, K., & Fei-Fei, L. (20–25 June 2009). ImageNet: A large-scale hierarchical image database. In *Proceedings of the 2009 IEEE Conference on Computer Vision and Pattern Recognition, Miami, FL, USA* (pp. 248–255).
86. Zhang, X., Qiao, Y., Meng, F., Fan, C., & Zhang, M. (2018). Identification of maize leaf diseases using improved deep convolutional neural networks. *IEEE Access*, *6*, 30370–30377.
87. Salimans, T., Goodfellow, I., Zaremba, W., Cheung, V., Radford, A., & Chen, X. (2016). Improved techniques for training GANs. *arXiv*, arXiv:1606.03498.
88. Liu, B., Zhang, Y., He, D., & Li, Y. (2018). Identification of apple leaf diseases based on deep convolutional neural networks. *Symmetry*, *10*, 11.
89. Ramcharan, A., McCloskey, P., Baranowski, K., Mbilinyi, N., Mrisho, L., Ndalahwa, M., Legg, J., & Hughes, D. P. (2019). A mobile-based deep learning model for cassava disease diagnosis. *Frontiers in Plant Science*, *10*, 272.
90. Lin, T. Y., Maire, M., Belongie, S., Hays, J., Perona, P., Ramanan, D., Dollár, P., & Zitnick, C. L. (2014). Microsoft COCO: Common objects in context. In *European Conference on Computer Vision* (pp. 740–755). Springer: Berlin, Germany.
91. Toda, Y., & Okura, F. (2019). How convolutional neural networks diagnose plant disease. *Plant Phenomics*, *2019*, 9237136.
92. Szegedy, C., Vanhoucke, V., Ioffe, S., Shlens, J., & Wojna, Z. (27–30 June 2016). Rethinking the inception architecture for computer vision. In *Proceedings of the IEEE Conference on Computer Vision and Pattern Recognition, LasVegas, NV, USA* (pp. 2818–2826).
93. Lu, Y., Yi, S., Zeng, N., Liu, Y., & Zhang, Y. (2017). Identification of rice diseases using deep convolutional neural networks. *Neurocomputing*, *267*, 378–384.
94. Iandola, F. N., Han, S., Moskewicz, M. W., Ashraf, K., Dally, W. J., & Keutzer, K. (2016). SqueezeNet: AlexNet-level accuracy with 50x fewer parameters and <0. 5MB model size. *arXiv*, arXiv:1602.07360.
95. Durmus¸, H., Günes¸, E. O., & Kırcı, M. (7–10 August 2017). Disease detection on the leaves of the tomato plants by using deep learning. In *Proceedings of the 2017 6th International Conference on Agro-Geoinformatics, Fairfax, VA, USA* (pp. 1–5).
96. Hu, G., Yang, X., Zhang, Y., & Wan, M. (2019). Identification of tea leaf diseases by using an improved deep convolutional neural network. *Sustainable Computing: Informatics and Systems*, *24*, 100353.
97. LeCun, Y., Bottou, L., Bengio, Y., & Haffner, P. (1998). Gradient-based learning applied to document recognition. *Proceedings of the IEEE*, *86*(11), 2278–2324.
98. Bi, C., Wang, J., Duan, Y., Fu, B., Kang, J. R., & Shi, Y. (2020). MobileNet based apple leaf diseases identification. *Mobile Networks and Applications*, *27*, 172–180.
99. Atila, Ü., Uçar, M., Akyol, K., & Uçar, E. (2021). Plant leaf disease classification using EfficientNet deep learning model. *Ecological Informatics*, *61*, 101182.
100. Tuncer, A. (2021). Cost-optimized hybrid convolutional neural networks for detection of plant leaf diseases. *Journal of Ambient Intelligence and Humanized Computing*, *12*, 8625–8636.
101. Chohan, M., Khan, A., Chohan, R., Katpar, S. H., & Mahar, M. S. (2020). Plant disease detection using deep learning. *International Journal of Recent Technology and Engineering*, *9*(1), 909–914.

102. Szegedy, C., Liu, W., Jia, Y., Sermanet, P., Reed, S., Anguelov, D., Erhan, D., Vanhoucke, V., & Rabinovich, A. (7–12 June 2015). Going deeper with convolutions. In *Proceedings of the IEEE Conference on Computer Vision and Pattern Recognition, Boston, MA, USA* (pp. 1–9).
103. Jadhav, S. B., Udupi, V. R., & Patil, S. B. (2020). Identification of plant diseases using convolutional neural networks. *International Journal of Information Technology, 13*, 2461–2470.
104. Arora, J., Agrawal, U., & Sharma, P. (2020). Classification of maize leaf diseases from healthy leaves using deep forest. *Journal of Artificial Intelligence and Systems, 2*, 14–26.
105. Sibiya, M., & Sumbwanyambe, M. (2019). A computational procedure for the recognition and classification of maize leaf diseases out of healthy leaves using convolutional neural networks. *AgriEngineering, 1*, 119–131.
106. Singh, U. P., Chouhan, S. S., Jain, S., & Jain, S. (2019). Multilayer convolution neural network for the classification of mango leaves infected by anthracnose disease. *IEEE Access, 7*, 43721–43729.
107. Türkogˇlu, M., & Hanbay, D. (2019). Plant disease and pest detection using deep learning-based features. *Turkish Journal of Electrical Engineering and Computer Sciences, 27*(3), 1636–1651.
108. Adedoja, A., Owolawi, P. A., & Mapayi, T. (5–6 August 2019). Deep learning based on NASnet for plant disease recognition using leave images. In *Proceedings of the 2019 International Conference on Advances in Big Data, Computing and Data Communication Systems (icABCD), Winterton, South Africa* (pp. 1–5).
109. Li, Y., Nie, J., & Chao, X. (2020). Do we really need deep CNN for plant diseases identification? *Computers and Electronics in Agriculture, 178*, 105803.
110. Zeng, W., & Li, M. (2020). Crop leaf disease recognition based on self-attention convolutional neural network. *Computers and Electronics in Agriculture, 172*, 105341.
111. Liu, T., Chen, W., Wu, W., Sun, C., Guo, W., & Zhu, X. (2016). Detection of aphids in wheat fields using a computer vision technique. *Biosystems Engineering, 141*, 82–93.
112. Barbedo, J. G. A., Koenigkan, L. V., Halfeld-Vieira, B. A., Costa, R. V., Nechet, K. L., Godoy, C. V., Junior, M. L., Patricio, F. R. A., Talamini, V., Chitarra, L. G., et al. (2018). Annotated plant pathology databases for image-based detection and recognition of diseases. *IEEE Latin America Transactions, 16*(6), 1749–1757.
113. Brahimi, M., Arsenovic, M., Laraba, S., Sladojevic, S., Boukhalfa, K., & Moussaoui, A. (2018). Deep learning for plant diseases: Detection and saliency map visualisation. In *Human and Machine Learning* (pp. 93–117). Springer: Berlin, Germany.
114. Hassan, S. M., Amitab, K., Jasinski, M., Leonowicz, Z., Jasinska, E., Novak, T., & Maji, A. K. (2022). A survey on different plant diseases detection using machine learning techniques. *Electronics, 11*(17), 2641.

CHAPTER 3

Machine, Deep, and Reinforcement Learning Algorithms and Applications

K. GOWRI[1] and B. L. SHIVAKUMAR[2]

[1]*Department of Computer Science, Sri Ramakrishna College of Arts & Science, Coimbatore, Tamil Nadu, India*

[2]*Sri Ramakrishna College of Arts & Science, Coimbatore, Tamil Nadu, India*

ABSTRACT

The investigation of technological techniques that allow a machine to acquire knowledge from and adjust to fresh data despite the need for human interaction is known as machine, deep, and reinforcement learning. The world of technology is rife with information in this industry 4.0 era. Examples of this information include cybersecurity, mobile, social media, business, Internet of Things, and health information. The key to developing smart analyses of these data and correspondingly clever and robotic applications is understanding artificial intelligence, and specifically machine learning (ML). There are several dissimilar kinds of ML algorithms in the field, including supervised, unsupervised, semi-supervised, and reinforcement learning. Furthermore, deep learning (DL), a subset of a larger class of ML techniques, is capable of large-scale, intelligent data analysis. Many tools and models are being launched for handling a large amount of data that is generated every day due to the growth of emerging technologies. Reinforcement learning is one part of ML that focuses on the idea of rewards and consequence. The above framework can be made more

Real-Time Artificial Intelligence (AI): Key Motivations, Technologies, Platforms, and Use Cases.
Pethuru Raj, B. Sundaravadivazhagan, V. Kavitha, B. Narendra Kumar Rao, & Hannah Vijaykumar (Eds.)

DOI: 10.1201/9781998511358-3

effective by using a DL strategy. In this study, we provide a systematic understanding of the machine, deep, and reinforcement learning techniques that may be used to improve an application's knowledge and capacities.

3.1 INTRODUCTION

Humans have utilized a wide variety of instruments since the beginning of time to complete a wide range of jobs. The individual's brain's capacity for invention led to the involvement of many devices. By permitting individuals to fulfill a variety of demands, such as computing, journey, industry, and building, these machines made life easier for humans.

The essential distinction among human beings and machines in carrying out their duties has been brought to light by intellect, notwithstanding the machine business's tremendous advancements. Humans employ their senses to take details from their environment, which is then analyzed by the brain to make appropriate decisions. In contrast, machines lack intelligence by design. Machines are not capable of making decisions.

Deep learning (DL) is a method of machine learning (ML) and artificial intelligence (AI) technique that uses certain human brain functions to make judgments with the goal of intimidating people and their behavior. This is a crucial component of data science, channeling forecasting based on methods that use data under statistical and forecasting conditions. In order for an organism to possess the capacity to gain knowledge, adapt, and behave like a human, it must be driven by enormous forces known as programming.

Figure 3.1 demonstrates how DL is a part of both "ML and AI" as a whole. Machine learning, also referred to as ML, automates the method of creating analytical models by learning from data or experience, while AI typically incorporates human intellect and behavior into algorithms or machines [1–2]. Furthermore, data-driven learning methods that compute using multilayer neural networks and processing are referred to as deep learning which alludes to the notion of several stages of computation with the goal of building a framework that is fueled by information.

Therefore, DL can be viewed as one of the fundamental technologies of AI, a frontier in AI that can be functional to the development of automated and intelligent systems. Most importantly, it pushes AI into a new realm called "Smarter_AI." Although DL can learn from data, it is closely related to "Data Analytics" [3]. For sophisticated analytics and prudent decision-making in a particular issue area, DL approaches may be essential. Overall, the method of extracting concepts or meaning through data is referred to as statistical analysis [4, 5]. DL technology has the ability to revolutionize the contemporary world, particularly with reference to a powerful computer engine. In addition, it can help accomplish Industry 4.0 objectives and facilitate technology-driven robotics, AI, and sophisticated technology in particular [6]. Next, the process of learning how to optimize a scalar reward or reinforcement signal by translating conditions to behaviors is known as reinforcement learning [7].

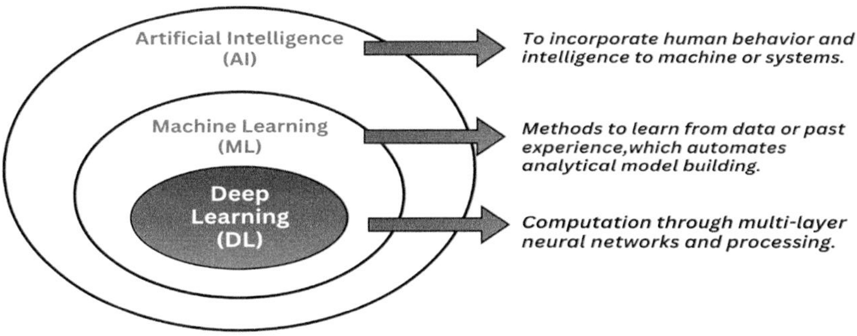

FIGURE 3.1 An example of where deep learning (DL) stands in relation to artificial intelligence (AI) and machine learning (ML).

3.2 INTRODUCTION TO MACHINE LEARNING (ML)

Common phrases like AI and ML have the potential to capture the interest of many scientists for varying lengths of time. Among the most incredible advances in computer engineering is the ability of machines to behave and think like people, which is a remarkable feat. Over time, various computations have been designed to make machines replicate human thought processes, and a wide range of coding languages have been used to carry out those programs [8]. The discipline of software engineering known as "ML" was developed from AI's consideration of design acknowledgment and cognitive theories of learning. ML aims to analyze trends and make recommendations to solve company issues rather than just training a model to foresee the amount based on previously labeled information [9]. People use AI and ML on either an unconscious or a conscious level pretty frequently these days. It is plausible that ML systems are using it in one way or another. Machine learning is a popular application where various algorithms generate understanding (which is essentially the same as what a human being does) of data. Virtual personal assistants (VAs) such as Siri, Alexa, Google, and GPS route administrators are some of the most famous instances of using data in some way from the human mind [10]. AI developments have made online shoppers "cleaver" and have also enhanced their lives. Systems are used by large shops and e-commerce sites, like Amazon, to select phrases that are closest to the online search goods. Therefore, the difficult, lengthy assignments are completed requiring human involvement. Usually, the shopping app or website makes a few suggestions that, in some manner or a different one, fit the user's preferences.

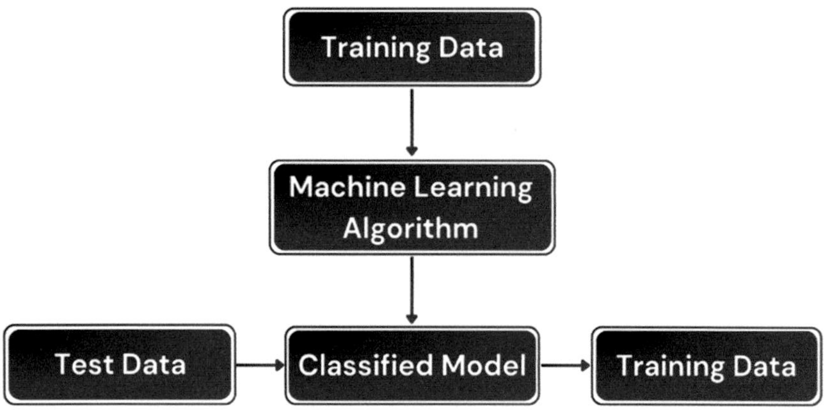

FIGURE 3.2 The overall machine learning method.

3.2.1 *MACHINE LEARNING OVERVIEW*

Arthur Samuel, a pioneer in AI and computer games, contributed to the evolution of machines in 1959 when he coined the term "ML." Prior to that, Turing and Champernowne developed the first computer-based chess game in 1948 by using paper and pencil. Subsequently, Dietrich Prinz unveiled his innovative chess-playing system in 1951. Christopher Strachey developed the first draughts method in 1952. The draught session proceeded at a good clip. Nilsson wrote a book on ML by using pattern categorization in the 1960s. Duda and Hart (1970) provided an explanation of the persistence of interest in pattern classification. In addition, neural networks on computer terminals started learning 40 characters in the year 1981. Neural network scientists Hinton et al. successively demonstrated the multilayered perceptron (MLP) concept with useful back propagation (BP) training in 1985 and 1986. As we get closer to the present, a new phase of neural networks known as DL has emerged. In 2005, researchers Andrew Ng et al. started the third wave of the neural network's development.

3.3 MACHINE LEARNING ALGORITHMS

The goal of ML, a level of AI, is to develop models and algorithms that let machines gain knowledge and experience. To find patterns, anticipate outcomes, and make decisions, ML algorithms use information rather than directly programming them to accomplish tasks. Allowing computers to learn and adapt without having to be specifically trained for every case is the core objective of ML.

Three main categories of ML exist.

3.3.1 SUPERVISED LEARNING

The majority of ML users employ the supervised learning approach. The term "Supervised Learning" has a foundational reason. Since an algorithm uses a set of training data to carry out its learning, it is known as supervised learning. Additionally, the procedure might be compared to a professor supervising the learning process while utilizing the training data set. The system already knows and has cached the right response. The instructor makes the necessary corrections after the algorithm assists in predicting the information used in the training process. The learning process only comes to finish when the algorithm performs to an adequate level.

Issues with supervised learning fall into two categories. Regression and classification issues are other subgroups of these supervised problems.

1. Classification Problems: A classification problem is the one that presents an outcome variable that only belongs to one or more

categories, like "red"/"blue," or it might vary between "disease" and "no disease."

2. Regression: Regression problems arise when an actual value, such as "dollars" or "weight," is the result of a factor.

There are certain issues that you might see in the data type. The common issues arise or are developed upon the classification and regression challenges. Time-series prediction and recommendation are two examples of prevalent challenges.

Some of the highly well-liked supervised ML algorithms are as follows:

1. DT—decision_trees
2. NB—Naïve_Bayes classification
3. SVM—support_vector_machines
4. RF—random_forest
5. LR problems—linear_regression
6. LSR—ordinary least-squares regression
7. LR—logistic regression
8. Ensemble methods

3.3.2 UNSUPERVISED LEARNING ALGORITHM

Unsupervised learning is a kind of ML in which the system learns through the data without the requirement for outside supervision. The technique must act on the unsupervised data without supervision in order for the unsupervised modeling to be trained using the unlabeled dataset, which is neither classified nor grouped. Without a predetermined output, the model searches through the massive volume of knowledge for insightful information in unsupervised learning.

They are employed in the resolution of the association and clustering issues. Thus, it can be further divided into two categories:

1. Clustering: Using this technique, items are grouped into clusters so that those who share the greatest number of parallels stay in that category and share little to none with those in other groups. By identifying patterns among the data objects, cluster analysis groups them into groups based on whether or not such patterns exist.
2. Association: To ascertain the relationships among parameters in an enormous database, a type of unsupervised learning known as

the relationship rule is applied. It determines the set of components that are present in this information set.

3.3.3 REINFORCEMENT LEARNING

Using reinforcement learning, an agent gains decision-making skills through engaging with its surroundings. For every action the agent does, it gets response in a way of incentives or punishments. The agent's objective is to optimize the incremental incentive as time passes by figuring out which behaviors result in desired results.

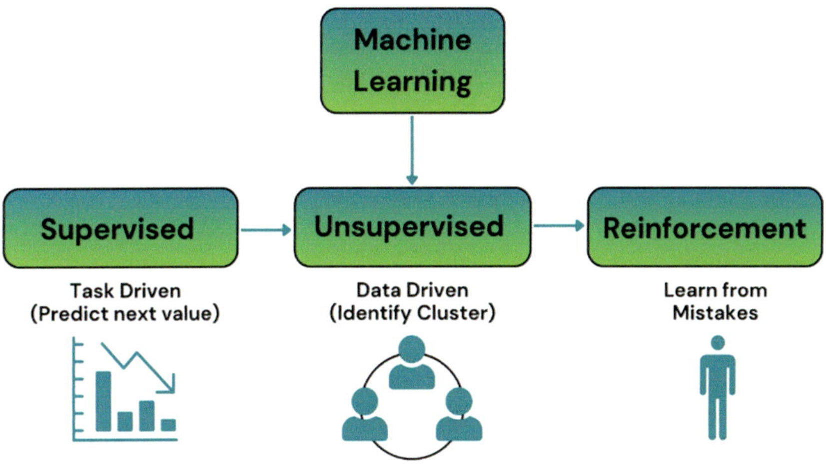

FIGURE 3.3 Three main categories of machine learning (ML).

3.4 APPLICATIONS OF MACHINE LEARNING

1. Image Recognition:

Image recognition is one of the many widely used computer applications of ML. It is used to identify objects, persons, places, digital pictures, and so on. Robotic buddy labeling recommendation is a typical use case for personal identification and recognition of pictures.

We may dynamically tag pals in posts on Facebook. Every time we post a photograph of ourselves with a friend on Facebook, we get a tagged recommendation with their name thanks to an ML facial identification and recognizing algorithm. Its foundation is the Facebook project "Deep Face," which aims to identify people and recognize faces in images.

2. Speech Recognition:

The "Search by voice" option that appears when we use Google falls within the speech recognition category and is a widely used ML application.

3. Traffic Prediction:

Maps by Google are our go-to resource when planning a trip; it not only provides the quickest path but also forecasts congestion.

It uses two approaches to estimate roadway circumstances, such as whether it is pure, moving gradually, or really overfilled:

- The car's location in real time using sensors and the Google Maps app.
- The average time was taken at the same time on previous days.

By using Google Map, all the users contribute to the app's improvement. To increase efficiency, it receives information supplied by the clients and transmits it back to its system.

4. Product Recommendations:

Machine learning is widely used by media and electronic commerce companies such as YouTube, Netflix, and more to provide invention recommendations to their users. As of ML, even though we only employ one browser to access the Internet, we begin to see ads for similar products when we search for anything on Amazon.

Facebook determines customer's preferences using a range of ML techniques, and then makes product recommendations that match those hobbies.

Simultaneously, Amazon uses ML to recommend television series, films, and various other kinds of entertainment.

5. Self-Driving Cars:

Among the many fascinating features of ML is automatic car driving. Machine learning is important to the operation of self-driving cars. Tesla, a well-known automaker, is occupied which creating autonomous cars. The method employed for instructing car models to identify objects and people while navigating is called unsupervised training.

6. Email Spam and Malware Filtering:

All new communications that arrive are directly characterized as junk mail regular, or essential. Machine learning is a technology that allows us to consistently accept important emails in our inbox with significant symbols and unwanted messages in their trash box.

7. Virtual Personal Assistant:

There are many virtual human assistants available to us, including Siri, Alexa, Cortana, and Google Assistant. As their name implies, they assist us in locating the data by means of our voice commands.

8. Online Fraud Detection:

Machine learning is improving the integrity as well as protection of our online buying by detecting unauthorized transactions. Every time we make a purchase through the Internet, there are several ways that an unauthorized purchase could happen, such as using false identities or accounts, as well as taking money while the payment is being made. Therefore, the Feed Through Neural Network helps us by identifying if the interaction is illegitimate or real to recognize this.

Every legitimate transaction's outputs are converted into an assortment of hashing values, which serve for the next round's input. By searching for a specific pattern that varies with each valid transaction, it detects fraud and improves the reliability of our electronic payments.

9. Stock Market Trading:

Investment in the stock market makes extensive use of ML. Machine learning leverages long instantaneous memories in neural network models—which react to changes in stock value—to forecast trends in the financial sector.

10. Medical Diagnosis:

In medicine, AI is cast-off to diagnose diseases. Because of this, health technology is proceeding swiftly and can now provide 3D images that show the exact location of tumors in the head. It facilitates the detection of brain cancers along with additional nervous system problems.

11. Automatic Language Translation:

Today, it is not a big deal if we are moved to an unacquainted location and do not voice the language; ML aids with it as well by translating texts onto languages we are familiar with. This capability, known as automated translation, is obtainable by Google's "Google Neural Machine_Translation" system. It uses neural ML to convert texts into an alternate language that we are acquainted with.

3.5 INTRODUCTION TO DEEP LEARNING

Deep neural networks are a particular type of ML that draw inspiration from the structure and operations of the human brain. Neural networks like these are made up of quite a lot of layers of networked nodes/neurons; this process inputs and acquires models of information at various complexity degrees. Deep learning methods are continuously constructed to traverse a few neural network layers.

3.5.1 DEEP LEARNING OVERVIEW

Machine learning's subclassification of deep neural networks: It is a network model in which the neurons are separated into input and output by layers and have several settings. DL adopts a technique based on neural network topologies, hence the term "deep neural networks." With DL, features can be automatically learned as well as represented at various tiers of a hierarchy. In summary, the entire DL network is utilized for the data selection and modification process, which gives it strength in contrast to conventional ML techniques. The lower layers handle input data in a straightforward way or discover the easy aspects; their result is then sent to the upper layers for learning complex features. Deep learning is therefore appropriate for handling more complicated and large-scale data [11].

The Turing machine built by connecting the neurons was demonstrated by McCulloch and Pitts [12]. Rosenblatt [13] showed in 1958 that if the perceptrons could represent the information they want to learn, they would intersect. The limitation of perceptrons, which is to stop experiments in neurons operating for at least 10 years, was first described by Minsky and Papert in 1969 [14]. After that, the back-propagation technique was introduced by Geoffrey Hinton et al. [15]. Then, in 1988, a hierarchical neural network known as Neocognitron [16] demonstrated proficiency in recognizing visual patterns. Additionally, Yan LeCun examined back propagation using CNN for document analysis in 1998 [17]. The training problem for DNNs was then resolved by the Hinton laboratory in 2006 [18, 19]. From 2012 to 2020, the development of a DL algorithm has accelerated in several domains.

3.6 DEEP LEARNING ALGORITHMS

3.6.1 CONVOLUTIONAL NEURAL NETWORKS (CNNs)

Convolutional neural networks, often referred to as ConvNets, mainly comprise of many layers and are utilized especially for recognizing objects and image processing. Originally known as LeNet, Yann LeCun developed it in 1998. It was planned to recognize numbers and zip code characters back then. CNNs are generally used in series projections, finding of anomalies, medical image processing, and imagery from satellites documentation.

In instruction to display convolutional processes, CNNs processes the information by hitting numerous layers and identifying attributes. Rectified linear units (ReLUs), which serve to fix the feature map, make up the convolutional layer [20]. The map features are corrected and added to the subsequent feedstuff by means of the pooling layer. In general, pooling is a downsampled sampling technique that lowers the feature map's extents. Eventually, the output is a set of 2-D arrays composed of a single long unremitting linear vector that has been flattened in the map. Using information received from the data pooling layer, the next layer, referred to as the fully connected layer, classifies the image and uses it to produce a flattened matrix.

3.6.2 LONG SHORT-TERM MEMORY NETWORKS (LSTMs)

Recurrent neural networks with long-term dependency variation built in are recognized as LSTMs. Its only behavior by definition is to memorize and recall historical facts for a longer duration. Because LSTMs can constrain recollection or preceding inputs, they are mostly used in time sequence predictions due to their ability to retain across time. Their structure is similar to a chain, with four interacting layers that communicate with each other in diverse ways. In Figure 3.5, this is where the analogy originates. In addition to their time sequence forecast applications, they can be utilized in the creation of music loops, speech recognition software, and clinical trials.

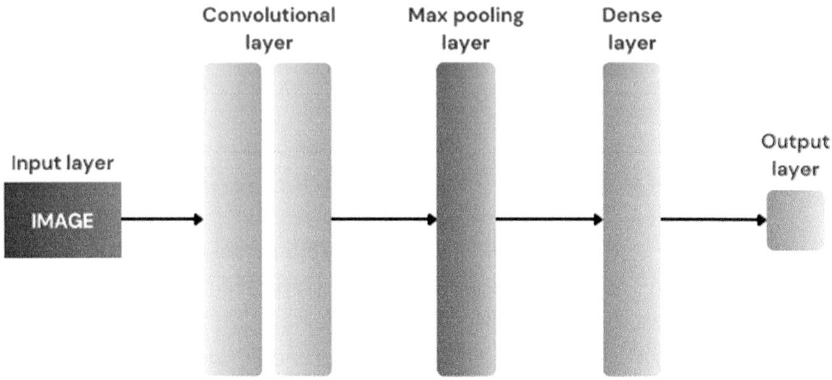

FIGURE 3.4 Structure of convolutional neural network (CNN).

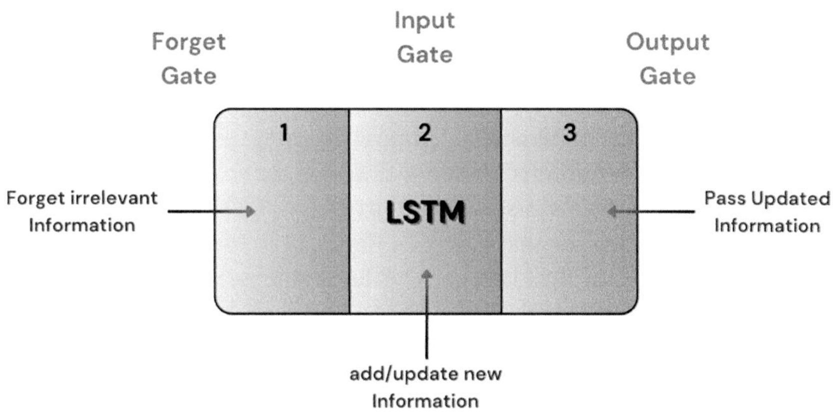

FIGURE 3.5 Structure of long short-term memory network (LSTM).

3.6.3 RECURRENT NEURAL NETWORKS (RNNs)

The data obtained from LSTMs can be employed as input for this stage of recurrent neural networks, or RNNs, which are composed of an ongoing sequence of indirect connectivity. These inputs are firmly ingrained in the system, and the long-term retention capacity of LSTMs allows these inputs to be provisionally stored in the interior memory. Because of this, RNNs operate under the synchronization phenomenon of LSTMs and depend on the inputs retained by LSTMs. RNNs are mainly used in time series analysis, captioning of images, data translation for devices, and written information recognition.

3.6.4 GENERATIVE ADVERSARIAL NETWORKS (GANs)

Deep learning techniques such as GANs are hand-me-down to create new data examples that correspond to the training set. A program that acquires to produce false information and a discriminator that alters by learning from this false data are the two main parts of a GAN. Since they are regularly employed to pretend lensing the gravitational dark substance and clarify celestial images, GANs have become increasingly popular over time. It is also utilized in video games to recreate 2D models in higher determinations, such as 4 K, in order to improve graphics for those textures. They are also utilized in the depiction of human faces, 3D objects, and lifelike cartoon characters.

In modeling, GANs produce and comprehend both fictitious and real data. The generator generates various types of fake data during the training process so that the system for discrimination can swiftly adapt and react to it as false data. These recognized findings are subsequently sent to GANs for informing. To see how it works, look at the graphic here.

3.6.5 RADIAL BASIS FUNCTION NETWORKS (RBFNs)

Neural networks with a feed-forward architecture that use radial functions as activation signals are known as RBFNs. They are primarily utilized for time-series estimation, regression testing, and classification. They are composed of three layers: the (i) input layer, (ii) hidden layer, and (iii) output layer.

By calculating the comparisons in the working out dataset, RBFNs accomplish these objectives. By comparing earlier data sets, they often have an input vector that feeds these data into the input layer, confirmative of the documentation and delivering findings. Specifically, the nodes in the input layer are effective at classifying the type of data because they have neurons that are sensitive to these signals. Although they are closely integrated with the input layer, neurons are initially found in the hidden layer. Gaussian transfer functions that are inversely proportional to the output's distance away from the neuron's center are present in the hidden layer. The radial-based data in the output layer is combined linearly, and output is produced by passing Gaussian functions through the neuron as parameters. Take a look at the provided graphic below to fully comprehend the procedure.

3.6.6 MULTILAYER PERCEPTRONS (MLPs)

The basis of DL technology is the MLP. It is a member of the class of feed_forward neural networks through different perceptron layers. These perceptrons contain a variety of activation mechanisms. The total number of connected input and output layers is the same for MLPs as well. Between these two layers is another layer that is kept hidden. The popular of the time, ML is utilized to make speech and image recognition tools as well as other interpretation software.

Information entering into the input layer is the first step in the MLP's operation. The layer's neurons make a chart to create a link that only drives in a particular way. It is exposed that the input layer and the hidden layer carry the same weight. Activation_purposes are used by MLPs to classify nodes that are ready to fire. ReLUs, sigmoid, and tanh function are some of these activation functions. The primary purpose of MLPs is to train models so that they can recognize the type of correlation that each layer is providing in directive to get the obligatory production from the provided data set. View the graphic below for a clearer understanding.

3.6.7 SELF-ORGANIZING MAPS (SOMs)

Teuvo Kohenen created SOMs in order to use artificial and self-organizing neural networks to see data and comprehend its features. The primary goal of SOMs is display information in order to address difficulties is on things that are not visible to humans. Since these data are typically high-dimensional, there is a lower likelihood of human error and involvement from humans.

SOMs select random vectors from the providing training data after resetting the weights of numerous systems to aid with data visualization. To understand connections, they look at every node to determine its relative weights. The best-matching unit (BMU) is the node that is selected as the winner. These victorious nodes are eventually originated by SOMs, but they progressively dissolve from the model vector. Hence, the greater the node's proximity to BMU, the greater the likelihood that it will notice the weight and perform extra tasks. Moreover, multiple iterations are carried out to guarantee that every node nearer BMU is overlooked.

3.6.8 DEEP BELIEF NETWORKS (DBNs)

Since DBNs include multiple layers of latent as well as stochastic factors they are referred to as models that generate. Since the hidden variable has 0 and 1 values, it is also known as a hidden unit. Because the RGM layers are piled up to create interaction with layers above and below, DBNs are employed in motional object detection and image and video classification applications.

The algorithms are used by Greedy power DBNs. The most popular method for using DBNs is the layer-to-layer strategy, which generates weights by leveraging a top-down method. DBNs employ a step-by-step Gibbs method of sampling on the upper, hidden two-layer. Next, these phases use a model based on the evolutionary method of sampling to extract an amount from the visible units. Using a bottom-up pass methodology, DBNs acquire knowledge from the values found in the hidden value of each layer.

3.6.9 RESTRICTED BOLTZMANN MACHINES (RBMs)

Geoffrey Hinton created RBMs, which are similar to stochastic neural networks in that they pick up knowledge based on the expected range of the data being supplied set. This approach is regarded as one of the fundamentals of DBNs and is mostly utilized in the domains of dimension reduction, regression and classification, and topic modeling. The two elements that make up an RBM are the visible layer and the concealed layer. These two layers are linked together by hidden units, and the nodes that produce the output are coupled to bias units. RBMs typically consist of two phases: the inward pass and the return pass.

RBMs work by taking in inputs and converting them to facts such that the inputs are stored in the frontward pass. Every input is given weight by RBMs, and the rearward pass uses these weights in order to further transform the inputs into recreated inputs. Subsequently, every one of the weights and the two translating inputs are concatenated. After that, those inputs are brought into the visible layer, where initiation takes place and easily reconstructable output is produced. Examine the image below to gain an understanding of this process.

3.6.10 AUTOENCODERS

A single kind of neural network known as an autoencoder finds that its inputs and results are typically the same. Its main goal when designing it was to address problems with unsupervised learning. Neural networks with extensive training that duplicate the information are called autoencoders. The input and output are typically the same because of this. They are employed to do tasks such as image processing, demography prediction, and identifying drugs.

An autoencoder consists of three components: (i) encoder, (ii) code, (iii) decoding unit. The structure of autoencoders allows them to receive signals and translate these into a multitude of conclusions. Replicating a starting point more accurately requires putting it upward. It does this by encrypting the graphic or input and decreasing its size. If the duplicate is not adequately viewed, it is passed to the neural network for confirmation. After that, the photo is known as a recovered image, and it is identical to the initial picture. To have a deeper comprehension of this intricate process, refer to the picture that is provided below.

3.7 APPLICATIONS OF DEEP LEARNING

1. Virtual Assistants:

Virtual assistants are cloud-based software that can understand voice orders in natural language and transmit out operations on the user's behalf. AI assistants like Google Assistant, Cortana, Siri, and Amazon Alexa are common examples. For them to function at their best, devices that are linked to the Internet are required. Based on the previous interaction, the assistant typically uses algorithms that use DL to deliver improved user knowledge each time a facility is agreed to it.

2. *Chatbots:*

Chatbots can quickly resolve client issues. An AI tool for text- or text-to-speech conversations on the internet is called a Chatbot. It can interact with humans and carry out human-like tasks. Chatbots are widely used for client instantaneous messaging, social media marketing, and customer service. It responds to human inputs automatically. It generates many kinds of response by using the DL and ML techniques.

3. *Medical Care:*

Deep learning is being used in the health care business. Deep learning has made it feasible to practice processors to aid with both computer-aided diagnosis and disease identification. Through the process of medical imaging, it is frequently utilized for drug development, medical research, and the identification of serious illnesses like cancer and retinopathy from diabetes.

4. *Amusement:*

To improve the user experience, productions like Netflix, Amazon, YouTube, and Spotify endorse related films, songs, and videos. It is all because of DL. Online streaming firms provide recommendations to customers based on their surfing history, interests, and activity to help them in selecting goods and services. Additionally, DL algorithms are utilized to mechanically produce subtitles and add sound to silent movies.

5. *Fake News Detection and News Aggregation:*

You can tailor news based on reader profiles thanks to DL. News content can be combined and filtered based on reader choices as well as social, regional, and financial variables. Classifiers that can identify and exclude skewed and fraudulent news from your feed are made possible in part by neural networks. They also alert you to potential privacy violations.

6. *Writing Music:*

A machine is capable of learning musical notation, patterns, and structures and can begin creating music on its own. Raw audio can be produced using reproductive copies based on DL, such as WaveNet. The Long Short Term Memory Network enables autonomous music generation. The toolbox for Music21 Python is used in computer-aided musicology. By studying music, creating music samples, and imparting the principles of musical theory, we may train a system to compose music.

7. Drawing Images:

Deep learning has significantly improved image colorization. The process of creating a colored image from a monochrome input is called image colorization. One example of a picture colorization model is ChromaGAN. An aggressive model that integrates a visual and semantic knowledge of colors and class patterns is used to frame a neural network that grows to colorize.

8. Automation:

Robotics that mimic human behavior is a major use of DL. Robots with DL capabilities use immediate information to identify obstacles and instantly plot their route. It can be utilized for managing inventories, product manufacturing, hospitals, factories, and warehouses, among other places.

The company's robots can reload a dishwasher, pick themselves up after falls, and perform other activities in addition to reacting to individuals pushing them over.

9. Picture Description:

Picture tagging is the method of writing an explanation for a picture. It makes use of machine vision to comprehend the subject matter of the image and a language model to translate the comprehension of the image into a suitable phrase order. An LSTM, a type of recurrent neural network, is utilized to combine the descriptions into a logical sentence. With Microsoft's caption bot, you may upload a picture or enter the URL of any image, and it will show the picture's text description. Caption AI is another such program that offers the ideal hashtags and caption for a photo.

10. Advertising:

Deep Learning in Marketing makes it possible to maximize the user encounter. Deep Learning enhances marketing initiatives and enables publishers and marketers make their ads more meaningful. It will allow ad networks to save expenses by bringing down the advertisement's cost per acquire from $60 to $30. Real-time ad auctioning, target display ads, and driven by data prediction advertising are all possible.

11. Natural Language Processing:

Natural language processing, or NLP, is a significant area where DL is demonstrating encouraging outcomes. It is the process that makes it possible for robots to learn and understand human language. But bear in mind that robots find it extremely challenging to comprehend human language. Not only are the alphabet and words a barrier to accurate machine comprehension or creation of human language, but context, accents, handwriting, and other elements also provide a barrier. Deep learning-based natural language processing (NLP) is addressing many of the problems related to human language comprehension by training computers (Autoencoders and Distributed Representation) on how to respond appropriately to language stimuli.

12. Visual Recognition:

Assume for the moment that you are looking at old pictures or memories. Some of these can be printed if you would like. The only way to accomplish this in the absence of metadata was by manual labor. Sorting them by date was your best option; however, sometimes, the downloaded photos did not include that metadata. Conversely, DL has simplified the task. It can be used to categorize images according to recognized locations, faces, a combination of people, events, dates, and other criteria. In order to identify features when looking for a certain picture in a library, sophisticated visual recognition algorithms with different levels of complexity are needed.

13. Fraud Detection:

Large organizations in the financial network industry are already experimenting with DL for fraud protection and detection, which is another appealing use. Predictive data analysis is one tool that PayPal, for instance, utilizes to identify and stop acts of fraud. The company stated that by employing the lengthy short-term recollection architecture of neural networks to analyze user behavior sequences, anomaly identification may be boosted by as much as 10%. Any financial company, health platform, financial app, or organization that collects and handles sensitive data needs to have durable detection of fraud methods. Deep learning has the potential to increase fraud's predictability and thus make it preventable.

14. Personalizations:

These days, every platform is trying to deploy Chatbots to give its users personalized, human-touch experiences. Deep learning is helping e-commerce giants like Amazon, E-Bay, and Alibaba to identify enormous revenue possibilities over the holiday season and to provide seamless, personalized experiences like product recommendations, personalized packaging, and discounts. Reconnaissance is carried out, even in more recent markets, by offering products, services, or strategies that are more likely to appeal to human psychology and support the expansion of micro-markets. Self-service online options are becoming more common, and trustworthy processes are bringing online services that were previously exclusively accessible offline.

15. Detecting Developmental Delay in Children:

Since early detection improves infants's prognoses, it is imperative that developmental abnormalities for kids be diagnosed as soon as possible. In the meantime, motor skill is taken into consideration in the early diagnosis of developmental disabilities since an increasing amount of research points to a connection among developmental delays and motor competency. However, casual surveys or questionnaires sent to parents are usually used to test motor abilities in order to diagnose developmental problems due to a shortage of professionals and time constraints. Deep learning technology is gradually making this possible. A computer system can identify speech, language, and cognition deficits before kindergarten was developed by researchers at the Institute of Health Professions at Massachusetts General Hospital and the Department of Computer Science and Artificial Intellect Laboratory at MIT.

16. Colorization of Photos in Black and White:

Image colorization is the method of taking grayscale photos as input and creating colorized images for output that correspond to the input's semantic colors and tones. This technique was previously performed by hand employing human labor due to the intricate nature of the work. However, in order to color the image in the same way as a human operator would, it is now applied to objects and their context inside the shot utilizing today's DL technology. In supervised layers, premium convolutional neural networks are used to replicate the image with the addition of color.

17. Adding Sounds to Silent Movies:

Audio effects that were not recorded during filming are often added to pictures to give them a more authentic sense. This is known as "Foley." University of Texas researchers employed DL to automate this process. The network of neurons was trained using 12 renowned movie scenes where Foley effects were frequently used by the directors. Their linear network generates the sound, while their neural network recognizes the tone that needs to be produced. To move from time synchronized pictures to sound creation—a whole different medium—they used neural network!

18. Automatic Machine Translation:

In the earlier few years, DL has transformed a number of disciplines. The field of machine translation has moved to using deep-learning neural-based techniques thanks to those developments, replacing more traditional methods like rule-based systems or statistical phrase-based techniques. Because of vast amounts of learning data and unmatched processing capacity, neural MT (NMT) models are able to utilize all information available anywhere in the source phrase and mechanically determine which piece is significant at which stage of synthesizing the production text. The main reason for the notable improvement in translation quality is the removal of earlier independent constraints. As the outcome, the level of distinction between neural and human translators was closed by neural translations.

19. Automatic Handwriting Generation:

One of the uses of DL is producing a fresh set of handwriting for a given corpus of words or phrases. The writing is successfully shown as a set of coordinates that a pen uses to create the samples. Pen movement and letter creation are found to be related, and more examples are created.

20. Automatic Game Playing:

Here, one learns from a corpus of text that has been produced word for word or character for character from fresh text. Spelling, punctuation, and even the style of the text in corpus phrases can be learned with DL algorithms. For text production, large recurrent neural networks are usually used to learn from objects in input word sequence. But, recently, LSTM recurrent neural networks have demonstrated amazing performance in this task by using a syllable-based model that generates just one character at a time.

21. Language Translations:

Technologies are paying close attention to computer translation. Translation quality has significantly increased as a result of this investment and current developments in DL. Google claims that switching to DL improved the precision of translation by 60% over the previous phrase-based method used in Google Translate. Nearly human precision translations into more than 100 languages are now possible in numerous languages thanks to Google and Microsoft.

22. Restoration of Pixels:

Before DL, movies could not be zoomed into above their true resolution. In order to infer a person's face from extremely poor quality pictures, scientists at Google Brain built a DL network in 2017.

23. Demographic and Election Predictions:

Gebru et al. explored the potential of 50 million Google Street View images to see what may be achieved with a DL network. The results were outstanding as usual. The computer gained the ability to identify and locate cars and their specifications. It was capable of recognizing the make, model, body style, and year of almost 22 million cars. Motivated by the success of these DL abilities, the research continued. It was demonstrated that the system could infer each location's demographics just from the composition of the cars there.

24. Deep Dreaming:

A study called DeepDream uses neural network-taught patterns to visualize data. DeepDream exaggerates and misinterprets every pattern it detects in an image, much like a toddler gazing at clouds and trying to make sense of seemingly random shapes.

3.8 INTRODUCTION TO REINFORCEMENT LEARNING

In addition to supervised learning and unsupervised learning, there is one more ML paradigm: reinforcement learning (RL). It makes use of agents that take on the role of human specialists in a given field. RL learns from knowledges by networking with the situation, observing, and reacting to outcomes; it does not require data with labels. As shown in Figure 3.6, RL can be described using the Markov Decision Process. Every situation is branded by a state that embodies the current conditions therein. The RL agent acts in the context, changing its existing state and creating a new one in the process. The agent is rewarded for its efforts. In order to enable the agent assess the completed action in the current situation and gain knowledge from circumstances, the (agent) manager obtains a positive recompense for good actions and a negative reward for poor acts [21].

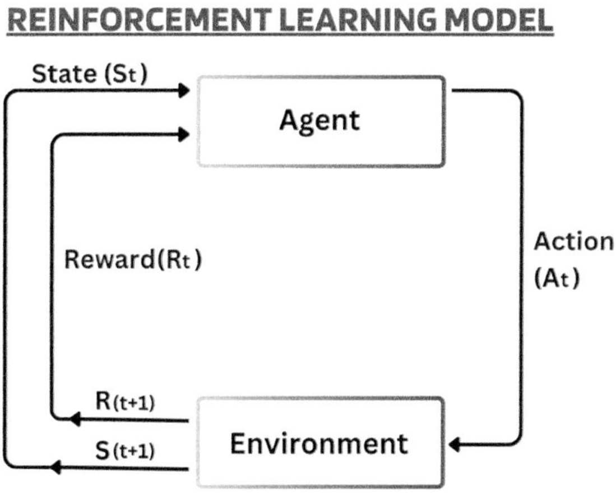

FIGURE 3.6 Reinforcement learning model.

Source: https://www.spiceworks.com/tech/artificial-intelligence/articles/what-is-reinforcement-learning/

Reinforcement learning has been one of the main engines behind recent advances in AI. Reinforcement learning can be defined as teaching a machine to behave as well as practicable by providing it with feedback on how it performed, which suggests determining a course of action that maximizes the anticipated benefit. To educate the agent for its actions, reinforcement learning algorithms use slightly varied techniques.

Nonetheless, the majority of RL algorithms use a comparable structure to choose the best values or policy, combining an on-policy or off-policy approach with the choice of a model-free or model-based method.

When using the techniques, one must grasp the common terminology used in all the RL methods.

1. Action (A): The agent's movements.
2. State (S): The state of the surroundings at the moment.
3. Reward (R): A return that the state sends back to assess the previous action.
4. Policy: The method the agent uses to decide what to do given the current situation.
5. Q-value: Also referred to as the action value, this term describes the agent's estimated quality of action in a given condition.

Remarkably, OpenAI advanced the reinforcement learning approach to AI by introducing reinforcement learning with human feedback (RLHF), which eventually resulted in the creation of ChatGPT. Proximal Policy Optimization (PPO) method, released by OpenAI in 2017, was used as the basis for RLHF storehouse.

3.8.1 OVERVIEW OF REINFORCEMENT LEARNING

The task of creating a controller to minimize an amount of the behavior of a dynamical system over time was first referred to as "optimal control" in the late 1950s. A solution to this issue was created in the middle of the 1950s by Richard Bellman and associates by expanding on a theory of Hamilton and Jacobi from the nineteenth century. By using the ideas of a value function, sometimes known as the "optimal return function," and the state of a dynamical system, this technique defines a practical equation that is now regularly mentioned to as the Bellman equation. Dynamic programming is the name given to the set of methods for resolution ideal controller problems by figuring out this calculation [21]. Markovian decision processes (MDPs) are a discontinuous stochastic form of the best control problem that remained also labeled by Bellman [22]. The policy iteration approach for MDPs was developed by Ron Howard [23]. These are all the fundamental components that support the theories and techniques of contemporary reinforcement learning.

3.9 REINFORCEMENT LEARNING ALGORITHMS

RL algorithm is a model-based algorithm; a predefined reinforcement learning model is present that gains knowledge from the initial state, actions, and state changes brought about by the actions. These kinds so save actions and status data for later use. Conversely, model-free algorithms do not require state and action data to be stored in storage because they work through trial and error techniques.

The following mathematical notations can aid in a better understanding of on-policy and off-policy methods:

The state is represented by the letter "s," the action is represented by the letter "a," and the chance of determining the reward is represented by the symbol "π." The Q(s, a) function understands and learns from states, actions, and state transitions, which is useful for the prediction process and provides future rewards to the agents.

Therefore, off-policy concentrates on learning [Q(s, a)] from random states and actions, whereas on-policy employs the Q(s, a) function to learn from current states and actions.

Furthermore, rather than depending solely on data pertaining to previous states, the Markov decision process stresses current state, which aids in predicting future states. This suggests that present states have a greater influence on the probability of future ones than does the process leading to current ones. Reinforcement learning relies heavily on the Markov property [24].

3.9.1 QLEARNING

First, Q-Learning is an off policy, model-free reinforcement learning algorithm created on the Bellman Equation. The method makes use of a lookup database called a Q-table, which holds the agent's estimated utility, or "quality," of doing a specific action in a particular condition. Through trial and error, the agent adjusts the Q-values in the table to maximize it, and finally, it converges to the best course of action.

Q-Learning does not require any storage capacity in order to store every potential arrangement of actions and states because it is a model-free approach. Moreover, unlike SARSA (explained below), this algorithm can choose the optimal successor action, which is decided by a different policy, because it does not adhere to a single policy. It also makes it possible to include additional factors.

$$Q(s,a) \leftarrow Q(s,a) + \alpha(r + \gamma\, max_{a}'(s',a') - Q(s,a))$$

The benefit of acting in state "s" is currently estimated as $Q(s, a)$.

The learning rate, or α, is a number between 0 and 1 that establishes how much value is given to the incoming information in relation to the current prediction.

The reward obtained in state "s" for action "a" is denoted by r.

The discount factor (γ), which ranges from 0 to 1, establishes the significance of benefits in the future.

Following action "a" in state "s," state "*s*"—next state.

The action chosen in state *s*' is *a*'.

3.9.2 SARSA

The main distinction between SARSA (State Action Reward State Action) and Q-Learning is that the former is an on-policy algorithm; for this reason, it is frequently referred to as "on-policy Q-learning." This suggests that, in contrast to the Q-learning algorithm, which is unconstrained by the next action, this algorithm derives the Q-value from the action carried out by the present rule.

The acronym SARSA represents a flowchart wherein the procedure begins in a state(S), does an action (A), and subsequently generates a reward (R). This modifies the value of the Q-function.

$$Q(s,a) \leftarrow Q(s,a) + \alpha(r + \gamma Q(s',a') — Q(s,a))$$

Those now-obtained Q-values are kept in a database, and the policy selects the one with the greatest values based on the state it is in at the time. This process repeats for the subsequent states. Q-values are continuously modified until an optimal policy is found. However, because it is limited by a single policy, Q-learning provides greater flexibility in terms of value selection.

Both Q-learning and SARSA are tabular techniques that do not scale well for large state and action spaces because of their high memory requirements and inability to visit every state and action during training. Neural networks can help with this.

3.9.3 DEEP Q NETWORK (DQN)

Using neural networks, DQN is an extension of Q-Learning that estimates the value of the Q functions. Because of this, it can overcome Q-learning's limitations, which include its inability to calculate values for states that are not visible. DeepMind even used DQN on the Atari game in 2013.

The Q-learning update equation is used to train the neural network:

$$Q\_\theta(s,a) \leftarrow Q\_\theta(s,a) + \alpha((r + \gamma\, max_a{}' Q_{tar}(s',a')) - Q\_\theta(s,a))$$

Researchers determine the best function after determining the optimal values, or the largest rewards, using the traditional Q-learning technique.

3.10 APPLICATIONS OF REINFORCEMENT LEARNING

1. Video Games:

Algorithms such as AlphaGo and Alpha Zero have achieved superhuman performance in board games like Go and Chess, demonstrating the successful application of reinforcement learning in this domain.

2. Robotics:

Robots may be trained to carry out intricate tasks and figure out the best control strategies in practical settings by using reinforcement learning.

3. Autonomous Systems:

In autonomous devices such as factory automation and unmanned drones, reinforcement learning is used to make decisions.

4. Resource Management:

In situations like energy management and traffic control, where the best possible resource allocation is needed, reinforcement learning is applied.

5. Finance:

To generate successful choices for investments, computerized trading and portfolio optimization use reinforcement learning.

3.11 CONCLUSION

In conclusion, using a variability of algorithms, ML techniques analyze and interpret data, draw conclusions from it, and then use that knowledge to make the best judgments. Deep learning relies on layers of artificial neural networks within the system. Comprehensive analyses of ML and DL are presented, along with the different uses they have. These days, ML is being used by everyone, in either a direct or indirect manner, from getting suggestions for goods when shopping online to updating photos on social media platforms. The evolution, salient characteristics, shared characteristics, and distinctions between reinforcement, DL and ML are also explained.

KEYWORDS

- **DL algorithms**
- **DL applications and reinforcement algorithms**
- **ML algorithms**
- **ML applications**
- **reinforcement learning applications**

REFERENCES

1. Liao, X., Ding, L., & Wang, Y. (2011). Secure machine learning a brief overview. In *2011 Fifth International Conference on Secure Software Integration and Reliability Improvement—Companion* (pp. 26–29).
2. Rosiene, J. A., & Pe Rosiene, C. (2020). SPAM: Simplifying python for approaching machine learning. In *2020 IEEE Frontiers in Education Conference (FIE)* (pp. 1–3).
3. Khalaf, M., et al. (2018). A data science methodology based on machine learning algorithms for flood severity prediction. In *2018 IEEE Congress on Evolutionary Computation (CEC)* (pp. 1–8).
4. Escudero, G., Marquez, L., & Rigau, G. (2000, Sept 22). A comparison between supervised learning algorithms for word sense disambiguation.
5. Sharma, N., Sharma, R., & Jindal, N. (2021, June). Machine learning and deep learning applications—A vision. *2*(1), 24–28.

6. Lei, Z., Shuai, W., & Bing, L. (2017). *Deep learning for sentiment analysis: A survey*. National Science Foundation (NSF), and by Huawei Technologies Co. Ltd.
7. McCulloch, W. S., & Pitts, W. (1943). A logical calculus of the ideas immanent in nervous activity. *Bulletin of Mathematical Biology, 5*, 115–133.
8. Rosenblatt, F. (1958). The perceptron: A probabilistic model for information storage and organization in the brain. *Psychological Review, 65* (6), 386–408.
9. Minsky, M., & Papert, S. A. (2017). *Perceptrons: An introduction to computational geometry*. MIT Press, Cambridge.
10. Ackley, D. H., Hinton, G. E., & Sejnowski, T. J. (1985). A learning algorithm for Boltzmann machines. *Cognitive Science, 9*, 147–169.
11. Fukushima, K. (1988). Neocognitron: A hierarchical neural network capable of visual pattern recognition. *Neural Network, 1*, 119–130.
12. LeCun, Y., Bottou, L., Bengio, Y., & Haffner, P. (1998). Gradient-based learning applied to document recognition. *Proceedings of the IEEE, 86*, 2278–2324.
13. Hinton, G. E., Osindero, S. et al. (2006). A fast learning algorithm for deep belief nets. *Neural Computers, 18*, 1527–1554.
14. Hinton, G. E., & Salakhutdinov, R. R. (2006). Reducing the dimensionality of data with neural networks. *Science, 313*, 504–507.
15. Sarker, I. H. Deep learning: A comprehensive overview on techniques. Taxonomy, Applications and Research Directions.
16. Sarker, I. H. (2021). Data science and analytics: An overview from data-driven smart computing, decision-making and applications perspective. *SN Computer Science*.
17. Sarker, I. H. (2021). Machine learning: Algorithms, real-world applications and research directions. *SN Computer Science, 2* (3), 1–21.
18. Sarker, I. H., & Furhad, M. H. (2021). Nowrozy. *SN Computer Science, 2* (3), 1–18.
19. Sarker, I. H., Hoque, M. M., & Uddin, M. K. (2021). Mobile data science and intelligent apps: Concepts, AI-based modeling and research directions. *Mobile Network Applications 26* (1), 285–303.
20. Sarker, I. H., Kayes, A. S. M., Badsha, S., Alqahtani, H., Watters, P., & Ng, A. (2020). Cybersecurity data science: An overview from machine learning perspective. *Journal of Big Data, 7*(1), 1–29.
21. Sutton, R. S. Introduction: The challenge of reinforcement learning. The Springer International Series in Engineering and Computer Science Book Series, Vol. 173 (SECS).
22. Gupta, G., & Katarya, R. (2021, 13 July). A study of deep reinforcement learning based recommender systems. In *2021 2nd International Conference on Secure Cyber Computing and Communications (ICSCCC)*.
23. Zhang, J., Hao, B., Chen, B., Li, C., Chen, H., & Sun, J. (2019). Hierarchical reinforcement learning for course recommendation in MOOCs. *Proceedings of the AAAI Conference on Artificial Intelligence, 33*, 435–442.
24. Al-Mahamid, F., & Grolinger, K. *Reinforcement learning algorithms: An overview and classification*. IEEE.

CHAPTER 4

AIOps-Powered Entrepreneurship: Harnessing AI for Operational Excellence and Rapid Growth

M. ASHOK KUMAR[1], ALIYU MOHAMMED[2], G. KALAIARASI[3], and B. SUNDARAVADIVAZHAGAN[4]

[1]*Department of Computer Science, Bluecrest University, Monrovia, Liberia*

[2]*Department of Management, Skyline University Nigeria, Kano, Nigeria*

[3]*Department of Commerce, Government of Arts and Science College for Women, Karimangalam, Dharmapuri, Tamil Nadu, India*

[4]*Department of Information Technology, University of Technology and Applied Sciences-Al Mussana, Oman*

ABSTRACT

In the trendy digital world, AIOps (Artificial Intelligence for IT Operations) has become a giant turnaround in business, helping operations to run more smoothly and start growing business quickly. This research explores how entrepreneurial ventures can leverage AI-driven automation and real-time decision making, coupled with predictive analytics. Yet, research on AIOps has closely aligned them to operational management, and on the other hand, research on AI and more broadly on entrepreneurship studies exists, but has covered them in separate areas. By filling this gap, the study analyzes from a theoretical perspective, the historical development,

Real-Time Artificial Intelligence (AI): Key Motivations, Technologies, Platforms, and Use Cases. Pethuru Raj, B. Sundaravadivazhagan, V. Kavitha, B. Narendra Kumar Rao, & Hannah Vijaykumar (Eds.)

DOI: 10.1201/9781998511358-4

and practical usage of AIOps created entrepreneurship. It also identifies major challenges, including the fact that frameworks for AIOps aren't well defined, there is conceptual ambiguity, and there is resistance from entrepreneurs to adopting the concept. This is to reinforce the significance of having an AIOps implementation strategy in place and offers strategic insights into combating the inefficiencies of the operations and the uncertainty of the market. This study bridges the theoretical and the practical gap and provides the entrepreneur with a complete framework for using AIOps for sustainable growth. The research extends our understanding of both academic discourse as well as real-world applications to help business leaders, policymakers, and researchers learn to foster innovation and resilience of the digital economy.

4.1 INTRODUCTION

In the dynamic intersection of artificial intelligence (AI) and operations (Ops), a transformative force has emerged, redefining entrepreneurship: "AIOps-Powered Entrepreneurship: Harnessing AI for Operational Excellence and Rapid Growth." This study explores the historical trajectory and global perspectives surrounding this crux, shedding light on the evolution of AI and Ops within the entrepreneurial landscape. In the globalized marketplace, where digital technologies burgeon, the imperative to integrate AI into operational strategies has become the hallmark of success for forward-thinking enterprises. Companies worldwide, in the pursuit of operational efficiency and competitive advantages, have increasingly turned to AIOps to optimize processes and make data-driven decisions [2]. Historical examinations of literature unveil a gradual shift in focus. Early works centered on standalone concepts of AI and operations. Over time, the integration of these domains garnered attention. However, a comprehensive exploration of AIOps in entrepreneurship remained notably limited in the existing literature [18]. The existing literature provided insights into AI and operational management independently. However, a synthesis of these domains within entrepreneurship was not extensively explored. The existing knowledge base primarily highlighted technological advancements and operational strategies in isolation, leaving a noticeable void in understanding the holistic impact of AIOps on entrepreneurial ventures. Addressing these gaps was imperative for several reasons. Businesses, operating in a competitive landscape, needed to harness the full potential of AI to ensure operational excellence and sustainable growth. Understanding the nuances of AIOps in entrepreneurship was pivotal for informed decision-making, strategic planning, and risk mitigation. Filling these knowledge gaps aimed to provide a comprehensive framework guiding entrepreneurs in navigating the complexities of AIOps adoption. The rationale behind this study lay in the recognition of the transformative potential of AIOps in entrepreneurship and the concurrent lack of comprehensive frameworks guiding its integration. Bridging this gap was critical for scholars, practitioners, and policymakers alike, facilitating the development of informed strategies and policies fostering innovation, efficiency, and growth in entrepreneurial ventures [4]. Recognizing the interdisciplinary nature of this study, it brought together insights from the realms of AI, operations management, and entrepreneurship. An interdisciplinary approach was crucial for crafting holistic and applicable solutions that accounted for the multifaceted challenges and opportunities arising at the intersection of these fields. This study delved into both theoretical issues and practical problems associated with AIOps-powered entrepreneurship. Theoretical issues involved conceptual frameworks, integration models, and the development of a unified theoretical foundation. Practical problems encompassed the challenges faced by entrepreneurs in implementing AIOps, potential barriers to adoption, and real-world examples of successful integration [30]. By addressing both theoretical and practical aspects, this research sought to offer a nuanced understanding that went beyond theoretical speculation. In subsequent sections, we embarked on a journey to explore the theoretical foundations, historical developments, and practical implications of AIOps-powered entrepreneurship. We believed that by doing this, we would significantly add to the body of research already in existence and deepen our understanding of how AI helps startups grow faster and streamline operations [14].

4.1.1 PROBLEM STATEMENT

The dynamic field of AIOps-powered entrepreneurship has a number of important issues and gaps that need to be carefully considered and resolved. First, the absence of well thought out AIOps frameworks indicates a glaring gap in the entrepreneurial space [22]. Due to this disparity, entrepreneurs are unable to fully capitalize on AI's revolutionary potential to enhance operational plans and promote growth. The dearth of specialized frameworks indicates the urgent need for comprehensive regulations that close the gap between AI capabilities and entrepreneurial practices. Second, in today's fast-paced business climate, achieving operational excellence a requirement for long-term success presents entrepreneurs with daunting hurdles [17]. These include the inability to get real-time information, the inability to swiftly adapt to rapidly changing market conditions, and inherent inefficiencies in operational processes. AIOps can help with these issues by facilitating data-driven decision-making and increasing productivity in entrepreneurial strategy. For AIOps to be implemented effectively, however, a detailed grasp of the unique obstacles and bottlenecks that prevent operational excellence must be achieved is necessary.

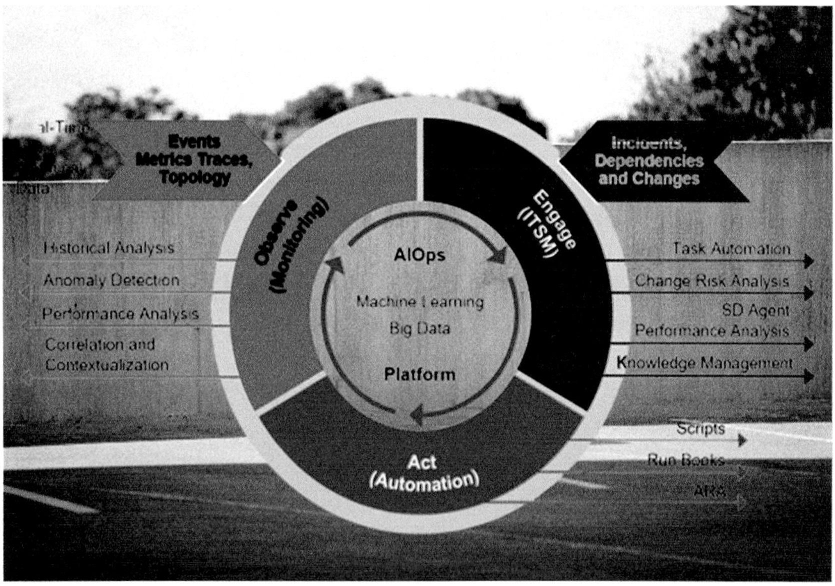

FIGURE 4.1 Artificial intelligence for IT operations (AIOps).
Source: Gartner, ID 378587.

Third, the rapid improvements in AI technologies are not keeping up with the conceptualization and integration of AIOps into entrepreneurial practices [7]. It can be difficult for entrepreneurs to integrate AIOps into their operations because there is a lack of clear direction and conceptual ambiguity. This conceptual and practical divide keeps AIOps from being effectively adopted and keeps entrepreneurs from realizing its disruptive potential. Therefore, it is imperative to have a clear conceptual framework and workable rules that make it easier to integrate AIOps into entrepreneurial endeavors. This research attempts to contribute to theoretical discourse as well as practical frameworks and insights in tackling these complex difficulties. The ultimate objective is to equip business owners with the skills and resources they need to successfully negotiate the challenges of adopting AIOps [18]. To further our understanding of this intersection and promote useful applications that promote operational excellence and quick growth, it is imperative that we do a thorough investigation of the difficulties associated with AIOps-powered entrepreneurship.

4.1.2 RATIONALE FOR A CONCEPTUAL EXPLORATION

The decision to embark on a conceptual exploration of AIOps in entrepreneurship stems from the intrinsic complexity of integrating AI into operational strategies [38]. Given the nascent stage of AIOps frameworks in entrepreneurial literature, a conceptual lens provides a structured and foundational approach to unravel the intricacies and potentialities of this intersection. This exploration seeks to build a theoretical foundation, offering insights into the conceptual underpinnings of AIOps integration and its implications for entrepreneurial practices.

4.1.3 OBJECTIVES OF THE CONCEPTUAL ANALYSIS

The primary objectives of the conceptual analysis are two-fold: first, to distill a clear and comprehensive conceptual framework for the integration of AIOps in entrepreneurship; and second, to elucidate the theoretical foundations that underpin the symbiotic relationship between AI and entrepreneurial processes. This analysis aims to synthesize the existing theoretical perspectives, identify gaps, and proposes a cohesive framework that facilitates the understanding and practical implementation of AIOps in entrepreneurial endeavors [11].

4.1.4 SIGNIFICANCE OF CONCEPTUALIZING AIOPS IN ENTREPRENEURSHIP

Conceptualizing AIOps within the entrepreneurial context holds immense significance for both academia and industry. From an academic standpoint, this conceptual exploration contributes to the advancement of theoretical discourse by bridging the gap between AI and entrepreneurship. It lays the groundwork for future empirical research and establishes a foundation for interdisciplinary scholarship [10]. From a practical perspective, the significance lies in providing entrepreneurs with a conceptual roadmap, guiding them through the integration of AIOps to achieve operational excellence and foster rapid growth. The outcomes of this conceptualization hold the potential to inform strategic decision-making, enhance risk management, and revolutionize entrepreneurial practices in the era of AI-driven innovation.

4.2 THEORETICAL FOUNDATIONS OF AIOPS

In unraveling the theoretical foundations of AIOps within the entrepreneurial sphere, this section delves into key aspects that define and contextualize this dynamic intersection.

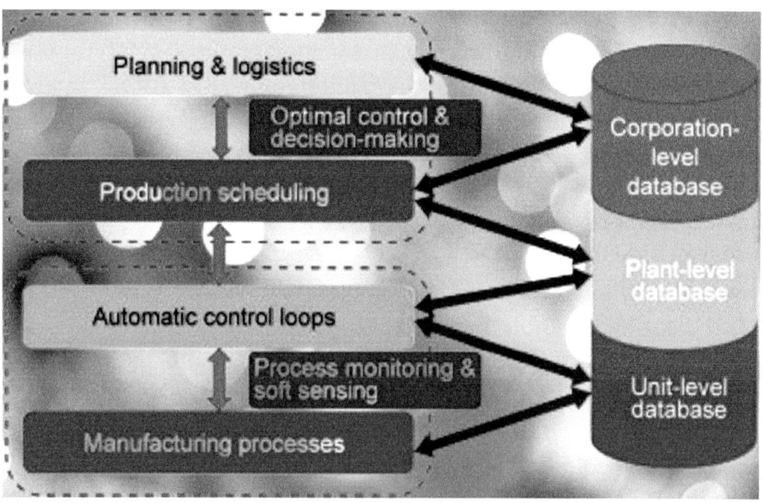

FIGURE 4.2 Data analytics and machine learning.

Source: Adapted from Ref. [40]

4.2.1 DEFINING AIOPS IN THE CONTEXT OF ENTREPRENEURSHIP

To commence, a clear definition of AIOps tailored to entrepreneurship is paramount. AIOps, in this context, refers to the seamless integration of AI into operational processes within entrepreneurial ventures. This definition aligns with the fundamental idea that AI technologies are not only tools for automation but strategic enablers that enhance decision-making, efficiency, and innovation in entrepreneurial operations [1].

4.2.2 THE EVOLUTION OF AIOPS CONCEPTS

Understanding the evolution of AIOps concepts is vital for contextualizing its current state within entrepreneurship. Originally, people may have considered AIOps as just an IT process, but today's applications and implications go all the way to business strategy [29]. In this context, it becomes clear that modern entrepreneurship is being adapted into the data-driven and AI-based formulation.

4.2.3 KEY COMPONENTS OF AIOPS IN ENTREPRENEURSHIP

For comprehensive evaluation of AIOps, it takes a look at its constituents as it relates to business. These include machine learning algorithms that combine with real-time monitoring systems and data analytics. All of these components contribute toward improving decisions, streamlining processes, and innovations in business projects [26]. Thus, a review of some basic parts of AIOps and their relevance to entrepreneurship becomes relevant. The following table lists the necessary components:

TABLE 4.1 Matrix 2: Key Components of AIOps Conceptual Framework.

Framework Components	Data Integration and Analysis	Machine Learning Algorithms	Real-Time Monitoring Systems
Unified AIOps framework	Yes	Yes	Yes
Collaborative AIOps model	Yes	Yes	Yes

Source: Researcher's Review, 2023.

This matrix outlines the key components of the AIOps conceptual framework. It visually indicates which components are present in each of the three AIOps models, providing a quick overview of the shared elements among them.

These matrices can be inserted into your paper as figures or tables, depending on the formatting guidelines of the journal. They offer a concise and structured representation of the concepts discussed in your research, making it easier for readers to grasp key information.

A comprehensive understanding of AIOps necessitates an exploration of its key components in the entrepreneurial context. The following table summarizes the key components:

TABLE 4.2 Key Components of AIOps in Entrepreneurship.

Key Components	Description
Machine Learning Algorithms	Integration of algorithms to facilitate learning, prediction, and decision-making.
Real-time Monitoring Systems	Implementation of systems providing continuous monitoring and analysis for proactive decision-making.

Source: Researcher's Review, 2023.

FIGURE 4.3 Stages of the AIOps framework.

Source: Adapted from https://vitria.com/resources/aiops-its-a-new-day-for-these-companies/

4.2.4 RELATIONSHIP BETWEEN AIOPS AND OPERATIONAL EXCELLENCE

It is quite complicated how AIOps contribute to achieving or rather contributing toward successful implementation of operational excellence.

AIOps works like a catalyst where process is automated and instant insights improve output. To succeed in a venture on the long run, operations excellence is important and requires cooperation of these components [33].

4.2.5 THEORETICAL PERSPECTIVES ON AIOPS-DRIVEN RAPID GROWTH

It is important to understand how AIOps affects entrepreneurial paths in order to understand the theories of AIOps-driven fast growth. This part examines the theories that underscore AIOps' contribution toward innovations, market agility, competitive advantage, and its being a driver of rapid growth. These include resource-based view and innovation diffusion theory which could provide a theoretical framework for analysis about revolutionary impact of AIOps [36].

This section provides definitions, traces the conceptual evolution of AIOps in entrepreneurship, clarifies key components, explores its relationship with operational excellence, and looks at theoretical perspectives on AIOps-driven rapid growth [39]. In summary, it lays the theoretical foundation for understanding AIOps in entrepreneurship. The incorporation of these theoretical underpinnings will guide further examinations and aid in the creation of a strong conceptual framework for entrepreneurship enabled by AIOps.

4.3 INTEGRATION OF AIOPS IN ENTREPRENEURIAL MODELS

The integration of AIOps into entrepreneurial models marks a pivotal exploration in leveraging AI for operational optimization and strategic advancement within the entrepreneurial landscape.

4.3.1 EXPLORING EXISTING ENTREPRENEURIAL MODELS

Entrepreneurial endeavors often draw on established models to guide their strategies. Understanding these models provides a foundation for integrating AIOps effectively.

Lean Startup: This model advocates for iterative development, customer feedback, and validated learning. Incorporating AIOps can enhance this approach by infusing data-driven decision-making into the iterative process, enabling faster learning cycles and improved adaptability [13].

Business Model Canvas: A visual representation detailing key components of a business, the Business Model Canvas serves as a strategic thinking tool. AIOps integration can provide each canvas element with a data-centric viewpoint, assisting entrepreneurs in coordinating their business components with current market dynamics and insights [16].

Effectuation Theory: This theory emphasizes forming alliances, utilizing already-existing resources, and adjusting to unanticipated events. By offering constant monitoring, AIOps integration can improve adaptive capacity and enable entrepreneurs to make data-driven decisions in response to the dynamic entrepreneurial landscape [5].

TABLE 4.3 Exploring Existing Entrepreneurial Models.

Entrepreneurial Model	Key Features	AIOps Integration Notes
Lean Startup	Iterative development, customer feedback, validated learning.	Infuse AIOps for data-driven iteration, faster learning cycles.
Business Model Canvas	Visual representation of key business components.	Incorporate AIOps for real-time insights, aligning with market dynamics.

Source: Researcher's Review, 2023.

4.3.2 CONCEPTUAL FRAMEWORKS FOR AIOPS INTEGRATION

Important elements that promote smooth adoption are included in a conceptual framework for incorporating AIOps into entrepreneurial models.

Data Collection and Analysis: The thorough gathering and examination of pertinent data is the first step in the AIOps integration process. This fundamental stage guarantees informed decision-making processes that are in line with the ever-changing landscape of entrepreneurship.

Machine Learning Integration: Predictive analytics and pattern recognition are made possible for entrepreneurial models by integrating machine learning algorithms. The capacity to foresee market trends and take proactive measures is improved by this integration.

Real-time Monitoring Systems: It is critical to put in place mechanisms for ongoing observation and evaluation. Real-time insights are made possible by AIOps, which helps business owners react quickly to changes and disruptions in their operational environment [23].

TABLE 4.4 Key Components of the AIOps Integration Framework.

AIOps Integration Framework	Key Components	Purpose
Data collection and analysis	Comprehensive data analysis for informed decision-making.	Foundation for data-driven strategies.
Real-time monitoring systems	Implement continuous monitoring and analysis.	Enable real-time insights and proactive decision-making.

Source: Researcher's Review, 2023.

4.3.3 IMPLICATIONS OF AIOPS ON ENTREPRENEURIAL DECISION-MAKING

Entrepreneurial decision-making processes are reshaped by the integration of AIOps, which brings new factors and considerations.

Enhanced Data-Driven Decisions: AIOps gives entrepreneurs access to real-time insights that support well-informed decisions based on extensive data analytics. This guarantees that choices are in line with the state of the market and new trends.

Improved Forecasting Accuracy: The incorporation of machine learning improves the precision of prediction models. More accurate forecasts will enable entrepreneurs to more confidently foresee changes in the market and adjust their strategies.

Increased Operational Efficiency: AIOps reduces decision-making latency and boosts overall efficiency by streamlining operational processes. Enhanced resource allocation and optimized workflows are advantageous for entrepreneurs [6]. This section clarifies the steps involved in integrating AIOps into entrepreneurial models, including understanding the implications for decision-making, conceptualizing integration strategies, and investigating current frameworks. Adopting AIOps can greatly enhance entrepreneurial activities and provide strategic advantages in a business environment that is becoming more and more dynamic.

4.4 CHALLENGES AND OPPORTUNITIES IN CONCEPTUALIZING AIOPS-POWERED ENTREPRENEURSHIP

Conceptualizing AIOps within the entrepreneurial domain presents a nuanced landscape that encompasses both challenges and promising opportunities for innovation. This section delves into the intricacies of identifying challenges, exploring avenues for innovation, and addressing ethical considerations in the context of AIOps-driven entrepreneurship.

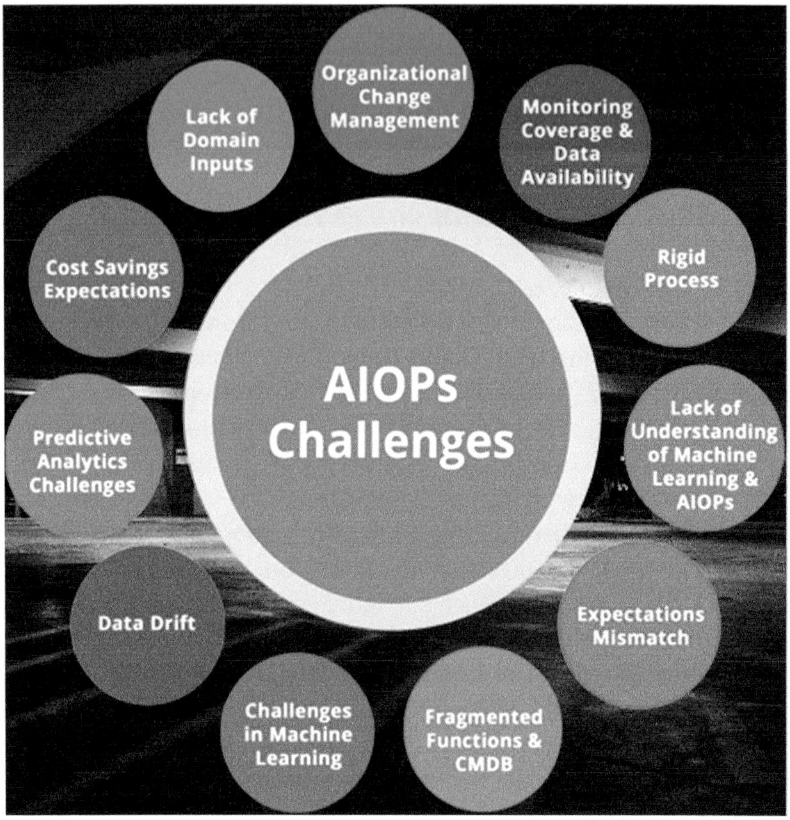

FIGURE 4.4 AIOPs challenges.
Source: Adapted from Ref. [31]

4.4.1 IDENTIFYING CHALLENGES IN AIOPS CONCEPTUALIZATION

4.4.1.1 LACK OF STANDARDIZED FRAMEWORKS

The lack of uniform frameworks regarding incorporating AIOps in the entrepreneurship approach constitutes a key problem [30]. Nevertheless, most frameworks are customized for particular sectors meaning that entrepreneurs need to adjust them so that they fit perfectly into their local settings. Because of this, it inhibits ease in which AIOps can be adapted, a critique is called for on some conception theories used.

4.4.1.2 COMPLEXITY IN IMPLEMENTATION

Integration of AIOps is by essence difficult because it merges various technological devices, including machine learning algorithms and real-time monitoring systems [8]. This complexity also poses difficulties for entrepreneurs who must ensure that their undertakings are compatible with current operational procedures and infrastructure. Properly implementing AIOps is not simple because one false move could hinder efficiency gains associated with it.

4.4.1.3 RESISTANCE TO CHANGE

However, organizational resistance to change remains a common obstacle in AIOps conception. The adoption of AI technologies may generate resistance by employees and stakeholders due to job loss concerns or unawareness of the advantages associated with these tools. Addressing this challenge involves strategic communication, training programs, and fostering a culture that embraces the transformative potential of AIOps.

4.4.2 OPPORTUNITIES FOR INNOVATION AND IMPROVEMENT

4.4.2.1 CUSTOMIZED AIOPS FRAMEWORKS

An opportunity lies in the development of customized AIOps frameworks tailored to the unique needs of entrepreneurial ventures [21]. By crafting frameworks that consider the specific challenges and operational contexts of startups and small enterprises, entrepreneurs can enhance the feasibility and effectiveness of AIOps integration. This customization fosters innovation in conceptualization methods and ensures a more seamless adoption process.

4.4.2.2 INTEGRATION WITH EMERGING TECHNOLOGIES

Opportunities for innovation emerge through the integration of AIOps with other emerging technologies such as blockchain and edge computing [27]. Collaborative integration opens avenues for enhanced security, decentralized decision-making, and improved real-time analytics. Entrepreneurs can capitalize on these synergies to create novel AIOps-powered entrepreneurial models that are at the forefront of technological advancement.

4.4.2.3 COLLABORATION AND KNOWLEDGE SHARING

Collaboration within the entrepreneurial community and knowledge sharing play pivotal roles in overcoming challenges and fostering innovation in AIOps conceptualization [7]. Entrepreneurial ecosystems can create platforms for sharing best practices, case studies, and success stories related to AIOps adoption. Collaborative efforts enhance the collective knowledge base, enabling entrepreneurs to navigate challenges more effectively.

4.4.3 ADDRESSING ETHICAL CONSIDERATIONS IN AIOPS-DRIVEN ENTREPRENEURSHIP

4.4.3.1 TRANSPARENCY AND EXPLAINABILITY

Ethical considerations in AIOps-driven entrepreneurship necessitate a commitment to transparency and explainability [31]. Entrepreneurs must prioritize developing AIOps systems that can articulate their decision-making processes in a comprehensible manner. This not only builds trust among stakeholders but also aligns with ethical principles of accountability.

4.4.3.2 FAIRNESS AND BIAS MITIGATION

Ensuring fairness and mitigating biases in AIOps algorithms is imperative. Entrepreneurs must actively address the potential biases embedded in training data and algorithms to prevent discriminatory outcomes [34].

Implementing fairness-aware machine learning techniques and regularly auditing AIOps models contribute to ethical practices in entrepreneurship.

4.4.3.3 DATA PRIVACY AND SECURITY

AIOps often involves the processing of vast amounts of sensitive data. Entrepreneurs must prioritize data privacy and security, implementing robust measures to safeguard against unauthorized access and potential breaches [35]. Respecting user privacy and complying with data protection regulations are fundamental ethical considerations in AIOps-driven entrepreneurship.

In conclusion, navigating the conceptualization of AIOps in entrepreneurship involves addressing challenges related to standardized frameworks, implementation complexity, and resistance to change. Simultaneously, entrepreneurs have the opportunity to innovate through customized frameworks, integration with emerging technologies, and collaborative knowledge-sharing efforts. Ethical considerations, including transparency, fairness, and data privacy, are integral components of responsible AIOps-driven entrepreneurship, ensuring alignment with ethical principles and societal values.

4.5 THE CONCEPTUAL FRAMEWORK FOR AIOPS-POWERED ENTREPRENEURSHIP

The conceptual framework for AIOps-powered entrepreneurship serves as a blueprint for understanding the dynamic interplay of components, interactions, and expected outcomes within the integration of AI into operational practices.

4.5.1 COMPONENTS OF THE FRAMEWORK

4.5.1.1 DATA INTEGRATION AND ANALYSIS

The foundation of the framework lies in robust data integration and analysis. This involves collecting diverse data sets relevant to entrepreneurial processes and applying advanced analytics to derive actionable insights [9].

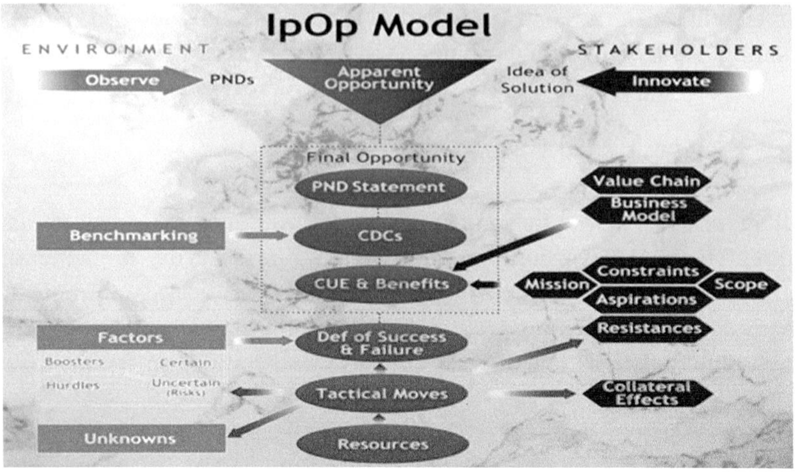

FIGURE 4.5 Entrepreneurship and start-up.

Source: Adapted from https://www.startup-book.com/2012/09/25/proven-tools-for-converting-your-projects-into-success-without-a-business-plan/

4.5.1.2 MACHINE LEARNING ALGORITHMS

Incorporating machine learning algorithms is pivotal for predictive analytics, pattern recognition, and decision support. These algorithms continuously learn from data patterns, enhancing their accuracy and effectiveness over time [3].

4.5.1.3 REAL-TIME MONITORING SYSTEMS

Real-time monitoring systems provide continuous surveillance of entrepreneurial operations. These systems facilitate proactive decision-making by offering insights into ongoing processes and potential areas for improvement [20].

4.5.2 INTERACTIONS AND RELATIONSHIPS

4.5.2.1 DATA–ALGORITHM INTERACTION

The interaction between data and machine learning algorithms is dynamic and iterative. The algorithms analyze the integrated data, generating patterns and insights that, in turn, inform further data collection and analysis [32].

4.5.2.2 HUMAN–AIOPS COLLABORATION

The framework emphasizes collaboration between human expertise and AIOps capabilities. Entrepreneurs and stakeholders work in tandem with AIOps systems, combining domain knowledge with AI-driven insights for more informed decision-making [28].

4.5.2.3 FEEDBACK LOOPS

Continuous feedback loops are integral for adaptability. Feedback from outcomes and decision effectiveness informs adjustments in data collection strategies, algorithm refinement, and overall framework enhancement [14].

4.5.3 EXPECTED OUTCOMES AND BENEFITS

4.5.3.1 ENHANCED DECISION-MAKING

The primary expected outcome is the enhancement of decision-making processes. AIOps empowers entrepreneurs with real-time, data-driven insights, leading to more informed and strategic decision-making [24].

4.5.3.2 OPERATIONAL EFFICIENCY

AIOps integration aims at optimizing operational efficiency by streamlining processes and reducing decision-making latency. This results in improved resource allocation and workflow optimization [37].

4.5.3.3 INNOVATION AND ADAPTATION

The framework anticipates fostering a culture of innovation and adaptation. AIOps-powered entrepreneurship encourages experimentation, agility, and the ability to adapt to rapidly changing market conditions [25].

4.5.4 ADAPTABILITY AND SCALABILITY

4.5.4.1 CONTINUOUS IMPROVEMENT MECHANISMS

Adaptability is ensured through continuous improvement mechanisms embedded within the framework. Regular assessments, feedback loops, and iterative adjustments contribute to the ongoing evolution of AIOps integration [19].

4.5.4.2 SCALABILITY FOR GROWTH

The framework is designed to be scalable, accommodating the growth trajectories of entrepreneurial ventures. As enterprises expand, the AIOps integration scales seamlessly to meet the evolving needs and complexities [29].

In summary, the conceptual framework for AIOps-powered entrepreneurship encompasses key components such as data integration, machine learning algorithms, and real-time monitoring systems. The interactions and relationships within the framework highlight the dynamic collaboration between data, algorithms, and human expertise. Expected outcomes include enhanced decision-making, operational efficiency, and a culture of innovation. The framework is designed for adaptability, incorporating continuous improvement mechanisms and scalability for sustained growth in entrepreneurial endeavors.

4.6 COMPARATIVE ANALYSIS OF CONCEPTUAL MODELS

The comparative analysis critically assesses three existing conceptual models in AIOps, considering their relevance and efficacy within the nuanced context of AIOps-powered entrepreneurship.

4.6.1 REVIEW OF EXISTING CONCEPTUAL MODELS IN AIOPS

4.6.1.1 MODEL A: THE UNIFIED AIOPS FRAMEWORK

Overview: Model A advocates for a unified approach, integrating machine learning, data analytics, and real-time monitoring within a cohesive framework [15].

Strengths:

Comprehensive Integration: Model A offers a holistic integration of AIOps components, aligning with the need for a unified and synergistic approach in entrepreneurial settings.

Operational Streamlining: The model promotes streamlined operations by consolidating diverse AIOps functionalities into a unified structure.

Weaknesses:

Potential Complexity: The comprehensive nature of the framework may introduce complexity, requiring careful consideration of its applicability to diverse entrepreneurial ventures.

4.6.1.2 MODEL B: THE ADAPTIVE AIOPS ECOSYSTEM

Overview: Model B emphasizes adaptability, creating an ecosystem that facilitates continuous learning and evolution of AIOps capabilities [27].

Strengths:

Continuous Learning: The adaptive ecosystem allows for continuous learning, essential for entrepreneurial ventures operating in dynamic environments.

Flexibility: Model B provides flexibility, accommodating changes in entrepreneurial strategies and adapting to evolving business landscapes.

Weaknesses:

Complex Implementation: The adaptive nature may introduce complexity in implementation, requiring a strategic approach to balance adaptability and operational efficiency.

4.6.1.3 MODEL C: THE COLLABORATIVE AIOPS MODEL

Overview: Model C underscores collaboration between AI systems and human expertise, recognizing the complementary nature of human intuition in decision-making [12].

Strengths:

Human-Centric Approach: The model incorporates a human-centric approach, acknowledging the unique strengths of human decision-makers alongside AIOps capabilities.

Collaborative Decision-Making: Encourages collaboration, leveraging the expertise of both humans and AI systems for more informed and nuanced decisions.

Weaknesses:

Potential Resistance: The collaborative nature may face resistance from stakeholders not accustomed to shared decision-making responsibilities.

4.6.2 COMPARATIVE EVALUATION OF AIOPS-POWERED ENTREPRENEURSHIP MODELS

4.6.2.1 INTEGRATION WITH ENTREPRENEURIAL PRACTICES

Model A: Demonstrates strong integration potential, aligning well with entrepreneurial processes and offering a unified framework.

Model B: Excels in adaptability, providing flexibility for seamless integration with evolving entrepreneurial practices.

Model C: Highlights collaboration, fostering synergy between AIOps and human-centric entrepreneurial strategies.

4.6.2.2 SCALABILITY AND FLEXIBILITY

Model A: Shows scalability through a diverse range of AIOps components, offering flexibility for customization.

Model B: Excels in scalability with an adaptive ecosystem, ensuring flexibility for growth and adaptation.

Model C: Prioritizes flexibility through collaborative approaches, allowing adaptability to varying entrepreneurial scenarios.

4.6.2.3 ETHICAL CONSIDERATIONS

Model A: Addresses ethical considerations through transparency and explainability in its unified framework.

Model B: Incorporates ethical considerations with continuous improvement mechanisms and fairness-aware algorithms.

Model C: Acknowledges ethical dimensions through its collaborative approach, promoting human-centric decision-making and ethical AI practices.

4.6.2.4 INNOVATION AND LEARNING

Model A: Demonstrates innovation through a unified structure, promoting continuous learning via machine learning integration.

Model B: Excels in innovation with an adaptive ecosystem, fostering a culture of experimentation and learning.

Model C: Emphasizes innovation through collaboration, harnessing the creative potential of human input alongside AIOps capabilities.

4.6.3 CONCLUSION

Each model presents distinct strengths and considerations. Model A offers a comprehensive and unified approach, Model B excels in adaptability and scalability, and Model C prioritizes collaboration and ethical dimensions. The choice among these models depends on the specific needs, scale, and ethical priorities of the entrepreneurial venture. The comparative analysis provides a nuanced understanding to guide entrepreneurs in selecting the most suitable AIOps-powered model for their operational framework.

TABLE 4.5 Matrix 1: Comparative Analysis of AIOps Models.

Model Components	Unified AIOps Framework	Adaptive AIOps Ecosystem	Collaborative AIOps Model
Integration with entrepreneurial practices	High	Moderate	High
Scalability and flexibility	High	High	Moderate

Source: Researcher's Review, 2023.

This matrix provides a representation of the comparative analysis of the three AIOps models, highlighting their strengths and characteristics in key areas such as integration with entrepreneurial practices, scalability, ethical considerations, and innovation.

4.7 DISCUSSION

4.7.1 SYNTHESIZING KEY CONCEPTS

Integration of AIOps: The conceptual framework established in this study revolves around the seamless integration of data analytics, machine learning, and real-time monitoring into the fabric of entrepreneurship. The interconnectedness of these components is designed to enhance operational efficiency, inform decision-making, and foster a culture of innovation.

Challenges and Opportunities: Identifying challenges in AIOps conceptualization, such as the absence of standardized frameworks and resistance to change, provides a foundation for innovation. Opportunities emerge through the customization of frameworks, integration with emerging technologies, and collaborative efforts within the entrepreneurial community.

Ethical Considerations: The discussion underscores the ethical dimensions of AIOps-driven entrepreneurship, emphasizing transparency, fairness, and data privacy. Addressing these ethical considerations is not only essential for responsible AI practices but also for building trust with stakeholders.

Conceptual Models: The comparative analysis of the existing conceptual models (Unified AIOps Framework, Adaptive AIOps Ecosystem, and Collaborative AIOps Model) illuminates their unique strengths. The choice among these models depends on factors such as integration with entrepreneurial practices, scalability, ethical considerations, and a focus on innovation.

4.7.2 IMPLICATIONS FOR FUTURE RESEARCH

Framework Refinement: Future research endeavors could focus on refining the conceptual framework through empirical testing and real-world applications. This involves assessing the practicality and effectiveness of the framework in various entrepreneurial contexts and identifying areas for refinement.

Dynamic Adaptability: Investigating the dynamic adaptability of AIOps-powered entrepreneurship models is a crucial avenue for future research. Understanding how these models respond to evolving business landscapes and emerging technologies will contribute to their sustained relevance and effectiveness.

User Acceptance: Exploring user acceptance and mitigating resistance to change represents a critical area for future research. Understanding the factors influencing the acceptance of AIOps technologies among entrepreneurs and stakeholders can inform strategies for successful implementation.

Long-Term Impact: Assessing the long-term impact of AIOps on entrepreneurship is an important area for exploration. Research can delve into sustained benefits, challenges faced over time, and the evolution of AIOps adoption in entrepreneurial practices.

4.7.3 THEORETICAL CONTRIBUTIONS TO THE FIELD

Unified AIOps Framework: The development of a unified AIOps framework contributes theoretically by providing a structured model for integrating AI technologies into entrepreneurship. This framework serves as a theoretical cornerstone, offering insights into the interconnectedness of AIOps components.

Adaptive Ecosystem Theory: It contributes to scientific reasoning by incorporating continuous learning and change into the notion of what AIOps means. For instance, this is a dynamic method which helps in adapting to changing entrepreneurial environments.

Collaborative Decision-Making: The theoretic discussion acknowledges that these two entities do not work in silicates rather they thrive on reciprocating efforts thereby forming a collaborated AIOps model. This framework recognizes that human knowledge is more essential than AI which enables teamwork decision-making.

Finally, in the Discussion section, the main concepts are summarized, it is outlined how future studies can be conducted based on this framework, and, finally, some theoretical perspectives are pointed out. This article serves as a point of reference for upcoming studies on the subject, as well as an addition to the already existing theory in this field.

4.8 CONCLUSION

4.8.1 RECAPITULATION OF KEY CONCEPTUAL INSIGHTS

Therefore, this study delves into entrepreneurship and AIOps concepts outlining a theoretical framework and comparison with other models. The key insights are as follows:

Integrated AIOps Framework: Incorporating these elements into a unified theoretical framework for entrepreneurial AIOps entails combining data analytics, machine learning, and continuous monitoring. Such integration is aimed at making processes more efficient, facilitating better decisions, and supporting innovation.

Challenges and Opportunities: Opportunities for innovation emerge from identified challenges (absence for standardized framework and resistant to change). These challenges can be addressed through customization of frameworks, collaboration, and integrating with upcoming technologies.

Ethical Considerations: Ethical considerations of the AIOps as an entrepreneurial endeavor are emphasized in the study, underlining the need for openness, equity, and protection of customer data. This is vital in responsible AI practices and in building trust with stakeholders.

Conceptual Models: The comparative analysis compares the various conceptual models, including Unified AIOps Framework, Adaptive AIOps Ecosystem, and Collaborative AIOps Model, by highlighting their strengths and feasibility in the entrepreneurial context.

4.8.2 PRACTICAL IMPLICATIONS FOR ENTREPRENEURS

Strategic Integration: The discussed idea could be helpful when entrepreneurs incorporate AIOps into their business strategy. It entails appreciating that data analytics, machine learning, and real-time monitoring can be used together for effective operational management.

Innovation and Customization: The study highlights opportunities for innovation through the customization of AIOps frameworks. Entrepreneurs are encouraged to tailor these frameworks to their specific operational contexts, fostering a culture of innovation within their ventures.

Ethical Guidelines: Practical implications include incorporating ethical guidelines in AIOps implementation. Entrepreneurs should prioritize transparency, fairness, and data privacy to ensure responsible and ethical AI practices, thereby gaining trust from stakeholders.

4.8.3 RECOMMENDATIONS FOR FUTURE CONCEPTUAL AND EMPIRICAL RESEARCH

Refinement of Frameworks: Future research should focus on the empirical testing and refinement of the conceptual framework. Assessing its practicality in diverse entrepreneurial contexts and identifying areas for improvement will contribute to its ongoing development.

User Acceptance Studies: Further studies on user acceptance and strategies to mitigate resistance to change are recommended. Understanding the factors influencing the acceptance of AIOps technologies among entrepreneurs and stakeholders will inform successful implementation.

Long-Term Impact Assessment: Research exploring the long-term impact of AIOps on entrepreneurship is essential. Examining sustained benefits, challenges faced over time, and the evolution of AIOps adoption in entrepreneurial practices will provide valuable insights.

Dynamic Adaptability: Investigations into the dynamic adaptability of AIOps-powered entrepreneurship models are crucial. Understanding how these models respond to evolving business landscapes and emerging technologies will contribute to their continued relevance.

In conclusion, this study offers practical implications for entrepreneurs looking to embrace AIOps in their ventures. Recommendations for future research focus on refining frameworks, understanding user acceptance, assessing long-term impacts, and exploring the dynamic adaptability of AIOps models. This collective effort aims to contribute to the ongoing advancement of AIOps in the entrepreneurial domain.

KEYWORDS

- **artificial intelligence**
- **AIOPs**
- **conceptual models**
- **entrepreneurship**
- **machine learning**

REFERENCES

1. Allioui, H., & Mourdi, Y. (2023). Unleashing the potential of AI: Investigating cutting-edge technologies that are transforming businesses. *International Journal of Computer Engineering and Data Science (IJCEDS), 3*(2), 1–12.
2. Atwal, H. (2019). *Practical DataOps: Delivering agile data science at scale*. Apress.

3. Bharadiya, J. P. (2023). Machine learning and AI in business intelligence: Trends and opportunities. *International Journal of Computer (IJC), 48*(1), 123–134.
4. Cao, Z., & Shi, X. (2021). A systematic literature review of entrepreneurial ecosystems in advanced and emerging economies. *Small Business Economics, 57*, 75–110.
5. Choudhuri, B., Srivastava, P. R., Gupta, S., Kumar, A., & Bag, S. (2021). Determinants of smart digital infrastructure diffusion for urban public services. *Journal of Global Information Management (JGIM), 29*(6), 1–27.
6. Dar, A. R., & Ravindran, D. (2019). Fog computing resource optimization: A review on current scenarios and resource management. *Baghdad Science Journal, 16*(2).
7. Das, S. D., & Bala, P. K. (2023). What drives MLOps adoption? An analysis using the TOE framework. *Journal of Decision Systems,* 1–37.
8. Deutsch, D. (2023). Machine learning operations–domain analysis, reference architecture, and example implementation.
9. Duggineni, S. (2023). Data analytics in modern business intelligence. *Journal of Marketing & Supply Chain Management, 114,* 2–4. DOI: 10.47363/JMSCM/2023.
10. Dupin, N. (2023). Interdisciplinarity as a political instrument of governance and its consequences for doctoral training.
11. El Namaki, M. S. S. (2023). *Neo strategic management: Conceptual and operational foundations of tomorrow's strategic thinking*. Springer Nature.
12. Eskandarivatannezhad, M. (2023). *Human-AI symbiosis: The best approach for AI implementation in business decision-making in complex systems* (Master's thesis, UIS).
13. Ganesan, K. (2022). *The business case for AI: A leader's guide to AI strategies, best practices & real-world applications*. Opinosis Analytics.
14. George, B., & Wooden, O. (2023). Managing the strategic transformation of higher education through artificial intelligence. *Administrative Sciences, 13*(9), 196.
15. Hadyaoui, A., & Cheniti-Belcadhi, L. (2023). Ontology-based group assessment analytics framework for performances prediction in project-based collaborative learning. *Smart Learning Environments, 10*(1), 43.
16. He, F., Shen, L., & Peng, X. (2021). Resource choreography in cyber-physical-social systems: Representation, modeling and execution. *IEEE Transactions on Services Computing.*
17. Holbeche, L. (2023). *The agile organization: How to build an engaged, innovative and resilient business*. Kogan Page Publishers.
18. Hua, Y. (2021). A systems approach to effective AIOps implementation (Doctoral dissertation, Massachusetts Institute of Technology).
19. Idowu, S. O. (2023). Towards next-gen machine learning asset management tools.
20. Kayode-Ajala, O. (2023). Applications of Cyber Threat Intelligence (CTI) in financial institutions and challenges in its adoption. *Applied Research in Artificial Intelligence and Cloud Computing, 6*(8), 1–21.
21. Khan, A. (2023). Harnessing the power of AI: A review of advancements in healthcare. *BULLET: Jurnal Multidisiplin Ilmu, 2*(3), 546–556.
22. Kreutzer, R. T. (2021). Toolbox für digital business. *Leadership, Geschäftsmodelle, Technologien.*
23. L'Esteve, R. C. (2023). Impacts of modern AI and ML trends. In *The cloud leader's handbook: Strategically innovate, transform, and scale organizations* (pp. 135–155). Apress.

24. Mahendra, A. (2023). Fundamental of AI startups. In *AI startup strategy: A blueprint to building successful artificial intelligence products from inception to exit* (pp. 1–22). Apress.
25. Onkamo, M., & Rahman, S. T. (2023). Artificial intelligence for it operations.
26. Patel, A. K. (2023). Technological innovations in solar heater materials and manufacturing. *United International Journal for Research & Technology, 4*(11).
27. Pendy, B. (2023). From traditional to tech-infused: The evolution of education. *BULLET: Jurnal Multidisiplin Ilmu, 2*(3), 767–777.
28. Pinheiro, H. (2021). *How to implement AI-driven businesses in communication service providers (CSPs)* (Doctoral dissertation, Universidade Católica Portuguesa).
29. Ramu, V. B., & Yeruva, A. R. (2023). Optimising AIOps system performance for e-commerce and online retail businesses with the ACF model. *International Journal of Intellectual Property Management, 13*(3–4), 412–429.
30. Ravindran, A. A. (2023). Internet-of-things edge computing systems for streaming video analytics: Trails behind and the paths ahead. *IoT, 4*(4), 486–513.
31. Sabharwal, N., & Bhardwaj, G. (2022). AIOps architecture and methodology. In *Hands-on AIOps: Best practices guide to implementing AIOps* (pp. 19–60). Apress.
32. Saggi, M. K., & Jain, S. (2018). A survey towards an integration of big data analytics to big insights for value-creation. *Information Processing & Management, 54*(5), 758–790.
33. Sawik, B. (2023). Space mission risk, sustainability and supply chain: Review, multi-objective optimization model and practical approach. *Sustainability, 15*(14), 11002.
34. Su, Y., & Liu, Z. (2023). A study of ChatGPT empowering college students' innovation and entrepreneurship education in the context of China. *International Journal of New Developments in Education, 5*(13), 1–7.
35. Telo, J. (2023). Smart city security threats and countermeasures in the context of emerging technologies. *International Journal of Intelligent Automation and Computing, 6*(1), 31–45.
36. VK, R. K., Ukko, J., Rantala, T., & Saunila, M. (2023). The value of novel technologies in context to performance measurement and management: A systematic review and future research directions. *Data and Information Management,* 100054.
37. Wadhwa, H., & Aron, R. (2023). Optimized task scheduling and preemption for distributed resource management in fog-assisted IoT environment. *The Journal of Supercomputing, 79*(2), 2212–2250.
38. Xu, J. (2022). The technical foundations of FinTech: ABCDI and more. *The Future and FinTech, ABCDI and Beyond.*
39. Yeruva, A. R., & Ramu, V. B. (2023). AIOps research innovations, performance impact and challenges faced. *International Journal of System of Systems Engineering, 13*(3), 229–247.
40. Shang, C., & You, F. (2019). Data analytics and machine learning for smart process manufacturing: Recent advances and perspectives in the big data era. *Engineering, 5*(6), 1010–1016.

CHAPTER 5

Leveraging Incremental Learning for Agile Entrepreneurship: A Practical Guide to Continuous Learning and Adaptation in Dynamic Markets

ALIYU MOHAMMED[1], M. ASHOK KUMAR[2], G. KALAIARASI[3], and B. SUNDARAVADIVAZHAGAN[4]

[1]*Department of Management, Skyline University Nigeria, Kano, Nigeria*

[2]*Department of Computer Science, Bluecrest University, Monrovia, Liberia*

[3]*Department of Commerce, Govt Arts and Science College for Women, Karimangalam, Dharmapuri, Tamilnadu, India*

[4]*Department of Information Technology, University of Technology and Applied Sciences-Al Mussana, Oman*

ABSTRACT

This study explores the terrain of agile entrepreneurship and provides an extensive manual for ongoing education and market adaptation. This qualitative research, which is based on Teece's dynamic capacities paradigm, attempts to integrate adaptability, continuous learning, and incremental learning. The main goal is to provide entrepreneurs the academic underpinnings and real-world experience they need to thrive in unpredictable situations. By combining these ideas into a coherent framework, the study finds and fills a sizable research vacuum in the

Real-Time Artificial Intelligence (AI): Key Motivations, Technologies, Platforms, and Use Cases.
Pethuru Raj, B. Sundaravadivazhagan, V. Kavitha, B. Narendra Kumar Rao, & Hannah Vijaykumar (Eds.)

DOI: 10.1201/9781998511358-5

literature. The study explores the connections between incremental learning and continuous learning by using qualitative approaches, such as case studies and literature reviews, and highlights the significance of adaptive decision-making frameworks. Important discoveries highlight the connections between these ideas and demonstrate the transformational potential of establishing a learning culture in businesses. Entrepreneurs may learn from the successful adaptation tactics exemplified by case studies of Netflix and Amazon. The study emphasizes how important it is for leaders to drive adaptable methods, overcome obstacles, and foster a resilient culture. Recommendations based on the results promote the use of technologies that facilitate continuous learning, the creation of a learning culture, and the use of agile approaches. The conclusion highlights how agile entrepreneurship is a comprehensive approach and stresses how firms must view adaptation as a lifelong process. This study bridges theoretical understandings with practical strategies to provide a more nuanced understanding of agile entrepreneurship. The study's suggestions provide as a guide for business owners navigating dynamic marketplaces, making sure their endeavors not only survive but flourish in a constantly changing environment.

5.1 INTRODUCTION

For long-term success, entrepreneurship in changing marketplaces requires constant learning and adaptability. This conceptual, qualitative research delves into the essence of using incremental learning for agile entrepreneurship and offers a thorough manual for ongoing adaptation in unstable settings. The worldwide landscape of entrepreneurship must be examined in order to fully understand the substance of this research. Because markets are changing so quickly, agility and flexibility have been more valued in recent years. Dynamic techniques are being considered in lieu of traditional approaches. This change is the result of an increasing understanding that entrepreneurship is a dynamic activity that calls for ongoing learning. Majority of the literature on entrepreneurship in the past has been on static models that highlight stability and planning. But the state of the global corporate environment has become more unstable, necessitating a paradigm change. This conceptual study's basis was laid by the literature's identification of crucial gaps in understanding the dynamic character of modern marketplaces. Closing these gaps is really important.

Adaptability is hampered by the present understanding of entrepreneurship, which lacks a sophisticated approach to ongoing learning. This work aims to close these gaps and offer theoretical guidance for entrepreneurs navigating changing marketplaces through the application of incremental learning.

The importance is in providing a theoretical framework for responsiveness and resilience, which are essential components of long-term success. This qualitative, conceptual study was justified by the necessity for a theoretical investigation into the incorporation of incremental learning into commercial activities. Our goal is to provide a theoretical framework that encapsulates the complex dynamics of entrepreneurial learning in a dynamic setting by utilizing a comprehensive qualitative methodology. Since entrepreneurship has many facets, this research takes an interdisciplinary approach. Our goal is to offer a comprehensive understanding by integrating knowledge from information technology, psychology, and business. This strategy is essential since entrepreneurship is intrinsically multidisciplinary and necessitates the synthesis of several viewpoints to effectively handle its intricate problems. The integration of incremental learning into entrepreneurial strategies, learning theories, and agile techniques are all incorporated into the theoretical framework. The mismatch between conventional static models and the dynamic character of markets leads to real-world issues. For instance, established planning models often fall short in capturing the agility required for success. By addressing these issues theoretically, this research contributes to a more nuanced understanding of entrepreneurship and offers theoretical solutions for real-world challenges.

5.2 STATEMENT OF THE PROBLEM

Entrepreneurship, as a dynamic and evolving field, faces a myriad of challenges in the contemporary business landscape. Traditional approaches to entrepreneurial endeavors often fall short in adapting to the rapid changes and uncertainties inherent in dynamic markets. This conceptual study seeks to address several key issues within the intersection of entrepreneurship, incremental learning, and agility. Historically, entrepreneurship literature has been anchored in static models that emphasize planning and stability. However, the existing paradigms inadequately address the dynamic and unpredictable nature of today's markets. There is a pressing need to identify and understand the gaps in current entrepreneurship models, particularly in relation to continuous learning and adaptability. Despite the acknowledged importance of continuous learning, there exists a significant gap in the theoretical foundations concerning the integration of incremental learning into entrepreneurial practices. The absence of a robust theoretical framework hampers our understanding of how entrepreneurs can effectively leverage incremental learning to navigate challenges in dynamic markets.

While theoretical insights are essential, the translation of these concepts into practical applications within entrepreneurial settings remains unclear. This study aims to bridge the gap between theoretical knowledge and practical implementation, providing a conceptual guide that entrepreneurs can use to enhance their adaptability and learning capabilities. The interdisciplinary nature of entrepreneurship calls for a holistic understanding that incorporates insights from various fields. Nevertheless, there is a dearth of thorough research on the ways in which an interdisciplinary approach integrating psychology, business, and IT might strengthen the theoretical underpinnings of agile entrepreneurship through incremental learning. There is a deficiency in qualitative investigation because the majority of the field's current study favors quantitative approaches. By using a qualitative research technique, this study seeks to close this gap by allowing for a thorough examination of the subtleties of incremental learning in agile entrepreneurship and capturing the complexity and depth of experiences that quantitative methods could miss. There is a pressing need for a conceptual framework that helps entrepreneurs adjust to risks gradually given the speed at which markets are changing. The lack of a framework of this kind makes it more difficult for entrepreneurs to successfully navigate and seize opportunities in dynamic marketplaces. This conceptual study seeks to close knowledge gaps in the field of agile entrepreneurship by offering theoretical insights and conceptual frameworks that address the difficulties that have been highlighted. This will provide the groundwork for future research and practical applications in the field.

5.3 OBJECTIVES OF THE STUDY

Providing a theoretical framework for utilizing incremental learning within the framework of agile entrepreneurship is the main objective of this research. The following objectives are intended to close the gaps and tackle the issues that have been noted in the field:

1. To find and examine gaps in the models of entrepreneurship that is already in use.
2. To create a theoretical framework for entrepreneurship's progressive learning.
3. To close the knowledge gap between abstract ideas and real-world implementation.
4. To explore the interdisciplinary approach in agile entrepreneurship.
5. To conduct a qualitative exploration of incremental learning in entrepreneurship.
6. To propose a conceptual framework for agile entrepreneurship.
7. To assess the urgency of a conceptual framework for agile entrepreneurship.

Through achieving these objectives, this study aims to contribute theoretical insights, frameworks, and recommendations that advance the understanding of agile entrepreneurship, providing valuable guidance for entrepreneurs operating in rapidly changing business environments.

5.4 LITERATURE REVIEW

The literature review critically examines the existing scholarship in three key areas: Agile Entrepreneurship, Incremental Learning in Dynamic Markets, and the Role of Continuous Learning in Entrepreneurship.

5.4.1 AGILE ENTREPRENEURSHIP: CONCEPTS AND DEFINITIONS

The landscape of entrepreneurship has evolved over time, reflecting the changing demands of markets. Scholars such as Mansoori and Lackeus [37] and Patz [49] have contributed significantly by introducing the concepts of effectuation and Lean Startup, respectively. These models emphasize agility, experimentation, and adaptability as core elements of entrepreneurial success. Originally developed for software development, the Agile Manifesto has been modified for use in entrepreneurship. The use of agile concepts in entrepreneurial settings, such as iterative development and consumer feedback, was made famous by Le [35].

Due to this adaptation, conventional entrepreneurial procedures have changed, encouraging a more flexible and quick-thinking methodology. Despite the growing popularity of agile entrepreneurship, criticisms have surfaced. Fowler [27] emphasize the difficulties in striking a balance between the requirement for strategic direction and mobility. Academics like Moran [46] stress the significance of contextual elements in judging whether agile tactics are acceptable.

5.4.2 INCREMENTAL LEARNING IN DYNAMIC MARKETS

Education ideas and cognitive psychology are the foundation of incremental learning. Foundational ideas are offered by Piaget's constructivist theory [50] and Kolb's experiential learning model [33]. Incremental and continuous adjustments which form the principles inherent in these thoughts are taken forward into organizational learning, according to Sessa and London [55]. Other scholars, such as Sheng and Chien [57], have even examined the value of incremental learning as it relates to organizational adaptation from a business perspective. The emphasis that their work lays is on iterative improvement of process and gradual development on accumulation of wisdom to sharpen organizational skills. This makes clear such an idea as technology supports incremental learning. Dynamic capabilities were proposed by Teece [60] by highlighting the importance of technology in making firms responsive to changing environments. Technology plays a vital role in machine learning and data analytics, as well as incremental learning within a business organization.

5.4.3 THE ROLE OF CONTINUOUS LEARNING IN ENTREPRENEURSHIP

Senge (1990) developed the concept of a learning organization emphasizing the critical role of ongoing learning in achieving organizational goals. Scholars such as Prieto [53] emphasize the significance of a learning mindset in promoting creativity and flexibility within the setting of entrepreneurship. In spite of its significance, ongoing education in entrepreneurship is fraught with difficulties. Integration of learning approaches might be hampered by obstacles including organizational inertia Mikalef et al. [42] and risk aversion Ganz [28]. These difficulties highlight the necessity for sophisticated strategies in encouraging an ongoing learning culture. The connection between continuous learning and agile approaches has been studied recently. A framework that combines agile methods with organizational learning processes is put out by Flumerfelt et al. [26]. The responsiveness of agile approaches and the sustainability of continuous learning are combined in this integration to provide a comprehensive strategy.

5.4.4 ADAPTATION STRATEGIES IN AGILE ENVIRONMENTS

In agile contexts, the idea of adaptability has evolved significantly. The foundation was established by academics who stressed the significance of flexibility in organizational tactics, such as Amburgey and Miner [2]. The advent of agile approaches, as explained by Mishra and Mishra [45], brought adaptable concepts into software development and project management. Building on this basis, Teece's [61] introduction of the dynamic capacities viewpoint has been crucial to our understanding of adaptation in dynamic situations. Teece contends that for businesses to prosper in quickly evolving marketplaces, they require dynamic capabilities, more especially, the capacity to perceive, seize, and adapt. This viewpoint, which emphasizes the value of responsiveness and ongoing improvement, is consistent with the agile mindset. Although there are several advantages to using agile adaptation tactics, researchers such as Brown and Eisenhardt [11] advise against oversimplifying. They emphasize that adaptability might vary depending on the situation and is not always a one-size-fits-all approach. In complex and unpredictable contexts, the difficulties of striking a balance between stability and flexibility in adaptation techniques are especially pertinent.

5.4.5 RELATED WORK IN THE FIELD

As a result of their recognition of the complexity of entrepreneurship, academics are using multidisciplinary methods more frequently. Reymen et al. [54] promotes the "effectuation" method, which emphasizes the value of integrating theoretical viewpoints to improve entrepreneurial decision-making. This approach relies on ideas from other disciplines. Information technology integration has emerged as a major issue in entrepreneurship studies. Experts such as Shane and Venkataraman [56] underscore how technology shapes chances for entrepreneurship. IT integration is essential in this study's environment for supporting agile techniques and incremental learning. Notwithstanding the progress made, there are still study gaps about the relationship between incremental learning, adaptability, and agile entrepreneurship. Miceli et al. [40], among other recent studies, have highlighted the need for more in-depth analyses of the ways in which businesses might use agility and technology to support long-term innovation and adaptability.

Adaptation strategies in agile environments provide an overview of the literature that clarifies the history of adaptive thinking, the significance of dynamic skills, and the difficulties that arise when adjusting. The expanding trend of interdisciplinary study in entrepreneurship and the integration of information technology is highlighted in the section on Related Work in the Field. It also highlights the gaps that currently exist, which this conceptual research seeks to fill, offering a thorough basis for the next theoretical investigation.

5.5 CONCEPTUAL FRAMEWORK

This study's conceptual framework offers an organized and thorough manual for comprehending how incremental learning fits within agile entrepreneurship. This theoretical investigation draws on basic concepts of incremental learning theories as well as agile approaches.

5.5.1 DEFINITION AND COMPONENTS OF THE CONCEPTUAL FRAMEWORK

The conception of this study can be thought of as systematic model incorporating incremental learning within agile entrepreneurship. For instance, Yin [64] and Miles and Huberman [43] argue that a conceptual framework is necessary to categorize in a complex reality. The components include:

1. **Agile Entrepreneurship Dimensions:** The framework incorporates flexible factors, continued feedback and iterative planning as dictated by Beck's agile approach concepts [8]. The total strength of entrepreneurial process depends on each element.

2. **Incremental Learning Strategies:** The approach that defines procedures of incremental learning is based on organizational learning theories and cognitive psychology [19]. These strategies include slow disclosure of information, trial and error learning, and small continual adjustments.
3. **Organizational Culture:** The model includes cultural factors which are aimed at fostering experimentation, willingness for taking risks, as well as having a learning mentality, taking into account the culture influence on learning and adapting process.

5.5.2 THE ROLE OF AGILE METHODOLOGIES IN THE FRAMEWORK

The framework requires agile approaches. Concepts of agility are embedded in the iterative and flexible nature of the entrepreneurial process. It outlines agile methods, such as Scrum, Kanban and Lean Startup highlighting the adaptable nature of these methods. The concept adhering to Ries' works provides emphasis on customer-centricity in an agile approach. Customer involvement and feedback loops as critical elements ensure entrepreneurial operations are consistent with shifting customer needs and market dynamics. Iterative planning approach by Highsmith (2002), which is used as a means of offering adaptation for agile approaches. It promotes continuous assessment and adjustment in any undertaking based on instant input and external dynamic environment.

5.5.3 THEORETICAL UNDERPINNINGS OF INCREMENTAL LEARNING

Cognitive and experience learning are core components of this framework, based on Piagets's constructivist theory [50] and Kolbs's experiential learning model [33]. Entrepreneurs are urged to gain knowledge from practical experience, reflect on these experiences, and modify their strategies accordingly. A fundamental component of the approach is incremental learning, as proposed by Darr and Kurtzberg (2000). The procedure entails the ongoing acquisition of information and the ensuing modification of entrepreneurial tactics in light of this expanding body of knowledge. The paradigm acknowledges the significance of technology in supporting incremental learning, which is consistent with Teece's [60] dynamic capacities approach. The framework incorporates digital technologies and information systems to facilitate data-driven decision-making and improve the learning process.

5.5.4 FRAMEWORK FOR LEVERAGING INCREMENTAL LEARNING IN AGILE ENTREPRENEURSHIP

Building on the groundwork established in the preceding sections, the framework's growth entails integrating insights from incremental learning theories and agile approaches. Important contributions from academics like Teece [60], and Beck [8] serve as the foundation for creating an extensive and practical manual for business owners. The framework organizes iterative planning, ongoing feedback, and adaptation as fundamental elements, incorporating agile entrepreneurial features. Drawing from the tenets of agile approaches, including Scrum and Lean Startup, the framework highlights the fluid character of entrepreneurial endeavors. This is consistent with the work of Beck [8], who support customer-centric techniques and iterative development. Incremental learning techniques, influenced by organizational learning theories and cognitive psychology, are fundamental to the framework. According to Darr and Kurtzberg (2000), these tactics include experience learning and tiny, ongoing modifications. The framework puts these tactics into practice and gives entrepreneurs helpful pointers on how to apply incremental learning in their business endeavors. The framework includes cultural components in recognition of the influence that organizational culture has on learning and adaptation [22]. It promotes a culture that emphasizes ongoing development, welcomes experimentation, and cultivates a growth mentality. This is consistent with the idea that gradual learning and agile approaches can only be supported by an encouraging corporate culture.

5.5.5 RELATIONSHIPS BETWEEN FRAMEWORK COMPONENTS

A key component of the system is the interaction between incremental learning and agile approaches. Incremental learning solutions are easier to apply when agile principles like continuous feedback loops and iterative planning are used. According to researchers like Beck [8], integrating these two elements improves the flexibility of entrepreneurial processes.

The framework places emphasis on the connection between learning methods and corporate culture. Staff members are more likely to experiment, share expertise, and learn new things when they work in a friendly environment. This is consistent with Edmondson and McManus's (2007) results of how corporate culture affects learning practices. The concept emphasizes how technology may facilitate incremental learning as well as agile techniques. Agile techniques are supported by digital tools and information systems, which also make knowledge acquisition easier. Technology's significance in improving adaptability is recognized by the framework based on Teece's [60] dynamic capabilities viewpoint. The framework suggests combining corporate culture, incremental learning, and agile approaches in a comprehensive way. It highlights how these elements are interrelated and how they work together to promote agile entrepreneurship. This integrated method is in line with the recent demands for more thorough frameworks in entrepreneurship research.

With input from several academics, the conceptual framework for utilizing incremental learning in agile entrepreneurship has been thoughtfully developed. Combining incremental learning tactics, organizational culture, technology, and agile methodologies into a logical framework gives entrepreneurs a useful tool to improve learning and adaptation in ever-changing industries.

5.6 THEORETICAL FRAMEWORK

The investigation of the incorporation of incremental learning into agile entrepreneurship is guided by the theoretical framework of this study. It uses fundamental theories from organizational learning, cognitive psychology, and agile techniques to give a strong foundation for comprehending and applying theoretical ideas in an entrepreneurial setting.

5.6.1 THEORETICAL FOUNDATIONS OF INCREMENTAL LEARNING

5.6.1.1 COGNITIVE PSYCHOLOGY AND LEARNING THEORIES

Learning theories and cognitive psychology are the theoretical basis of incremental learning. Academics like Piaget [50] and Kolb [33] have established the foundation for comprehending how people learn via experiences and introspective processes. The framework's use of incremental learning, which emphasizes ongoing, little alterations based on experiential learning, welcomes these cognitive processes.

5.6.1.2 ORGANIZATIONAL LEARNING THEORIES

Bergsteiner et al. [9] expand on the theoretical underpinnings by applying experiential and cognitive learning to organizational contexts. Their research highlights how crucial organizational learning is to promoting flexibility. To give a basis for comprehending how incremental learning might be implemented at the organizational level, the theoretical framework incorporates various ideas of organizational learning.

5.6.1.3 DYNAMIC CAPABILITIES PERSPECTIVE

Teece [60] provides more depth to the theoretical underpinnings with his dynamic capacities approach. The notion of dynamic capacities pertains to an organization's capacity to perceive, grasp, and adjust to evolving surroundings. This viewpoint is consistent with the incremental learning method since it acknowledges the necessity of ongoing modifications and changes in response to new information and market realities.

5.6.2 AGILE METHODOLOGIES IN ENTREPRENEURIAL SETTINGS

The Agile Manifesto [6, 7] serves as a foundational element of the theoretical framework that explains the function of agile approaches in entrepreneurial contexts. Principles like collaborating with customers, adapting to change, and providing workable solutions iteratively are outlined by Beck [8] and others. The framework's approach to agility in entrepreneurial processes is guided by these concepts. The Lean Startup technique by Ries (2011) makes a substantial contribution to the theoretical framework. Lean Startup principles prioritize iterative development and customer-centricity, as shown by the build-measure-learn feedback cycle. These ideas are incorporated into the framework to help business owners apply agile techniques that complement the Lean Startup methodology. Another important theoretical component was presented by Schwaber and Sutherland (2017), the Scrum framework. The framework facilitates the understanding of how certain agile approaches might be implemented in entrepreneurial situations by drawing on Scrum principles including sprint planning, daily stand-ups, and retrospective reviews. The Scrum framework's emphasis on collaboration and flexibility is consistent with the agile mentality as a whole. The critiques and modifications of agile approaches offered by academics like Chan and Thong [15] contribute to a more sophisticated understanding. The framework's recognition of difficulties and contextual factors in the use of agile approaches in entrepreneurship is informed by theoretical criticisms.

5.6.3 MODELS FOR CONTINUOUS LEARNING IN DYNAMIC MARKETS

The Learning Organization model developed by Peter Senge in 1990 provides the theoretical framework for continuous learning. Senge stresses the value of establishing environments in which employees are constantly learning new things in order to promote flexibility and creativity. The model offers a thorough understanding of how businesses may include continuous learning into their core operations by introducing ideas like shared vision, system thinking, mental models, personal mastery, and team learning. The foundation of the approach is the psychological safety research conducted by Edmondson [22]. In dynamic marketplaces, psychological safety is essential to ongoing learning because it creates a secure space where people feel comfortable taking chances with one another, exchanging ideas, and learning from mistakes. Edmondson's work advances our knowledge of how establishing psychologically secure environments promotes lifelong learning. Leveraging outside ideas and partnerships to promote innovation is a notion introduced by Chesbrough [16] in his open innovation paradigm. Open innovation helps companies in dynamic marketplaces learn continuously by actively seeking out outside expertise, ideas, and collaborations. In order to improve organizational learning, Chesbrough's model emphasizes the need of integrating external information sources. The necessity of a more integrated approach to continuous learning is emphasized by Aamodt [1]. Their study adds to the framework by recommending that learning models take into account external collaborations and networked learning in addition to internal processes. This viewpoint is consistent with the dynamic nature of interconnected marketplaces and the ongoing learning that businesses must do from both internal and external sources.

5.6.4 FRAMEWORK FOR ADAPTATION STRATEGIES

A paradigm for adaptation that highlights the significance of strategic adaptation for organizational survival is put out by Amburgey and Miner [2]. It broadens our understanding on adaptations through emphasizing that organizations must be prepared to change plans as the environment changes. The paradigm supplies essential frameworks necessary in understanding why firms react to environmental changes in a strategic manner. Brown and Eisenhardt [11] advance the notion of dynamic capabilities which can be conceived as a capability for integrated, developed, and configured within and across firm boundaries. This viewpoint improves our knowledge of how organizations might develop the adaptive skills necessary to recognize and react to changes in their surroundings. Creating adaptation strategies requires an understanding of the dynamic capacities framework. Adaptation techniques can benefit from Teece's [60] dynamic capacities approach, which was previously explored in relation to incremental learning. Teece's research highlights that dynamic skills entail proactive environmental shaping and influence in addition to change adaptation. By emphasizing the strategic nature of adaptation and the requirement that firms take the initiative to actively shape their competitive environment, this viewpoint adds to the framework. The approach is enhanced by Chesbrough [17] and Shane and Venkataraman [56], who emphasize the part that innovation and technology play in adaptation tactics. According to their research, companies may strategically adjust by utilizing new technology and pursuing ongoing innovation. This viewpoint is consistent with the dynamic character of markets, where changes are frequently driven by technology.

To summarize, Senge, Edmondson, Chesbrough, West, Bogers, Amburgey, Miner, Brown, Eisenhardt, and Teece are just a few of the models and viewpoints that are integrated into the theoretical framework for continuous learning and adaptation tactics. This thorough synthesis offers a solid framework for comprehending how businesses might promote ongoing learning and make strategic adjustments in response to changing market conditions. Every model offers distinct perspectives to the overall framework, augmenting its all-encompassingness and relevance.

5.7 METHODOLOGY OF THE STUDY

This study conducts a conceptual investigation on utilizing incremental learning for agile entrepreneurship using a qualitative research technique. To fully explore the conceptual elements, theoretical foundations, and frameworks of the study issue, the qualitative technique is selected. A thorough explanation of the process is provided below:

5.7.1 RESEARCH DESIGN

The conceptual character of the study design is centered on integrating current theoretical frameworks, models, and viewpoints about incremental learning and agile entrepreneurship. Without gathering actual data, this method enables a thorough investigation of the conceptual underpinnings.

5.7.2 DATA COLLECTION

5.7.2.1 LITERATURE REVIEW

For this conceptual research, an extensive literature review serves as the main source of data. A thorough assessment and analysis of scholarly works, books, and pertinent publications on incremental learning, entrepreneurship, agile approaches, and related subjects was conducted. The literature study offers a wealth of frameworks and theoretical views from different academics.

5.7.2.2 DOCUMENT ANALYSIS

Analyses were conducted on documents, including reports, case studies, and academic publications, that show how agile approaches and incremental learning are applied in actual entrepreneurial contexts.

By examining real-world instances and experiences that have been described in the literature, document analysis improves conceptual comprehension.

5.7.3 DATA ANALYSIS

5.7.3.1 THEMATIC ANALYSIS

In order to find and classify important themes, concepts, and frameworks pertaining to incremental learning and agile entrepreneurship, thematic analysis was utilized. This methodical technique facilitates the identification of significant trends and the arrangement of the large body of theoretical data into logical topics.

5.7.3.2 FRAMEWORK SYNTHESIS

The process of data analysis entails combining the recognized themes and theoretical frameworks to create an all-encompassing conceptual framework. The synthesis process aims to integrate various perspectives, models, and theories into a cohesive guide that can inform the application of incremental learning in agile entrepreneurship.

5.7.4 SIGNIFICANCE OF THE STUDY

This qualitative conceptual study holds significance in contributing to the theoretical foundations of agile entrepreneurship. The synthesized conceptual framework provides entrepreneurs, researchers, and educators with valuable insights for navigating dynamic markets through incremental learning strategies. The study aims to bridge theoretical gaps and stimulate further research in the evolving field of agile entrepreneurship.

In conclusion, the methodology of this study employs a qualitative research design, relying on literature review, document analysis, thematic analysis, and framework synthesis. Ethical considerations revolve around proper citation, and the limitations center on the scope of literature and subjective interpretation. The significance lies in the contribution to theoretical understanding and the potential impact on entrepreneurial practices.

5.8 INCREMENTAL LEARNING IN AGILE ENTREPRENEURSHIP

5.8.1 DEFINITION AND CHARACTERISTICS

5.8.1.1 DEFINITION OF INCREMENTAL LEARNING IN AGILE ENTREPRENEURSHIP

Incremental learning in the context of agile entrepreneurship refers to an adaptive and continuous process where individuals and organizations accumulate knowledge, make small adjustments, and iteratively improve their strategies based on real-time feedback and experiences. It aligns with the principles of agile methodologies, emphasizing flexibility, responsiveness, and a dynamic approach to problem-solving [61].

5.8.1.2 CHARACTERISTICS OF INCREMENTAL LEARNING IN AGILE ENTREPRENEURSHIP

1. **Iterative Progress:** Incremental learning involves iterative cycles of planning, execution, and reflection. This iterative nature allows for constant adjustments and improvements.
2. **Customer-Centric Approach:** A key characteristic is the focus on customer needs and feedback. Incremental learning ensures that entrepreneurial efforts align with evolving customer preferences and market demands.
3. **Adaptability:** The adaptability of strategies based on real-time insights and changing market conditions is a fundamental characteristic. This adaptability is crucial for navigating uncertainties in dynamic markets.

5.8.2 IMPLEMENTATION CHALLENGES

One challenge in implementing incremental learning in agile entrepreneurship is striking a balance between agility and strategic direction. Scholars like Gerster et al. [29] emphasize the importance of maintaining a strategic focus while embracing agile practices. Finding this equilibrium requires careful consideration of organizational goals and market dynamics.

Resistance to change poses another challenge. Incremental learning often necessitates a cultural shift within organizations. Employees may resist continuous adjustments and a more flexible approach. A supportive company culture and efficient change management techniques are necessary to overcome this reluctance [34]. One recurrent difficulty is the appropriate integration of technology. Even if technology makes learning more gradual, businesses still have to get beyond issues like data security worries and the requirement for technology infrastructure [21]. To ensure successful deployment, technology must be integrated seamlessly.

5.8.3 SUCCESS STORIES AND CASE STUDIES

Airbnb offers an inspiring tale of success. Airbnb used incremental learning through ongoing experimentation and quick platform changes in response to user input by implementing Lean Startup concepts. This strategy was crucial to Airbnb's explosive expansion and market dominance [48]. The agile transformation of Spotify is a notable case study. The organization implemented agile approaches, such as Scrum and Kanban, to improve adaptability and agility. Spotify developed its development procedures, organizational structure, and product features incrementally, which helped the company succeed in the fiercely competitive music streaming market [24].

5.8.4 BEST PRACTICES

One of the finest practices in incremental learning is to promote cross-functional collaboration. Scrum and other agile approaches place a strong emphasis on interdisciplinary teams cooperating. This strategy promotes a comprehensive approach to problem-solving by ensuring a diversity of viewpoints and levels of experience [47]. It is essential to implement continual feedback loops. Feedback from stakeholders, consumers, and team members on a regular basis enables prompt modifications and enhancements. This best practice underpins incremental learning's iterative character and is the basis of agile techniques [39]. Empowering teams via agile leadership is crucial. In order to foster an atmosphere that encourages experimenting, taking risks, and learning from mistakes, leaders are essential. Agile leaders enable teams to take initiative and create an environment that supports incremental learning [32].

To summarize, the process of incremental learning in agile entrepreneurship is iterative and adaptable, marked by constant modifications, a focus on the client, and flexibility. Overcoming opposition to change and striking a balance between agility and direction are two implementation problems. Incremental learning is successful, as shown by success stories from businesses like Airbnb and Spotify. Other best practices include agile leadership, cross-functional cooperation, and constant feedback loops.

5.9 CONTINUOUS LEARNING IN DYNAMIC MARKETS

5.9.1 IMPORTANCE AND NEED

Dynamic markets are characterized by quick changes and uncertainty, it is imperative that learning be done continuously in these contexts. Teece [60] emphasized that in order for an organization to remain competitive, it must constantly learn new things, adjust its tactics, and react quickly to opportunities and dangers as they arise. The significance is in using innovation to obtain a competitive advantage. According to Bikse et al. [10], ongoing education encourages innovation, which helps start-ups keep ahead of the market by consistently launching new goods, services, or business strategies. Continual education aids in making smart decisions. According to Constantinides and Fountain [18], businesses that practice continuous learning are better able to make decisions based on the most recent developments in technology, consumer behavior, and market trends.

5.9.2 LEARNING MECHANISMS AND PLATFORMS

One important learning tool is the establishment of corporate training programs. According to Senge in 1990, these initiatives foster a culture of learning inside the company and provide staff members the abilities and information they need to adjust to shifting market conditions. Using online learning environments is a modern technique. According to Andreatos [4], platforms such as Udemy, Coursera, and LinkedIn Learning provide chances for workers and business owners to learn new skills, remain current with market trends, and partake in ongoing education. It is vital to cultivate networks of collaborative learning. Pittaway et al. [51] emphasize the value of networks that enable business owners to work together with colleagues in the field, exchange knowledge, and jointly overcome the difficulties presented by changing markets.

5.9.3 IMPACT ON ENTREPRENEURIAL VENTURES

The agility and adaptability of entrepreneurial endeavors are improved by ongoing learning. According to Beck [8], continuous learning is iterative and fits well with agile approaches because it allows company plans to be quickly adjusted in response to changes in the market. In dynamic markets, continuous learning contributes to improved decision-making under uncertainty. According to McGrath and MacMillan [38], firms that practice continuous learning are better able to make strategic choices in erratic situations, which lessens the negative effects of uncertainty on business endeavors. Continuous learning fuels innovation and product development. Shane and Venkataraman [56] highlight that entrepreneurial ventures engaged in continuous learning are more likely to identify new market opportunities, develop innovative solutions, and bring novel products or services to market.

5.9.4 OVERCOMING CHALLENGES

Developing an organizational learning culture is essential to overcoming obstacles. Edmondson [23] on psychological safety emphasizes how crucial it is to provide a work atmosphere in which staff members may freely exchange ideas, own up to their mistakes, and participate in ongoing education without worrying about the consequences. Keeping up with the fast advancements in technology is a problem for entrepreneurial endeavors. In order to get around this, Fitzgerald et al. [25] advises creating a culture that values technology, making use of digital tools, and keeping up with developments in technology so that they may be strategically incorporated into daily operations. It is essential to develop resilience via education. According to Bauman and Lucy [5], entrepreneurial endeavors may overcome obstacles by encouraging a resilient mentality, seeing failures as teaching opportunities, and utilizing setbacks as a platform for ongoing development.

In conclusion, adapting to changing market conditions, obtaining a competitive advantage, and formulating wise strategic choices all depend on ongoing learning. Online resources, collaborative networks, and organizational training programs are examples of learning methods. Innovation, better decision-making, and more agility are all impacted in entrepreneurial endeavors. Developing a learning culture, adjusting to technology advancements, and strengthening resilience via education are all necessary for overcoming obstacles.

5.10 ADAPTATION STRATEGIES FOR DYNAMIC MARKETS

5.10.1 FRAMEWORKS FOR ADAPTIVE DECISION-MAKING

Snowden and Boone [59] proposed the Cynefin framework, which provides a formal method for making decisions in dynamic markets. It classifies scenarios into chaotic, complicated, complicated, and simple domains and offers advice on the best course of action for each. This framework helps business owners identify the type of obstacles they face and make appropriate strategy decisions. According to Tremblay [62], the OODA loop is a paradigm for decision-making that emphasizes quick iterations and was created by military strategist John Boyd. Its application to business is keeping a close eye on the surroundings, adapting to changes, making choices fast, and following through on them. When navigating volatile marketplaces, entrepreneurs can effectively use this iterative cycle. Teece's [60] theory on dynamic capacities offers a theoretical foundation for comprehending how companies might adjust to evolving market circumstances. In order to adapt to changing circumstances, it highlights a company's capacity to integrate, develop, and restructure internal and external competences. This approach helps business owners build the skills necessary for strategic adaptability.

5.10.2 CASE STUDIES ON SUCCESSFUL ADAPTATION

The case of Netflix's transition from a DVD rental model to a streaming platform illustrates successful adaptation. By recognizing the shift in consumer preferences, Netflix strategically adapted its business model.

The company's agility in embracing technological changes and understanding market trends contributed to its continued success [41]. Amazon's evolution from an e-commerce giant to a leading provider of cloud services exemplifies successful adaptation. DaSilva et al. [20] detail how Amazon strategically diversified its business, leveraging its technological infrastructure to offer cloud computing services through Amazon Web Services (AWS). This adaptive strategy has contributed significantly to Amazon's resilience and growth.

5.10.3 CHALLENGES IN IMPLEMENTING ADAPTIVE STRATEGIES

Implementing adaptive strategies often faces resistance to change within organizations. Miller [44] highlight that employees may resist the alterations to existing processes or structures required for adaptation. It will need good change management and communication techniques to overcome this obstacle. Because dynamic markets are by their very nature unpredictable, entrepreneurs tend to be risk averse. Greenhalgh et al. [31] explores how adopting adaptive techniques might be impeded by fear of the unknown and failure. To overcome this obstacle, entrepreneurs must cultivate a culture that accepts measured risks and sees setbacks as chances for growth and development.

5.10.4 EVALUATING THE EFFECTIVENESS OF ADAPTATION

Developing key performance indicators (KPIs) is a necessary step in assessing how effective an adaptation is. According to Shane and Venkataraman [56], metrics including time to market, customer satisfaction, and revenue growth are crucial factors. These KPIs offer concrete indicators of how successfully adaptable initiatives are meeting consumer needs and enhancing overall company performance. Evaluating the efficacy of adaptation requires ongoing learning from iterations and feedback. The significance of feedback loops is underscored by the agile approaches, as noted by Sjödin et al. [58]. To improve their adaptive strategy, entrepreneurs should proactively seek out and consider input from stakeholders, consumers, and the market.

In conclusion, adaptive decision-making frameworks, case studies of successful adaptation, implementation obstacles, and techniques for assessing efficacy are all part of adaptation strategies for dynamic markets. OODA loop, dynamic capabilities framework, and Cynefin framework provide organized methods for making decisions. Successful adaptation is demonstrated by case studies of Netflix and Amazon; yet, obstacles include uncertainty and resistance to change. Establishing KPIs and promoting an iterative, continuous learning culture are essential to evaluating effectiveness.

5.11 IMPLEMENTATION GUIDELINES

5.11.1 INTEGRATING INCREMENTAL LEARNING, CONTINUOUS LEARNING, AND ADAPTATION

Acknowledging the connections between incremental and continuous learning is a necessary step toward integrating the two ideas. Because it focuses on making modest, ongoing changes based on experiences, incremental learning fits in well with continuous learning's iterative structure [63]. The integration guarantees that learning is not only ongoing but also applied in tiny, manageable increments, which improves the flexibility of entrepreneurial endeavors. Integrating decision-making that is adaptive is essential. The learning process should make use of the decision-making frameworks that were previously addressed, such as the OODA loop and the Cynefin framework. This guarantees that learning is an active process of observation, orientation, decision-making, and action in response to changing market conditions rather than merely a passive activity, American Economic Association, Royal Economic Society, and Simon [3]. Establishing a culture of learning inside the company is essential. This entails creating an atmosphere where workers of all levels are motivated to keep learning, trying new things, and adapting. Senge's (1990) Learning Organization model offers valuable perspectives on how companies may cultivate an environment that prioritizes and encourages ongoing learning on both an individual and organizational level.

5.11.2 PRACTICAL STEPS FOR ENTREPRENEURS

In order to remain up to date with market trends, competition activities, and technical breakthroughs, entrepreneurs need regularly undertake environmental scans. According to Porter and Heppelmann [52], this continual observation helps business owners foresee changes in the industry and serves as a basis for lifelong learning. Putting agile approaches like Scrum and Kanban into practice is a practical step toward enabling incremental learning and adaptive decision-making [36]. Agile approaches offer a methodical way to manage projects, with a focus on adaptability, teamwork, and iterative development. Encouragement of cross-functional collaboration is necessary to promote information exchange and a diversity of viewpoints. This entails dismantling organizational silos and encouraging cooperation among people with various specialties and skill sets. According to Schwaber and Majchrzak et al. (2012), cross-functional teams support both ongoing learning and comprehensive problem-solving.

5.11.3 TOOLS AND TECHNOLOGIES SUPPORTING IMPLEMENTATION

Agile methodology adoption is supported by using project management tools like Jira, Trello, or Asana. These solutions offer a digital workspace for managing assignments, monitoring development, and encouraging teamwork [14]. Tools for business intelligence and data analytics, such as Tableau or Power BI, help to draw conclusions from data. By giving business owners useful data on consumer behavior, industry trends, and the success of their plans, these solutions promote lifelong learning [13]. Continuous learning initiatives inside enterprises are supported by using learning management systems (LMSs). Learning management system platforms, like Moodle or Canvas, provide a single location for managing courses, training materials, and evaluations, which helps create a culture of learning.

5.11.4 MEASURING SUCCESS AND ITERATIVE IMPROVEMENTS

KPIs should be established by entrepreneurs in accordance with their flexible company plans and objectives. Metrics including time to market, revenue growth, and customer satisfaction may be included in KPIs [56]. A quantifiable framework for evaluating the effectiveness of incremental learning and continual adaptation is provided by clear KPIs. Iterative improvements require regular reviews of progress and retrospective analysis. Regular retrospectives, where teams evaluate what went well, what may be better, and how to adjust going forward, are crucial to agile approaches, such as Scrum Gonçalves and Linders [30]. This procedure guarantees an ongoing cycle of improvement and learning. It is important to establish feedback systems from all stakeholders, such as consumers and staff. Iterative improvements are guided by actively seeking and implementing feedback, which yields significant insights into the efficacy of adopted solutions [12]. It also promotes a mindset that is receptive to criticism and never stops learning.

In conclusion, combining these ideas, creating a learning culture, and putting agile entrepreneurship into practice are all necessary for achieving incremental learning, continuous learning, and adaptability. To assist with implementation, entrepreneurs may make use of technology and resources like LMSs and project management tools. Setting KPIs, carrying out routine reviews, and incorporating input for iterative changes are all necessary for measuring performance.

5.12 CASE STUDIES

Case Study 1: GE Healthcare

In this regard, GE Healthcare is internationally renowned as one of the world's leading medical technologies and healthcare services providers. The company has always been innovative, but now faces stiffening competition from newer and smarter start-ups. In addition, GE Healthcare has adopted the principle of incremental learning and agile enterprise. On incremental learning; an example is, for instance when developing new medical devices at GE Healthcare. This means that currently the company no longer uses the waterfall development model where, products are planned, developed, and then delivered according to schedule. On the contrary, GE Healthcare adopts an iterative approach by developing products incrementally and having them tested in stages together with customers. The organization therefore takes advantage of this approach that enables it to react to consumer demands in a timely manner.

Agile concepts have also been adopted by GE Healthcare organization structure. The company has put in place cross-functional teams charged with overseeing the whole product development cycle from conceptualization to market introduction. Such teams are autonomous and can try new things.

Due to this, there is now a culture of creativity and entrepreneurship within the corporation. That is why such companies like GE Healthcare have been focusing on incremental learning and agile entrepreneurship which enabled them to introduce new products and services faster and cheaper. It is also evident that through this strategy, the company has enhanced its customer satisfaction as well as the market share.

Case Study 2: Netflix

This is among the leading streaming entertainment companies with over 222 million subscribers spread around the globe. In business, it has been a pioneer in exploiting data and analytics. Netflix is renowned for its experimental nature. One of the strategies that Netflix has employed while coming up new product is incremental learning. The company would not have been guessing what exactly should be produced with the old tradition guts. On the other hand, Netflix targets customers' habits that are observed through customer viewing data for any tendencies and what customers need. This has helped the firm come up with quality original content.

Similar to the company behind Netflix, its software is also agile-based. The company uses the on-demand continuous delivery method that entails deployment of new codes in the production many times on daily basis. Human writer, the firm has an advantage by addressing bugs and coming up with additional features promptly, hence remaining before its competitors. Growth of the firm is rapid due to the inculcation of small-scale learning as well as Agile Entrepreneurship. Furthermore, the business has been able to improve on its customer satisfaction and profitability.

Case Study 3: Airbnb

AirBnb is a renowned accommodation service that enables individuals offer their house and flats to visitors. Disruptions in the hospitality industry are what the firm entails growth at an incremental learning rate using the agile entrepreneurship model. For instance, Airbnb is applying incremental learning when it comes to launching of new products or services into the marketplace. Traditional product roadmaps where features are developed linearly is no longer what the company now depends on. In contrast, Airbnb is more flexible and experiment with its design iteratively. The company will learn from its customers, making it quicker to build better products.

Its organizational structure has also incorporated Airbnb principles of agility. The firm established multifaceted teams in charge of different sections of business including marketing, product development, and customer service. These small teams have sufficient power to choose and implement innovations. It has brought about creativity and spirit of self-reliance in the organization. Due to its emphasis on incremental learning and agile enterprise, Airbnb has grown into a renowned international player in the hotel sector. This has also led to a higher level of customer satisfaction while at the same boosting business profits.

5.13 DISCUSSION

5.13.1 SYNTHESIS OF FINDINGS

The integration of results indicates how incremental learning, continuous learning, and adaptability are related to each other in the context of agile entrepreneurship. Combining these ideas creates a comprehensive strategy for negotiating dynamic marketplaces. Iterative adjustments describe incremental learning, which fits in well with continuous learning to form a synergistic connection that improves adaptability. The utilization of adaptive decision-making frameworks serves to reinforce the interrelatedness of these ideas by giving business owners a methodical way to make strategic modifications based on current information. The synopsis emphasizes the role that dynamic talents play in propelling agile entrepreneurship. A fundamental idea that supports an organization's capacity to integrate, develop, and reconfigure competences in response to dynamic markets is Teece's [60] dynamic capabilities framework. In line with the ideas of incremental and continuous learning, this framework provides a theoretical foundation for comprehending how businesses might promote flexibility and continual progress.

5.13.2 IMPLICATIONS FOR THEORY AND PRACTICE

Several theoretical ramifications for the discipline are highlighted in the discussion. First off, conventional notions of linear business models are called into question by the combination of incremental learning, continuous learning, and adaptability. The OODA loop and the Cynefin framework are two examples of theoretical frameworks that help us understand decision-making processes in dynamic contexts more deeply. The synthesis advocates for a shift in theoretical frameworks to take into account the iterative and dynamic character of entrepreneurial initiatives. From a practical standpoint, the conversation focuses on ideas that entrepreneurs and organizational leaders may implement. Senge's (1990) proposal for the creation of a learning culture appears to be a workable approach to encouraging ongoing learning and flexibility in businesses. The use of agile approaches, bolstered by tools and technology, provides a concrete means of converting abstract ideas into routine business operations. The summary highlights how organizational leaders must welcome change, encourage cross-functional cooperation, and aggressively seek out input in order to facilitate incremental changes.

5.13.3 CONTRIBUTIONS TO THE FIELD

By combining pre-existing concepts and models into a whole manual for agile entrepreneurship, this study contributes theoretically. A comprehensive approach to managing dynamic markets is provided by using decision-making frameworks in conjunction with the integration of incremental learning, continuous learning, and adaptability. The framework of dynamic capacities is notable for its ability to unite diverse theoretical ideas and offer a theoretical basis for comprehending how businesses might prosper in an unpredictable environment. The study makes a useful contribution in that it offers entrepreneurs actionable advice and methods for putting adaptive tactics into practice. The emphasis on establishing a learning culture, utilizing agile methodologies, and leveraging tools and technologies offers tangible insights for organizational leaders. The case studies of Netflix and Amazon serve as practical examples of successful adaptation, providing real-world illustrations of the concepts discussed in the study.

5.13.4 FUTURE RESEARCH DIRECTIONS

Future research directions should explore the context-specific applications of the synthesized concepts. Different industries and organizational sizes may require tailored approaches to agile entrepreneurship. Investigating how the integrated framework applies to specific contexts can provide nuanced insights for practitioners. The study suggests that organizational leaders play a pivotal role in driving adaptive strategies. Future research should delve deeper into the role of leadership in fostering a learning culture, overcoming resistance to change, and shaping organizational resilience. Exploring leadership styles and behaviors that align with agile entrepreneurship can contribute to leadership theory. Long-term outcomes of agile entrepreneurship strategies should be a focal point for future research. Understanding how organizations sustain success and adaptability over time, considering economic downturns or market disruptions, can provide valuable insights. Longitudinal studies tracking the application of adaptive strategies can contribute to the field's understanding of sustained organizational resilience.

In conclusion, the discussion synthesizes the findings, highlighting the integration of incremental learning, continuous learning, and adaptation in agile entrepreneurship. The implications for theory and practice emphasize the need for a shift in theoretical perspectives and actionable strategies for organizational leaders. Contributions to the field include theoretical insights and practical guidelines. Future research directions call for exploring context-specific applications, examining the role of leadership, and investigating long-term outcomes of agile entrepreneurship strategies.

5.14 CONCLUSION

In the dynamic symphony of agile entrepreneurship, this study orchestrates a harmonious blend of incremental learning, continuous adaptation, and dynamic capabilities. Beyond a mere juxtaposition of concepts, it forms an integrated framework that transcends conventional business strategies. Dynamic capabilities, as Teece [60] envisioned, emerge as the cornerstone, guiding organizations to integrate, build, and reconfigure competencies. Practically, the study illuminates the path for entrepreneurs, emphasizing the establishment of a learning culture through agile methodologies. Netflix and Amazon's case studies exemplify how strategic adaptation, fueled by continuous learning, propels organizational success. Leadership stands as the linchpin, steering organizations through the fog of change. Visionary leaders, adept at overcoming resistance and embracing collaboration, sculpt a legacy of adaptability. Success metrics and iterative improvements form a symphony, ensuring ventures not only survive but flourish.

As we close this chapter, it is not an end but a prelude to continuous discovery. The implications for theory and practice invite a journey into uncharted territories, where each step signifies a lesson learned and a challenge conquered. In the grand tapestry of agile entrepreneurship, this study stands as a guidepost an enduring testament to the boundless potential in the intersection of learning, adaptation, and the unwavering spirit of entrepreneurship. May it inspire those venturing into the unknown, where each step is a commencement of a continuous journey toward entrepreneurial excellence.

KEYWORDS

- **agile entrepreneurship**
- **continuous learning**
- **dynamic capabilities**
- **incremental learning**
- **leadership**

REFERENCES

1. Aamodt, A. (1991). *A knowledge-intensive, integrated approach to problem-solving and sustained learning* (Doctoral dissertation, Universitetet i Trondheim (Norway)).
2. Amburgey, T. L., & Miner, A. S. (1992). Strategic momentum: The effects of repetitive, positional, and contextual momentum on merger activity. *Strategic Management Journal*, *13*(5), 335-348.
3. American Economic Association, Royal Economic Society, & Simon, H. A. (1966). *Theories of decision-making in economics and behavioural science* (pp. 1-28). Palgrave Macmillan UK.
4. Andreatos, A. S. (2015). The use of MOOCs in the Continuing Education of Individuals and Organisations. In *Furthering Higher Education Possibilities through Massive Open Online Courses* (pp. 49-79). IGI Global.
5. Bauman, A., & Lucy, C. (2021). Enhancing entrepreneurial education: Developing competencies for success. *The International Journal of Management Education*, *19*(1), 100293.
6. Beck, K., Beedle, M., Van Bennekum, A., Cockburn, A., Cunningham, W., Fowler, M., ... & Thomas, D. (2001). The agile manifesto.
7. Beck, K., Beedle, M., Van Bennekum, A., Cockburn, A., Cunningham, W., Fowler, M., ... & Thomas, D. (2001). Manifesto for agile software development.

8. Beck, U. (2000). The cosmopolitan perspective: sociology of the second age of modernity. *The British journal of sociology*, *51*(1), 79-105.
9. Bergsteiner, H., Avery, G. C., & Neumann, R. (2010). Kolb's experiential learning model: critique from a modelling perspective. *Studies in Continuing Education*, *32*(1), 29-46.
10. Bikse, V., Lusena–Ezera, I., & Rivza, B. (2018). Innovative start-ups: challenges and development opportunities in Latvia. *International Journal of Innovation Science*, *10*(2), 261-273.
11. Brown, S. L., & Eisenhardt, K. M. (1998). *Competing on the edge: Strategy as structured chaos*. Harvard Business Press.
12. Bryk, A. S., Gomez, L. M., Grunow, A., & LeMahieu, P. G. (2015). *Learning to improve: How America's schools can get better at getting better*. Harvard Education Press.
13. Bughin, J., Chui, M., & Manyika, J. (2010). Clouds, big data, and smart assets: Ten tech-enabled business trends to watch. *McKinsey quarterly*, *56*(1), 75-86.
14. Cascio, W. F. (2000). Managing a virtual workplace. *Academy of Management Perspectives*, *14*(3), 81-90.
15. Chan, F. K., & Thong, J. Y. (2009). Acceptance of agile methodologies: A critical review and conceptual framework. *Decision support systems*, *46*(4), 803-814.
16. Chesbrough, H. (2003). The logic of open innovation: managing intellectual property. *California management review*, *45*(3), 33-58.
17. Chesbrough, H. (2010). Business model innovation: opportunities and barriers. *Long range planning*, *43*(2-3), 354-363.
18. Chesky, B., & Gebbia, J. (2016). How Airbnb's Agile Culture Fuels Innovation. *Harvard Business Review*, November-December 2016.
19. Constantinides, E., & Fountain, S. J. (2008). Web 2.0: Conceptual foundations and marketing issues. *Journal of direct, data and digital marketing practice*, *9*, 231-244.
20. Crossan, M. M., Lane, H. W., White, R. E., & Djurfeldt, L. (1995). Organizational learning: Dimensions for a theory. *The international journal of organizational analysis*, *3*(4), 337-360.
21. DaSilva, C. M., Trkman, P., Desouza, K., & Lindič, J. (2013). Disruptive technologies: a business model perspective on cloud computing. *Technology Analysis & Strategic Management*, *25*(10), 1161-1173.
22. Davenport, T. H. (2018). GE Healthcare Embraces Incremental Learning and Agile Entrepreneurship. *Harvard Business Review*, July-August 2018.
23. Duncan, N. B. (1995). Capturing flexibility of information technology infrastructure: A study of resource characteristics and their measure. *Journal of management information systems*, *12*(2), 37-57.
24. Edmondson, A. C. (2012). *Teaming: How organizations learn, innovate, and compete in the knowledge economy*. John Wiley & Sons.
25. Edmondson, A. C. (2018). *The fearless organization: Creating psychological safety in the workplace for learning, innovation, and growth*. John Wiley & Sons.
26. Eisenhardt, K. M., & Brown, S. L. (1998). Competing on the edge: Strategy as structured chaos. *Long range planning*, *31*(5), 786-789.
27. Eriksson, M., Fleischer, R., Johansson, A., Snickars, P., & Vonderau, P. (2019). *Spotify teardown: Inside the black box of streaming music*. Mit Press.

28. Fillion, G., Koffi, V., & Ekionea, J. P. B. (2015). Peter Senge's learning organization: A critical view and the addition of some new concepts to actualize theory and practice. *Journal of Organizational Culture, Communications and Conflict*, *19*(3), 73.
29. Fitzgerald, M., Kruschwitz, N., Bonnet, D., & Welch, M. (2014). Embracing digital technology: A new strategic imperative. *MIT sloan management review*, *55*(2), 1.
30. Flumerfelt, S., Bella Siriban-Manalang, A., & Kahlen, F. J. (2012). Are agile and lean manufacturing systems employing sustainability, complexity and organizational learning?. *The Learning Organization*, *19*(3), 238-247.
31. Fosnot, C. T., & Perry, R. S. (1996). Constructivism: A psychological theory of learning. *Constructivism: Theory, perspectives, and practice*, *2*(1), 8-33.
32. Fowler, A. (Ed.). (2013). *Striking a balance: A guide to enhancing the effectiveness of non-governmental organisations in international development*. Routledge.
33. Frederiksen, D. L., & Brem, A. (2017). How do entrepreneurs think they create value? A scientific reflection of Eric Ries' Lean Startup approach. *International Entrepreneurship and Management Journal*, *13*, 169-189.
34. Ganz, S. C. (2018). Ignorant decision making and educated inertia: Some political pathologies of organizational learning. *Organization Science*, *29*(1), 39-57.
35. Gephart, M. A., Marsick, V. J., Van Buren, M. E., Spiro, M. S., & Senge, P. (1996). Learning organizations come alive. *Training & Development*, *50*(12), 34-46.
36. Gerster, D., Dremel, C., & Kelker, P. (2018). " Agile meets non-agile": Implications of adopting agile practices at enterprises.
37. Gonçalves, L., & Linders, B. (2015). *Getting value out of agile retrospectives: a toolbox of retrospective exercises*. Ben Linders Publishing.
38. Grebić, B., & Stojanović, A. (2021). Application of the Scrum framework on projects in IT sector. *Eur. Proj. Manag. J*, *11*(2), 37-46.
39. Greenhalgh, T., Wherton, J., Papoutsi, C., Lynch, J., Hughes, G., Hinder, S., ... & Shaw, S. (2017). Beyond adoption: a new framework for theorizing and evaluating nonadoption, abandonment, and challenges to the scale-up, spread, and sustainability of health and care technologies. *Journal of medical Internet research*, *19*(11), e8775.
40. Hansen, J. Ø., Jensen, A., & Nguyen, N. (2020). The responsible learning organization: Can Senge (1990) teach organizations how to become responsible innovators?. *The learning organization*, *27*(1), 65-74.
41. Hewlin, T. (2018). How Netflix Uses Data and Analytics to Drive Innovation. *Harvard Business Review*, May-June 2018.
42. Joiner, B. (2009). Creating a culture of agile leaders: A developmental approach. *People and Strategy*, *32*(4), 28.
43. Kolb, B. (1984). Functions of the frontal cortex of the rat: a comparative review. *Brain research reviews*, *8*(1), 65-98.
44. Kotter, J. P., & Schlesinger, L. A. (1979). Choosing strategies for change.
45. Le, D. (2019). *Agile practices adoption with Lean in growing entrepreneur companies* (Master's thesis, D. Le).
46. Majchrzak, A., More, P. H., & Faraj, S. (2012). Transcending knowledge differences in cross-functional teams. *Organization science*, *23*(4), 951-970.
47. Malakar, S. (2021). *Agile Methodologies In-Depth: Delivering Proven Agile, SCRUM and Kanban Practices for High-Quality Business Demands (English Edition)*. BPB Publications.

48. Mansoori, Y., & Lackeus, M. (2020). Comparing effectuation to discovery-driven planning, prescriptive entrepreneurship, business planning, lean startup, and design thinking. *Small Business Economics*, *54*, 791-818.
49. McGrath, R. G., & MacMillan, I. C. (2000). *The entrepreneurial mindset: Strategies for continuously creating opportunity in an age of uncertainty* (Vol. 284). Harvard Business Press.
50. Meso, P., & Jain, R. (2006). Agile software development: adaptive systems principles and best practices. Information systems management, 23(3), 19-30.
51. Miceli, A., Hagen, B., Riccardi, M. P., Sotti, F., & Settembre-Blundo, D. (2021). Thriving, not just surviving in changing times: How sustainability, agility and digitalization intertwine with organizational resilience. *Sustainability*, *13*(4), 2052.
52. Mier, J., & Kohli, A. K. (2021). Netflix: reinvention across multiple time periods, reflections and directions for future research. *AMS Review*, *11*, 194-205.
53. Mikalef, P., van de Wetering, R., & Krogstie, J. (2021). Building dynamic capabilities by leveraging big data analytics: The role of organizational inertia. *Information & Management*, *58*(6), 103412.
54. Miles, M. B., & Huberman, A. M. (1994). *Qualitative data analysis: An expanded sourcebook*. sage.
55. Miller, D. (1982). Evolution and revolution: A quantum view of structural change in organizations. Journal of Management studies, 19(2), 131-151.
56. Mishra, D., & Mishra, A. (2011). Complex software project development: agile methods adoption. *Journal of Software Maintenance and Evolution: Research and Practice*, *23*(8), 549-564.
57. Moran, A. (2015). Managing agile. *Strategy, implementation, organisation and People. cham: Springer*.
58. Nelson, L. M. (2013). Collaborative problem solving. In *Instructional-design theories and models* (pp. 241-267). Routledge.
59. Oskam, J., & Boswijk, A. (2016). Airbnb: the future of networked hospitality businesses. *Journal of tourism futures*, *2*(1), 22-42.
60. Patz, M. (2013). *Lean Startup: adding an experimental learning perspective to the entrepreneural process* (Master's thesis, University of Twente).
61. Piaget, J. (1972). Development and learning. *Reading in child behavior and development*, 38-46.
62. Pittaway, L., Robertson, M., Munir, K., Denyer, D., & Neely, A. (2004). Networking and innovation: a systematic review of the evidence. *International journal of management reviews*, *5*(3-4), 137-168.
63. Porter, M. E., & Heppelmann, J. E. (2015). How smart, connected products are transforming companies. *Harvard business review*, *93*(10), 96-114.
64. Prieto, L. C. (2012). Knowledge is power? An inquiry into knowledge management, its effects on individual creativity, and the moderating role of an entrepreneurial mindset. *Academy of Strategic Management Journal*, *11*(1), 43.
65. Reymen, I. M., Andries, P., Berends, H., Mauer, R., Stephan, U., & Van Burg, E. (2015). Understanding dynamics of strategic decision making in venture creation: a process study of effectuation and causation. *Strategic entrepreneurship journal*, *9*(4), 351-379.

66. Sessa, V. I., & London, M. (2015). *Continuous learning in organizations: Individual, group, and organizational perspectives*. Psychology Press.
67. Shane, S., & Venkataraman, S. (2000). The promise of entrepreneurship as a field of research. *Academy of management review*, *25*(1), 217-226.
68. Shane, S., & Venkataraman, S. (2000). The promise of entrepreneurship as a field of research. *Academy of management review*, *25*(1), 217-226.
69. Sheng, M. L., & Chien, I. (2016). Rethinking organizational learning orientation on radical and incremental innovation in high-tech firms. *Journal of Business Research*, *69*(6), 2302-2308.
70. Siakas, K. V., & Siakas, E. (2007). The agile professional culture: A source of agile quality. *Software Process: Improvement and Practice*, *12*(6), 597-610.
71. Sjödin, D., Parida, V., Palmié, M., & Wincent, J. (2021). How AI capabilities enable business model innovation: Scaling AI through co-evolutionary processes and feedback loops. *Journal of Business Research, 134*, 574-587.
72. Snowden, D. J., & Boone, M. E. (2007). A leader's framework for decision making. *Harvard business review*, *85*(11), 68.
73. Teece, D. J. (2007). Explicating dynamic capabilities: the nature and microfoundations of (sustainable) enterprise performance. *Strategic management journal*, *28*(13), 1319-1350.
74. Tolfo, C., Wazlawick, R. S., Ferreira, M. G. G., & Forcellini, F. A. (2018). Agile practices and the promotion of entrepreneurial skills in software development. *Journal of Software: Evolution and Process*, *30*(9), e1945.
75. Tremblay Jr, M. P. (2015). Shaping and adapting: Unlocking the power of Colonel John Boyd's OODA Loop. *Master of Military Studies, United States Marine Corps, Command and Staff College, Quantico, Virginia*.
76. Turesky, E. F., & Gallagher, D. (2011). Know thyself: Coaching for leadership using Kolb's experiential learning theory. *The Coaching Psychologist*, *7*(1), 5-14.
77. Wong, B. M., & Headrick, L. A. (2021). Application of continuous quality improvement to medical education. *Medical Education*, *55*(1), 72-81.
78. Yin, R. K. (2003). Designing case studies. Qualitative research methods, 5(14), 359-386.

CHAPTER 6

Development of an Artificial Intelligence-Based Breast Cancer (BC) Prediction in Healthcare Systems

M. ASHOK KUMAR[1], ALIYU MOHAMMED[2], S. VEERAMA DEVI[3], and V. KAVITHA[4]

[1]*Department of Computer Science, Bluecrest University, Monrovia, Liberia*

[2]*Department of Management, Skyline University Nigeria, Kano, Nigeria*

[3]*Department of Data Science, Islamiah Women's Arts and Science College, Vaniyambadi, Tamilnadu, India*

[4]*Department of Computer Science with Cognitive Systems, Sri Ramakrishna College of Arts & Science, Coimbatore, Tamil Nadu, India*

ABSTRACT

Artificial intelligence (AI) helps to identify and detect the image of breast masses, segment breast masses, determine breast density, and determine cancer risk. One of the most crucial stages in computer-aided diagnosis is the detection of breast masses since they are a frequent finding in breast cancer (BC) patients. According to cancer statistics, BC is presently the most prevalent kind of cancer among women globally. It is one of the most common causes of mortality in women and grows in the breast tissue.

The second most common cause of mortality worldwide and one of the worst diseases is BC. If this cancer is discovered when it is in its early stages, it may be cured. 93% or more of BC patients survive

Real-Time Artificial Intelligence (AI): Key Motivations, Technologies, Platforms, and Use Cases. Pethuru Raj, B. Sundaravadivazhagan, V. Kavitha, B. Narendra Kumar Rao, & Hannah Vijaykumar (Eds.)

DOI: 10.1201/9781998511358-6

if the disease is discovered early. BC may be categorized in a variety of ways and has many different subtypes. It is easy to get perplexed. A type of BC is when certain cells in the breast become cancerous. Patients with BC may develop malignant or benign tumors. Malignant tumors are risky due to their more rapid development than benign ones. To treat a patient with BC effectively, early tumor type identification is crucial. The major goals of this chapter are to describe the dataset and assess how effectively various machine learning techniques predict BC. To categorize tumors into benign and malignant types, support vector machine, logistic regression, K-nearest neighbors, decision tree, Naive Bayes, and random forest classifiers have been used. In this chapter, the optimum algorithm is determined by calculating and comparing the accuracy of each method.

6.1 INTRODUCTION

One kind of cancer that affects the breast is breast cancer (BC). Cell growth or division that is uncontrolled can result in cancer. An X-ray typically reveals a tumor composed of BC cells. BC represents one of the most common illnesses that causes mortality in women. BC, being one of the most widespread diseases in women, affects 10% of women at various times in their lives and constantly has a relatively high prevalence and mortality rate. After lung cancer, it is the second most widespread cause of mortality for women. 25% of all malignancies and 12% of all new cases in women are BC. It is possible to identify BC by classifying tumors. There are two different types of tumors: malignant and benign, as shown in BC cases. Malignant tumors multiply faster than benign tumors do. To differentiate between various tumors, doctors need a precise diagnostic approach.

However, it can often be quite difficult to differentiate one tumor from another, even for experts. Therefore, identifying the type of malignancy requires a reliable automatic diagnostic system. Patients are often put to a flurry of tests, such as ultrasound, biopsy, and mammography, to ascertain the presence of BC based on the many forms of BC symptoms. The most typical of these methods is biopsy, which requires taking a sample of tissues or cells for identification. Images of cells and tissues taken at a microscopic scale can be used to quantify numerical properties including radius, roughness, perimeter, and area. Since early BC diagnosis lets patients to receive quick clinical therapy, it can considerably improve prognosis and survival chances. Accurate BC detection and patient classification into kinds of malignant or benign patients are crucial areas of research. Several methods for predicting BC have been discovered in recent years. Recent literature uses a classification technique called support vector machines (SVM).

The implementation of artificial intelligence (AI) to the medical sector, notably in the field of breast cancer prediction is discussed below. The following list outlines how and where AI can be used:

1. **Early Detection:** For the purpose of detecting breast cancer early, AI can be used to analyze mammograms and other medical pictures. Imaging patterns and anomalies that human radiologists might not see can be found by machine learning systems. The likelihood of a favorable outcome from treatment can be greatly increased by early identification.
2. **Risk Assessment:** A patient's medical history, genetics, and lifestyle characteristics can be used by AI to estimate their chance of acquiring breast cancer. It can offer tailored risk evaluations, enabling healthcare professionals to propose specialized prevention and screening techniques.
3. **Data Analysis:** AI can process and analyze vast amounts of patient data, including electronic health records, pathology reports, and genomic information. This can help healthcare providers make more informed decisions about diagnosis and treatment plans.
4. **Treatment Recommendations:** Oncologists can choose the best course of treatment for patients with breast cancer with the help of AI-powered tools. AI can recommend the optimal course of action based on unique patient features by examining treatment outcomes from a vast pool of patient data.
5. **Patient Monitoring:** Oncologists can choose the best therapy options for patients with breast cancer with the help of AI-powered tools. AI is able to recommend the optimal course of action for each patient based on their unique characteristics by analyzing the results of treatment from a huge pool of patient data.
6. **Resource Optimization:** By anticipating patient volumes, resource needs, and appointment availability, AI helps optimize healthcare resources. By doing this, healthcare facilities are guaranteed to

be ready to meet the demand for services related to breast cancer screening and treatment.

7. **Telemedicine and Remote Monitoring:** AI can help with telemedicine consultations and remote patient monitoring for those with breast cancer. Access to care is improved, especially in distant or disadvantaged locations, because patients can receive follow-up care and consultations without needing to make frequent in-person visits.
8. **Research and Drug Discovery:** AI's ability to analyze large datasets and find possible treatment candidates can hasten breast cancer research. Additionally, it can assist with planning clinical trials and predicting how patients would react to novel treatments.
9. **Patient Education and Support:** Virtual assistants and chatbots powered by AI can answer frequent queries from patients regarding breast cancer, treatment, and lifestyle modifications while also offering information, support, and guidance.
10. **Policy and Decision-Making:** AI can help administrators and policymakers in the healthcare industry make data-driven decisions on breast cancer screening programs, budget allocation, and healthcare policy creation.

In conclusion, AI can be used in the early detection, therapy, and beyond stages of breast cancer care. It is an effective tool for enhancing breast cancer prognosis, diagnosis, therapy, and patient outcomes within the healthcare system due to its capacity to handle and analyze big datasets, spot trends, and offer personalized insights.

6.1.1 OBJECTIVES

1. To examine the dataset for BC in order to categorize the features.
2. Discovering relationships between the values.
3. Training and testing portions of the dataset should be separated.
4. Using SVM to analyze the dataset will result in respectable accuracy.

The rest of the chapter are as follows: Section 6.2 discusses the literature study needed for BC classification. SVM is given in Section 6.3. Section 6.4 displays the discussion for the results. Section 6.5 shows the summary of this chapter.

6.2 LITERATURE REVIEW

The background research for BC classification is displayed in Table 6.1. It discusses the strategies and algorithms used to categorize the dataset as well as how to do so.

Most of the literature reviews are using many algorithms such as RF, LR, DT, NB, XGBoost, and LightGBM. From this table it is observed that RF, LR, DT, NB, XGBoost, and LightGBM cannot achieve more accuracy than any other algorithm but the SVM Algorithm to produce highest accuracy.

6.3 SUPPORT VECTOR MACHINE

In this section, the recommended system's methodology will be discussed, beginning with the correlation of the BC dataset and concluding with the categorization idea. According to this concept, the dataset should be split into training and test sets before an SVM classifier is used. Figure 6.1 displays the potential techniques.

The SVM methodology, a supervised machine learning method, has shown excellent performance in tackling classification issues utilizing a dataset of BC patients. SVM distinguishes between two classes once the input data have been mathematically transformed into a high-dimensional space, in contrast to LR, which relies on a pre-existing model to forecast whether or not a binary event will occur by fitting data to a logistic curve. SVM is data-driven and model-free, which accounts for this. This method has been used to build automated cancer classification and improve tumor classification algorithms in the BC dataset.

It shows the possibility of SVM as a method for dividing people into groups based on their tumor status. The goal was to create an SVM-based method to separate benign from malignant tumors in a dataset of BC cases. The factors that were utilized to produce the SVM model are shown.

6.3.1 METHODOLOGY

The SVM is one of the most fruitful supervised machine learning methods. This method of supervised machine learning is utilized for both classification and regression. SVM may categorize data points even if they are not

TABLE 6.1 Background Study for BC Classification.

Year	Author	Method	Dataset	Outcome
2020	Lakshmitha and Hayum [7]	SVM	Handwritten digits images	Detection
2021	Ara et al. [8]	Random Forest (RF), Logistic Regression (LR), Decision Tree (DT), Naive Bayes (NB), SVM and K-Nearest Neighbors (KNN).	WBCD dataset	Classification
2021	Abd-Elnaby et al. [5]	KNN, NB, SVM, K-means	Microarray data	Classification
2021	Lomboy and Hernandez [4]	SVM	BC Wisconsin (diagnostic) dataset	Classification
2021	Deepika et al. [2]	SVM	Synthetic data	Prediction
2021	Hasan et al. [9]	SVM, KNN, RF	WDBC dataset	Classification
2022	Egwom et al. [6]	SVM	Wisconsin Prognostic BC dataset	Classification
2022	Elkorany et al. [1]	SVM	WDBC dataset	Optimization
2022	Michael et al. [3]	K-NN, SVM, RF, XGBoost, and LightGBM	Synthetic data	Classification

otherwise linearly separable by translating the data into a high-dimensional feature space. Look at Figure 6.2, where a decision boundary or hyperplane is used to divide two distinct categories: SVM is a technique used to address classification challenges in machine learning. A point can be assigned to one of two classes in a two-dimensional problem space depending on the values of the two dimensions (let us say, independent variables), X1 and X2.

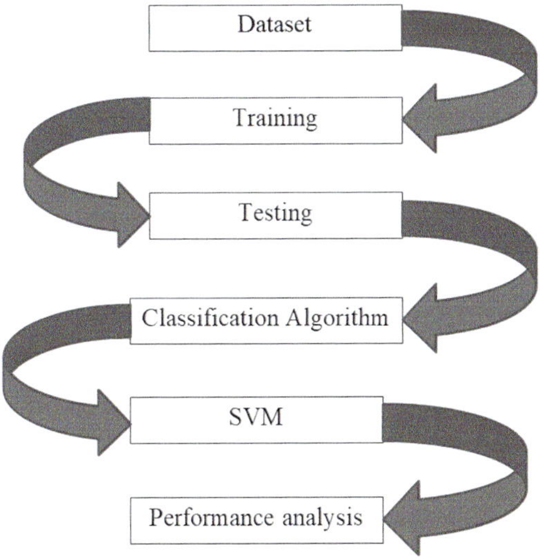

FIGURE 6.1 Potential techniques.

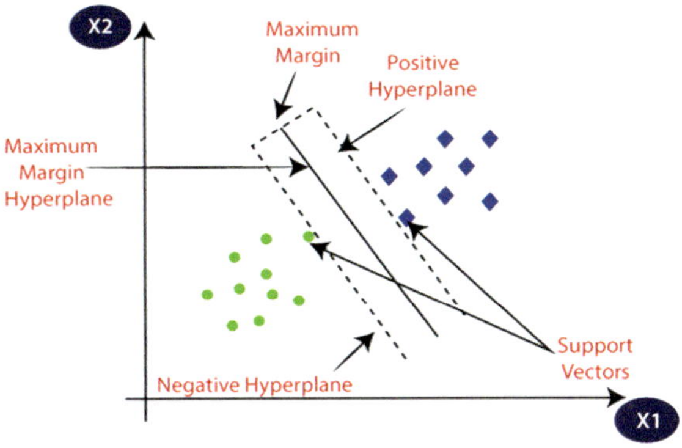

FIGURE 6.2 Visualize two distinct hyperplanes.

Source: Reprinted with permission from Ref. [10]. Copyright © 2023, The Author(s), under exclusive licence to Springer Nature Singapore Pte Ltd

6.3.2 DRAWBACKS OF SVM

1. The SVM approach is not well suited to large amount of datasets.
2. SVM performs poorly when the target classes overlap and there is greater noise in the dataset.
3. The performance of SVM is bad if there are few characteristics per data point than there for the training data samples.

Several issues with SVM were discussed in this chapter. One significant drawback of SVMs is their sensitivity to the kernel function and hyperparameters that are selected, which might have an impact on their performance. We undertake a process of hyperparameter tuning and kernel selection to find the best configuration for our particular dataset and situation in order to mitigate this. The feature selection methods and data pretreatment approaches minimize dimensionality and improve computational efficiency because SVMs often have scaling issues when working with huge datasets. In order to increase the robustness and generalization capabilities of SVMs, it also added ensemble learning techniques like bagging and boosting. Together, these approaches allowed us to maximize the benefits of SVMs while minimizing their drawbacks, ultimately producing more accurate and dependable research outcomes.

6.4 RESULTS AND DISCUSSION

The results obtained from applying classification to the datasets are shown in this section. Many studies have been conducted utilizing various techniques to enhance prediction performance for medical categorization. The suggested model's classification accuracy was calculated using a confusion matrix. The first classification challenge involved examining how feature reduction procedures affected the ability to anticipate BC cases. No features were derived from the datasets used in this experiment. The experimental results in light of the metrics are used for the model. Without feature reduction, the SVM results are shown while computing the mean and median for missing values. The process for importing libraries is shown in Figure 6.3. How to determine the dataset's form is shown in Figure 6.4. Figure 6.5 describes to check if there are any nulls in any column.

6.4.1 IMPORTING PACKAGES

Figure 6.3 shows importing libraries.

```
import numpy as np
import pandas as pd

import matplotlib.pyplot as plt
import seaborn as sns
import os

cancer=pd.read_csv("data.csv")
cancer.head()
```

	id	diagnosis	radius_mean	texture_mean	perimeter_mean	area_mean	smoothness_mean	compactness_mean	concavity_mean	concave points_mean	...	rac
0	842302	1	17.99	10.38	122.80	1001.0	0.11840	0.27760	0.3001	0.14710	...	
1	842517	1	20.57	17.77	132.90	1326.0	0.08474	0.07864	0.0869	0.07017	...	
2	84300903	1	19.69	21.25	130.00	1203.0	0.10960	0.15990	0.1974	0.12790	...	
3	84348301	1	11.42	20.38	77.58	386.1	0.14250	0.28390	0.2414	0.10520	...	
4	84358402	1	20.29	14.34	135.10	1297.0	0.10030	0.13280	0.1980	0.10430	...	

5 rows × 32 columns

FIGURE 6.3 Importing libraries.

6.4.2 SHAPE OF DATASET

Figure 6.4 shows shape of dataset.

```
cancer.shape
```

```
(569, 32)
```

```
cancer.tail()
```

	id	diagnosis	radius_mean	texture_mean	perimeter_mean	area_mean	smoothness_mean	compactness_mean	concavity_mean	concave points_mean	...	rac
564	926424	1	21.56	22.39	142.00	1479.0	0.11100	0.11590	0.24390	0.13890	...	
565	926682	1	20.13	28.25	131.20	1261.0	0.09780	0.10340	0.14400	0.09791	...	
566	926954	1	16.60	28.08	108.30	858.1	0.08455	0.10230	0.09251	0.05302	...	
567	927241	1	20.60	29.33	140.10	1265.0	0.11780	0.27700	0.35140	0.15200	...	
568	92751	0	7.76	24.54	47.92	181.0	0.05263	0.04362	0.00000	0.00000	...	

5 rows × 32 columns

FIGURE 6.4 Shape of dataset.

6.4.3 NULL VALUES

Figure 6.5 shows null values of the dataset.

```
cancer.isnull().sum()

id                         0
diagnosis                  0
radius_mean                0
texture_mean               0
perimeter_mean             0
area_mean                  0
smoothness_mean            0
compactness_mean           0
concavity_mean             0
concave points_mean        0
symmetry_mean              0
fractal_dimension_mean     0
radius_se                  0
texture_se                 0
perimeter_se               0
area_se                    0
smoothness_se              0
compactness_se             0
concavity_se               0
concave points_se          0
symmetry_se                0
fractal_dimension_se       0
radius_worst               0
```

FIGURE 6.5 Null values of the dataset.

6.4.4 PAIR PLOT

Figure 6.6 shows the Looking at the Visualization of important features in relation to target variable diagnosis to see on which features it is more related. Visualize the diagnosis of malignant and benign based of its count is shown in Figure 6.7. Figure 6.8 shows visualize the area mean and smoothness mean of the dataset.

```
sns.pairplot(cancer, hue = 'diagnosis', vars = ['radius_mean', 'texture_mean', 'area_mean', 'perimeter_mean', 'smoothness_mean']

<seaborn.axisgrid.PairGrid at 0x2a164150fd0>
```

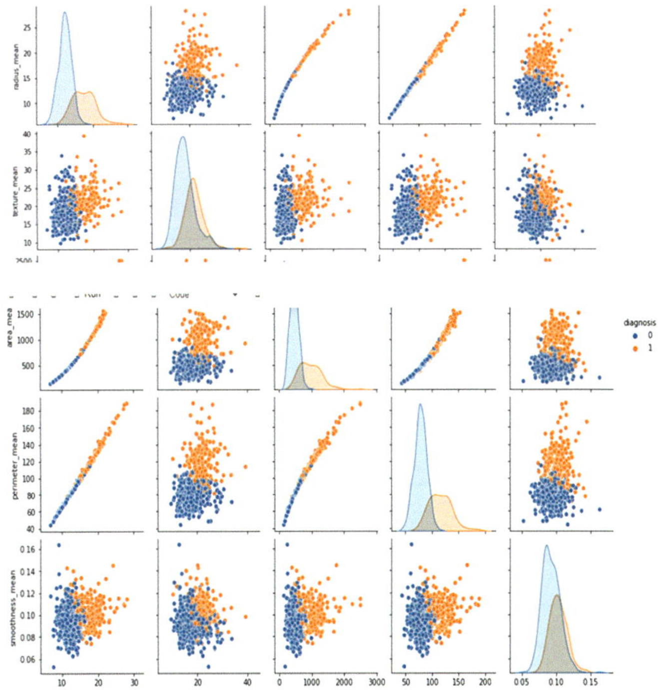

FIGURE 6.6 Pair plot.

6.4.5 *COUNT PLOT*

Figure 6.7 shows count plot.

```
sns.countplot(cancer['diagnosis'], label = "Count")

C:\Users\MATHU MATHI\anaconda3\lib\site-packages\seaborn\_decorators.py:36: FutureWarning: Pass the following variable as a key
word arg: x. From version 0.12, the only valid positional argument will be `data`, and passing other arguments without an expli
cit keyword will result in an error or misinterpretation.
  warnings.warn(

<AxesSubplot:xlabel='diagnosis', ylabel='count'>
```

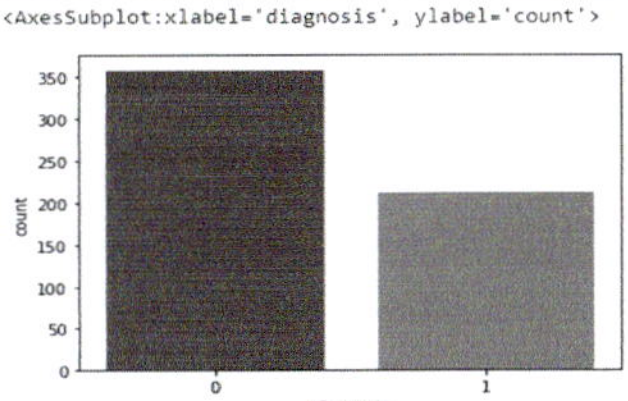

FIGURE 6.7 Count plot.

6.4.6 SCATTER PLOT

Figure 6.7 shows scatter plot or diagnosis.

```
sns.scatterplot(x = 'area_mean', y = 'smoothness_mean', hue = 'diagnosis', data = cancer)
```

```
<AxesSubplot:xlabel='area_mean', ylabel='smoothness_mean'>
```

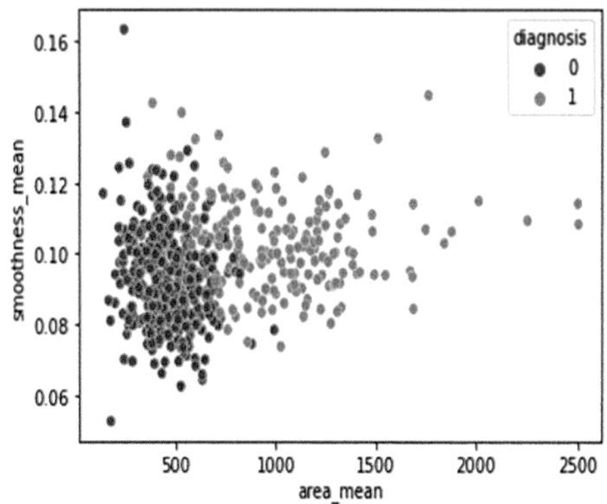

FIGURE 6.8 Scatter plot or diagnosis.

6.4.7 KDEPLOT 1

Figure 6.9 shows Kdeplot 1.

```
fig = sns.FacetGrid(cancer, hue="diagnosis",aspect=4)

fig.map(sns.kdeplot,'smoothness_mean',shade= True)

oldest = cancer['smoothness_mean'].max()

fig.set(xlim=(0,oldest))

fig.add_legend()
```

```
<seaborn.axisgrid.FacetGrid at 0x2a16a8f59a0>
```

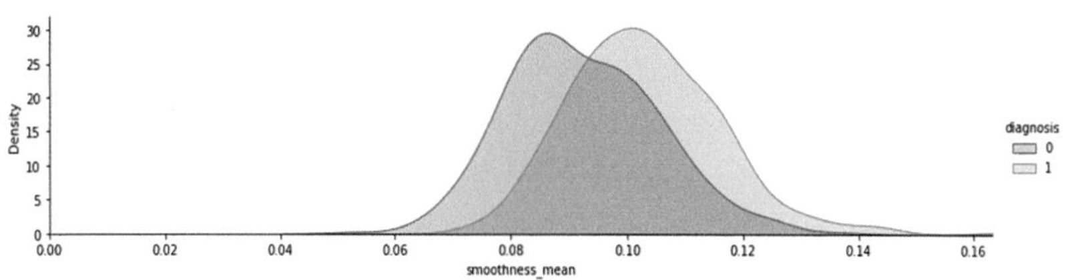

FIGURE 6.9 Kdeplot 1.

6.4.8 KEDPLOT 2

```
fig = sns.FacetGrid(cancer, hue="diagnosis",aspect=4)
fig.map(sns.kdeplot,'texture_mean',shade= True)
oldest = cancer['texture_mean'].max()
fig.set(xlim=(0,oldest))
fig.add_legend()
<seaborn.axisgrid.FacetGrid at 0x2a16a9bc4f0>
```

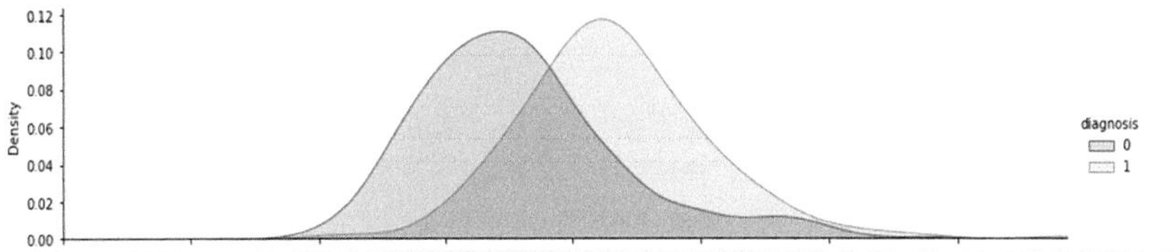

FIGURE 6.10 Kdeplot 2.

Figure 6.9 shows the map to plot all the possible kdeplots for the column of smoothness_mean by the hue choice. Map to plot all the possible kdeplots for column of texture_mean by the hue choice shown in Figure 6.10. Figure 6.11 shows in map to plot all the possible kdeplots for columns of area_mean by the hue choice. Figure 6.12 visualizes the axis for texture_mean and area_mean of hue diagnosis.

6.4.9 KEDPLOT 3

Figure 6.11 shows Kedplot 3.

```
fig = sns.FacetGrid(cancer, hue="diagnosis",aspect=4)
fig.map(sns.kdeplot,'area_mean',shade= True)
oldest = cancer['area_mean'].max()
fig.set(xlim=(0,oldest))
fig.add_legend()
<seaborn.axisgrid.FacetGrid at 0x2a16aa3c550>
```

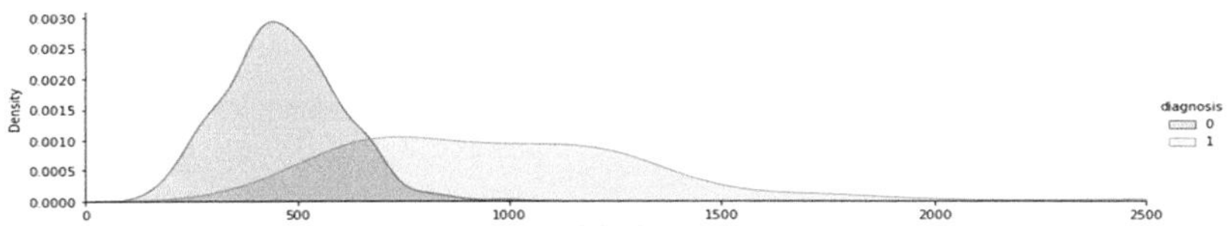

FIGURE 6.11 Kedplot 3.

6.4.10 FACTOR PLOT

Figure 6.12 shows factor plot.
sns.factorplot ('texture_mean', 'area_mean', hue= 'diagnosis',data =cancer)

```
<seaborn.axisgrid.FacetGrid at 0x2a16a9d3880>
```

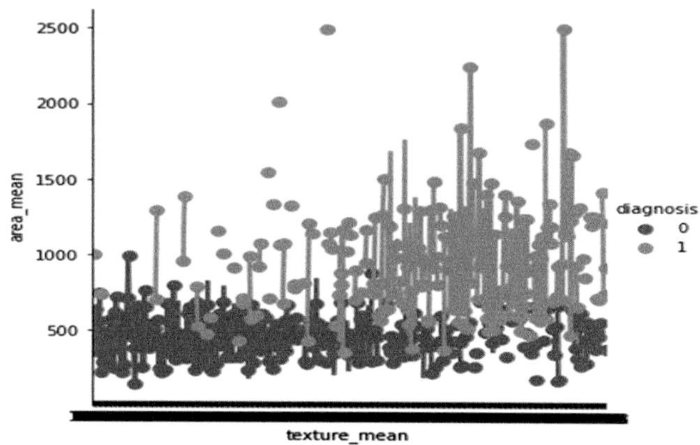

FIGURE 6.12 Factor plot.

6.4.11 SCATTERPLOT FOR CONCAVITY_SE

Figure 6.13 shows scatterplot for concavity_se.

```
sns.scatterplot('concavity_se', 'radius_mean', hue ='diagnosis', data = df_cancer)
```

```
<matplotlib.axes._subplots.AxesSubplot at 0x7f0050ad5908>
```

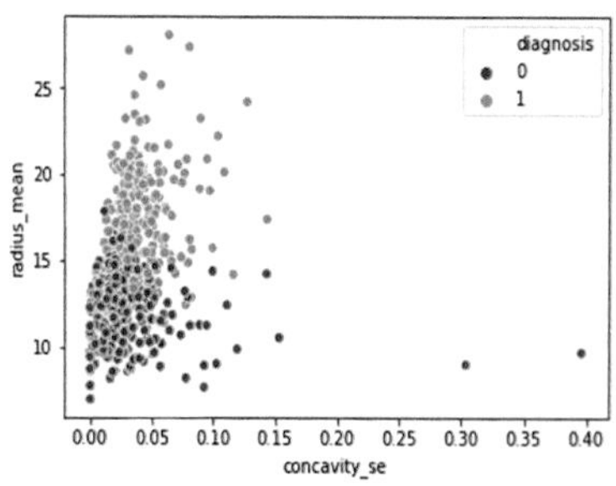

FIGURE 6.13 Scatterplot for concavity_se.

6.4.12 SCATTERPLOT FOR COMPACTNESS_SE

```
sns.scatterplot('compactness_se', 'radius_mean', hue ='diagnosis', data = df_cancer)
```

```
<matplotlib.axes._subplots.AxesSubplot at 0x7f00519ebb00>
```

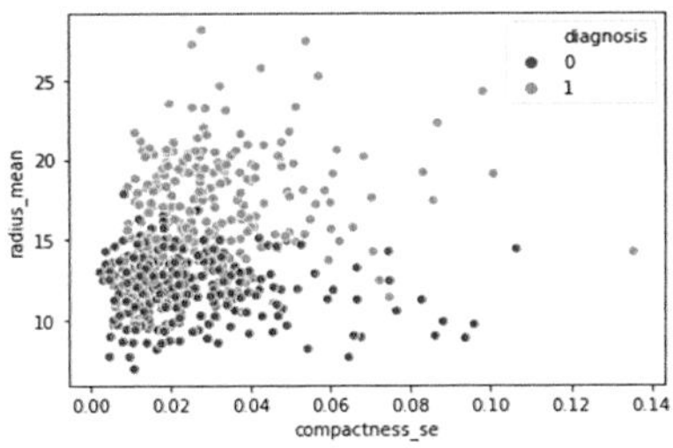

FIGURE 6.14 Scatterplot for compactness_se.

Figure 6.13 visualizes the axis of concavities and radius_mean in hue diagnosis. To visualize the axis of compactness and radius_mean in hue diagnosis in Figure 6.14.

6.4.13 HEAT MAP

plt.figure (figsize =(24,20))
sns.heatmap (cancer.corr(), annot=**True**)

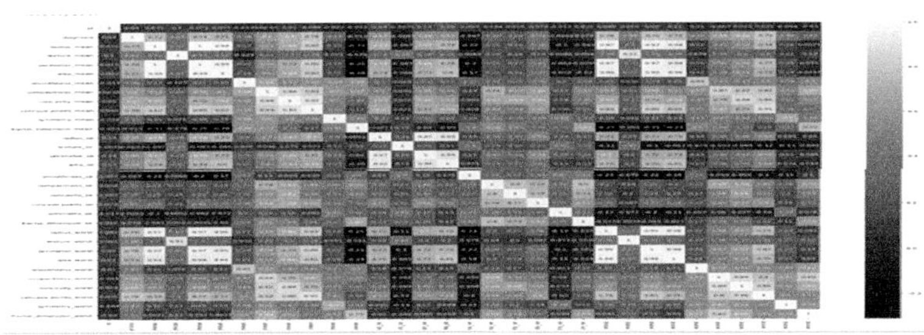

FIGURE 6.15 Heat map.

6.4.14 REMOVING UNWANTED COLUMNS

Figure 6.15 visualizes the correlation value of this dataset. The coding to remove unwanted columns is shown in Figure 6.16.

```
unwantedcolumnlist=["diagnosis","id"]
X = cancer.drop(unwantedcolumnlist,axis=1)
y = cancer['diagnosis']
```

FIGURE 6.16 Removing unwanted columns.

6.4.15 TRAINING AND TESTING

```
from sklearn.model_selection import train_test_split

X_train, X_test, y_train, y_test = train_test_split(X, y, test_size = 0.20, random_state=5)
```

FIGURE 6.17 Training and testing.

Figure 6.17 describes the coding of how to split training and testing dataset using sklearn to X_train, X_test,y_train,y_test. The coding of applying SVM Algorithm is shown in Figure 6.18. Figure 6.19 visualizes the heat map of confusion matrix by applying SVM algorithms.

6.4.16 SUPPORT VECTOR MACHINE

Figure 6.18 shows support vector machine.

```
from sklearn.svm import SVC
from sklearn.metrics import classification_report, confusion_matrix

svc_model = SVC()
svc_model.fit(X_train, y_train)

SVC()

y_predict = svc_model.predict(X_test)
cm = confusion_matrix(y_test, y_predict)
cm

array([[66,  0],
       [ 7, 41]], dtype=int64)
```

FIGURE 6.18 Support vector machine.

6.4.17 CONFUSION MATRIX CODING

Figure 6.19 shows confusion matrix.

```
sns.heatmap(cm, annot=True)
<AxesSubplot:>
```

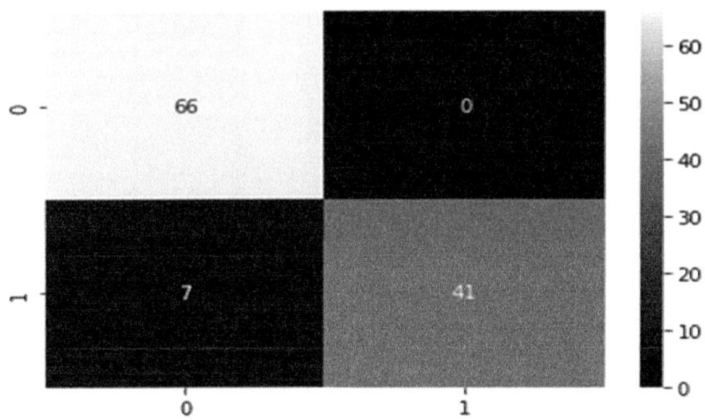

FIGURE 6.19 Confusion matrix.

6.4.18 *ASSESSMENT METRICS FOR SVM*

```
print(classification_report(y_test, y_predict))

              precision    recall  f1-score   support

           0       0.90      1.00      0.95        66
           1       1.00      0.85      0.92        48

    accuracy                           0.94       114
   macro avg       0.95      0.93      0.94       114
weighted avg       0.94      0.94      0.94       114
```

FIGURE 6.20 Assessment metrics for SVM.

Figure 6.20 shows the coding of printing evaluation metrics. Figure 6.21 discusses the training and testing.

6.4.19 *TRAINING AND TESTING*

min_train=X_train.min()
min_train

```
compactness_mean             0.019380
concavity_mean               0.000000
concave points_mean          0.000000
symmetry_mean                0.106000
fractal_dimension_mean       0.049960
radius_se                    0.111500
texture_se                   0.362100
perimeter_se                 0.757000
area_se                      6.802000
smoothness_se                0.001713
compactness_se               0.002252
concavity_se                 0.000000
concave points_se            0.000000
symmetry_se                  0.007882
fractal_dimension_se         0.000950
radius_worst                 7.930000
texture_worst               12.020000
perimeter_worst             50.410000
area_worst                 185.200000
smoothness_worst             0.071170
compactness_worst            0.027290
concavity_worst              0.000000
concave points_worst         0.000000
symmetry_worst               0.156500
fractal_dimension_worst      0.055040
dtype: float64
```

FIGURE 6.21 Training and testing.

6.4.20 RANGE OF TRAINING

range_train= (X_train-min_train).max()
range_train

```
compactness_mean              0.326020
concavity_mean                0.426800
concave points_mean           0.201200
symmetry_mean                 0.198000
fractal_dimension_mean        0.045790
radius_se                     2.761500
texture_se                    4.522900
perimeter_se                 21.223000
area_se                     518.798000
smoothness_se                 0.029417
compactness_se                0.133148
concavity_se                  0.396000
concave points_se             0.052790
symmetry_se                   0.071068
fractal_dimension_se          0.028890
radius_worst                 25.190000
texture_worst                37.520000
perimeter_worst             170.390000
area_worst                 3246.800000
smoothness_worst              0.129430
compactness_worst             1.030710
concavity_worst               1.105000
concave points_worst          0.291000
symmetry_worst                0.420900
fractal_dimension_worst       0.152460
dtype: float64
```

FIGURE 6.22 Range of training.

6.4.21 TRAINING DATA HEAD

Figure 6.23 shows training data head.

```
X_train_scaled = (X_train - min_train)/range_train
X_train_scaled.head()
```

	radius_mean	texture_mean	perimeter_mean	area_mean	smoothness_mean	compactness_mean	concavity_mean	concave points_mean	symmetry_mean	fractal_d
306	0.294335	0.206628	0.278350	0.167183	0.293220	0.101620	0.003423	0.016208	0.288889	
410	0.207251	0.265810	0.198328	0.108809	0.324546	0.103521	0.065206	0.104374	0.273232	
197	0.525297	0.410213	0.508673	0.373806	0.190304	0.205632	0.258435	0.287177	0.358586	
376	0.169861	0.355428	0.182157	0.082700	0.343956	0.449727	0.534208	0.295278	0.569697	
244	0.587770	0.466351	0.589524	0.429421	0.452018	0.418441	0.480084	0.441650	0.463636	

5 rows × 30 columns

FIGURE 6.23 Training data head.

6.4.22 SCATTER PLOT FOR AREA_MEAN AND SMOOTHNESS_MEAN

```
sns.scatterplot(x = X_train['area_mean'], y = X_train['smoothness_mean'], hue = y_train)
```

<AxesSubplot:xlabel='area_mean', ylabel='smoothness_mean'>

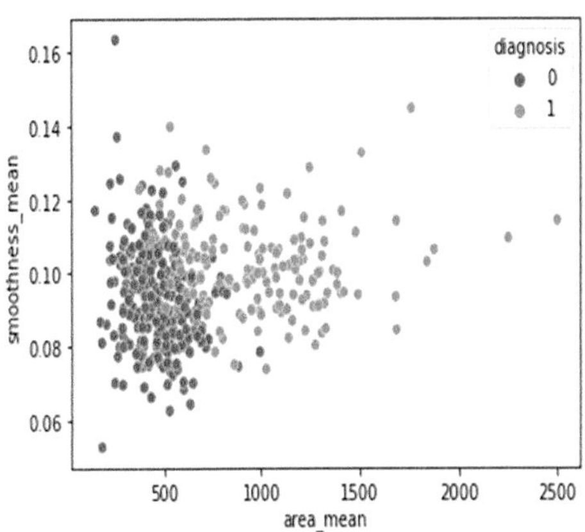

FIGURE 6.24 Scatter plot for area_mean and smoothness_mean.

The training dataset's head is depicted in Figure 6.23. The area mean and smoothness mean axes in hue diagnosis are depicted in Figure 6.24.

6.4.23 SCATTER PLOT FOR SCALED AREA_MEAN AND SCALED SMOOTHNESS_MEAN

Figure 6.25 shows scatter plot for scaled area_mean and scaled smoothness_mean.

```
sns.scatterplot(x = X_train_scaled['area_mean'], y = X_train_scaled['smoothness_mean'], hue = y_train)
```

<AxesSubplot:xlabel='area_mean', ylabel='smoothness_mean'>

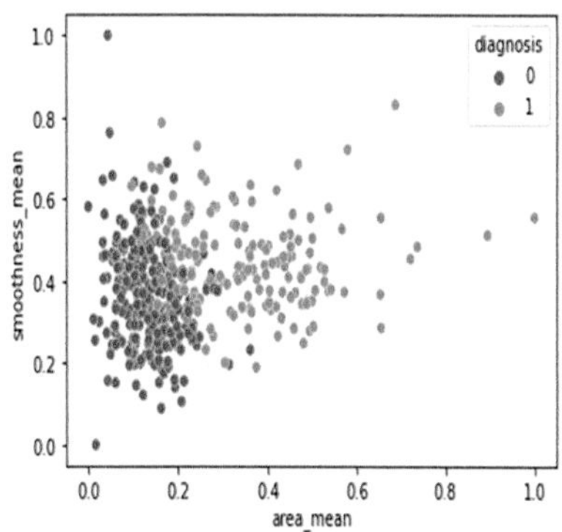

FIGURE 6.25 Scatter plot for scaled area_mean and scaled smoothness_mean.

6.4.24 CODING FOR SVM

```
min_test = X_test.min()
range_test = (X_test - min_test).max()
X_test_scaled = (X_test - min_test)/range_test
from sklearn.svm import SVC
from sklearn.metrics import classification_report, confusion_matrix

svc_model = SVC()
svc_model.fit(X_train_scaled, y_train)
```

SVC()

FIGURE 6.26 Coding for SVM.

The scaled area mean and scaled smoothness mean axes in hue diagnosis are shown in Figure 6.25. The SVM algorithm's code is seen in Figure 6.26.

6.4.25 CONFUSION MATRIX FOR SVM

Figure 6.27 shows confusion matrix for SVM.

```
y_predict = svc_model.predict(X_test_scaled)
cm = confusion_matrix(y_test, y_predict)

sns.heatmap(cm,annot=True,fmt="d")
```

```
<AxesSubplot:>
```

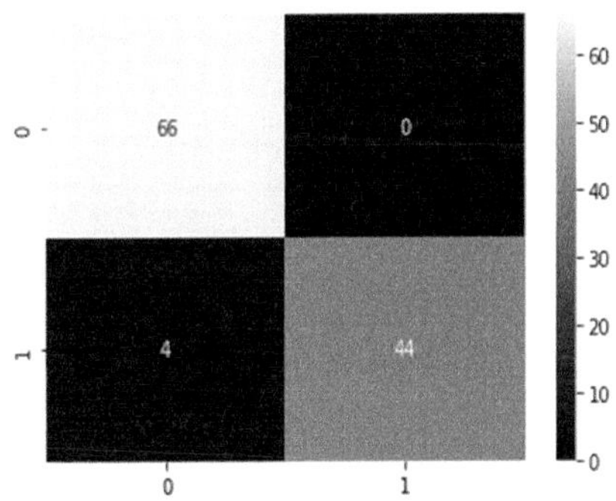

FIGURE 6.27 Confusion matrix for SVM.

6.4.26 ASSESSMENT METRICS FOR SVM

```
print(classification_report(y_test,y_predict))
```

```
              precision    recall  f1-score   support

           0       0.94      1.00      0.97        66
           1       1.00      0.92      0.96        48

    accuracy                           0.96       114
   macro avg       0.97      0.96      0.96       114
weighted avg       0.97      0.96      0.96       114
```

FIGURE 6.28 Assessment metrics for SVM.

Figure 6.27 depicts the SVM's confusion matrix. The coding for producing assessment metrics is seen in Figure 6.28. Figure 6.29 displays the SVM grid search CV. Figure 6.30 depicts the SVM's confusion matrix. The printing outcome of the assessment metrics is shown in Figure 6.31.

6.4.27 GRID SEARCH CV FOR SVM

param_grid= {'C': [0.1, 1, 10, 100], 'gamma': [1, 0.1, 0.01, 0.001], 'kernel': ['rbf']}
fromsklearn.model_selection**import**GridSearchCV
grid=GridSearchCV(SVC(),param_grid,refit=**True**,verbose=4)
grid.fit(X_train_scaled,y_train)

```
Fitting 5 folds for each of 16 candidates, totalling 80 fits
[CV 1/5] END ........C=0.1, gamma=1, kernel=rbf;, score=1.000 total time=   0.0s
[CV 2/5] END ........C=0.1, gamma=1, kernel=rbf;, score=0.945 total time=   0.0s
[CV 3/5] END ........C=0.1, gamma=1, kernel=rbf;, score=0.912 total time=   0.0s
[CV 4/5] END ........C=0.1, gamma=1, kernel=rbf;, score=0.956 total time=   0.0s
[CV 5/5] END ........C=0.1, gamma=1, kernel=rbf;, score=0.934 total time=   0.0s
[CV 1/5] END ......C=0.1, gamma=0.1, kernel=rbf;, score=0.945 total time=   0.0s
[CV 2/5] END ......C=0.1, gamma=0.1, kernel=rbf;, score=0.901 total time=   0.0s
[CV 3/5] END ......C=0.1, gamma=0.1, kernel=rbf;, score=0.890 total time=   0.0s
[CV 4/5] END ......C=0.1, gamma=0.1, kernel=rbf;, score=0.923 total time=   0.0s
[CV 5/5] END ......C=0.1, gamma=0.1, kernel=rbf;, score=0.868 total time=   0.0s
[CV 1/5] END .....C=0.1, gamma=0.01, kernel=rbf;, score=0.648 total time=   0.0s
[CV 2/5] END .....C=0.1, gamma=0.01, kernel=rbf;, score=0.637 total time=   0.0s
[CV 3/5] END .....C=0.1, gamma=0.01, kernel=rbf;, score=0.637 total time=   0.0s
[CV 4/5] END .....C=0.1, gamma=0.01, kernel=rbf;, score=0.637 total time=   0.0s
[CV 5/5] END .....C=0.1, gamma=0.01, kernel=rbf;, score=0.637 total time=   0.0s
[CV 1/5] END ....C=0.1, gamma=0.001, kernel=rbf;, score=0.648 total time=   0.0s
[CV 2/5] END ....C=0.1, gamma=0.001, kernel=rbf;, score=0.637 total time=   0.0s
[CV 3/5] END ....C=0.1, gamma=0.001, kernel=rbf;, score=0.637 total time=   0.0s
[CV 4/5] END ....C=0.1, gamma=0.001, kernel=rbf;, score=0.637 total time=   0.0s
[CV 5/5] END ....C=0.1, gamma=0.001, kernel=rbf;, score=0.637 total time=   0.0s
[CV 1/5] END ..........C=1, gamma=1, kernel=rbf;, score=1.000 total time=   0.0s
[CV 2/5] END ..........C=1, gamma=1, kernel=rbf;, score=0.956 total time=   0.0s
[CV 3/5] END ..........C=1, gamma=1, kernel=rbf;, score=0.967 total time=   0.0s
[CV 4/5] END ..........C=1, gamma=1, kernel=rbf;, score=1.000 total time=   0.0s
[CV 5/5] END ..........C=1, gamma=1, kernel=rbf;, score=0.967 total time=   0.0s

[CV 1/5] END ........C=1, gamma=0.1, kernel=rbf;, score=0.989 total time=   0.0s
[CV 2/5] END ........C=1, gamma=0.1, kernel=rbf;, score=0.945 total time=   0.0s
[CV 3/5] END ........C=1, gamma=0.1, kernel=rbf;, score=0.923 total time=   0.0s
[CV 4/5] END ........C=1, gamma=0.1, kernel=rbf;, score=0.967 total time=   0.0s
[CV 5/5] END ........C=1, gamma=0.1, kernel=rbf;, score=0.934 total time=   0.0s
[CV 1/5] END .......C=1, gamma=0.01, kernel=rbf;, score=0.945 total time=   0.0s
[CV 2/5] END .......C=1, gamma=0.01, kernel=rbf;, score=0.901 total time=   0.0s
[CV 3/5] END .......C=1, gamma=0.01, kernel=rbf;, score=0.879 total time=   0.0s
[CV 4/5] END .......C=1, gamma=0.01, kernel=rbf;, score=0.923 total time=   0.0s
[CV 5/5] END .......C=1, gamma=0.01, kernel=rbf;, score=0.868 total time=   0.0s
[CV 1/5] END ......C=1, gamma=0.001, kernel=rbf;, score=0.648 total time=   0.0s
[CV 2/5] END ......C=1, gamma=0.001, kernel=rbf;, score=0.637 total time=   0.0s
[CV 3/5] END ......C=1, gamma=0.001, kernel=rbf;, score=0.637 total time=   0.0s
[CV 4/5] END ......C=1, gamma=0.001, kernel=rbf;, score=0.637 total time=   0.0s
[CV 5/5] END ......C=1, gamma=0.001, kernel=rbf;, score=0.637 total time=   0.0s
[CV 1/5] END .........C=10, gamma=1, kernel=rbf;, score=1.000 total time=   0.0s
[CV 2/5] END .........C=10, gamma=1, kernel=rbf;, score=0.967 total time=   0.0s
[CV 3/5] END .........C=10, gamma=1, kernel=rbf;, score=0.956 total time=   0.0s
[CV 4/5] END .........C=10, gamma=1, kernel=rbf;, score=1.000 total time=   0.0s
[CV 5/5] END .........C=10, gamma=1, kernel=rbf;, score=0.956 total time=   0.0s
[CV 1/5] END .......C=10, gamma=0.1, kernel=rbf;, score=1.000 total time=   0.0s
[CV 2/5] END .......C=10, gamma=0.1, kernel=rbf;, score=0.967 total time=   0.0s
[CV 3/5] END .......C=10, gamma=0.1, kernel=rbf;, score=0.967 total time=   0.0s
[CV 4/5] END .......C=10, gamma=0.1, kernel=rbf;, score=0.989 total time=   0.0s
[CV 5/5] END .......C=10, gamma=0.1, kernel=rbf;, score=0.945 total time=   0.0s
```

```
[CV 5/5] END .....C=10, gamma=0.001, kernel=rbf;, score=0.879 total time=   0.0s
[CV 1/5] END ........C=100, gamma=1, kernel=rbf;, score=0.956 total time=   0.0s
[CV 2/5] END ........C=100, gamma=1, kernel=rbf;, score=0.956 total time=   0.0s
[CV 3/5] END ........C=100, gamma=1, kernel=rbf;, score=0.945 total time=   0.0s
[CV 4/5] END ........C=100, gamma=1, kernel=rbf;, score=0.989 total time=   0.0s
[CV 5/5] END ........C=100, gamma=1, kernel=rbf;, score=0.967 total time=   0.0s
[CV 1/5] END ......C=100, gamma=0.1, kernel=rbf;, score=1.000 total time=   0.0s
[CV 2/5] END ......C=100, gamma=0.1, kernel=rbf;, score=0.967 total time=   0.0s
[CV 3/5] END ......C=100, gamma=0.1, kernel=rbf;, score=0.945 total time=   0.0s
[CV 4/5] END ......C=100, gamma=0.1, kernel=rbf;, score=1.000 total time=   0.0s
[CV 5/5] END ......C=100, gamma=0.1, kernel=rbf;, score=0.956 total time=   0.0s
[CV 1/5] END .....C=100, gamma=0.01, kernel=rbf;, score=1.000 total time=   0.0s
[CV 2/5] END .....C=100, gamma=0.01, kernel=rbf;, score=0.967 total time=   0.0s
[CV 3/5] END .....C=100, gamma=0.01, kernel=rbf;, score=0.967 total time=   0.0s
[CV 4/5] END .....C=100, gamma=0.01, kernel=rbf;, score=0.989 total time=   0.0s
[CV 5/5] END .....C=100, gamma=0.01, kernel=rbf;, score=0.945 total time=   0.0s
[CV 1/5] END ....C=100, gamma=0.001, kernel=rbf;, score=0.989 total time=   0.0s
[CV 2/5] END ....C=100, gamma=0.001, kernel=rbf;, score=0.945 total time=   0.0s
[CV 3/5] END ....C=100, gamma=0.001, kernel=rbf;, score=0.923 total time=   0.0s
[CV 4/5] END ....C=100, gamma=0.001, kernel=rbf;, score=0.967 total time=   0.0s
[CV 5/5] END ....C=100, gamma=0.001, kernel=rbf;, score=0.934 total time=   0.0s

GridSearchCV(estimator=SVC(),
             param_grid={'C': [0.1, 1, 10, 100], 'gamma': [1, 0.1, 0.01, 0.001],
                         'kernel': ['rbf']},
             verbose=4)
```

FIGURE 6.29 Grid search CV for SVM.

6.4.28 *CONFUSION MATRIX FOR SVM*

Figure 6.30 shows confusion matrix for SVM.

```
grid.best_params_
{'C': 10, 'gamma': 0.1, 'kernel': 'rbf'}
grid.best_estimator_

SVC(C=1, gamma=1)

grid_predictions = grid.predict(X_test_scaled)
cm = confusion_matrix(y_test, grid_predictions)
sns.heatmap(cm,annot=True,fmt="d")

<AxesSubplot:>
```

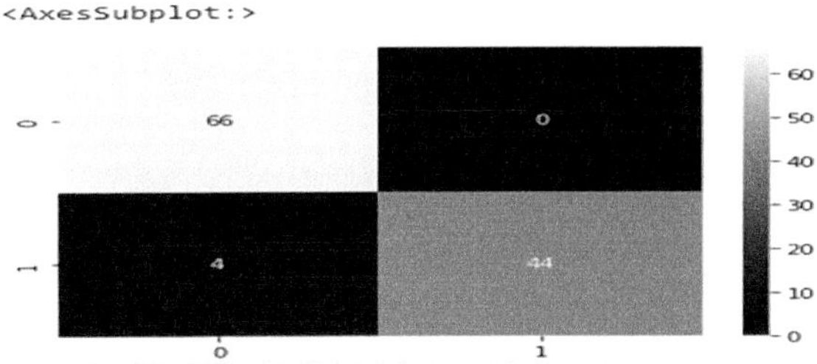

FIGURE 6.30 Confusion matrix for SVM.

6.4.29 *EVALUATION METRICS*

Figure 6.31 shows evaluation metrics.

```
print(classification_report(y_test,grid_predictions))

              precision    recall  f1-score   support

           0       0.94      1.00      0.97        66
           1       1.00      0.92      0.96        48

    accuracy                           0.96       114
   macro avg       0.97      0.96      0.96       114
weighted avg       0.97      0.96      0.96       114
```

FIGURE 6.31 Evaluation metrics.

6.5 CONCLUSION

BC is among the terrible illnesses that currently afflict women. Any all-encompassing BC-preventive approach must include a plan for BC detection and treatment. We used the Kaggle BC Dataset and SVM algorithms to analyze the effectiveness and application of various approaches and identify the strategy that could distinguish malignant from benign BC with the best degree of accuracy. The associations between the different characteristics in the dataset have been looked at for feature selection. With a 97% accuracy rate, SVM offers the study's greatest level of accuracy. This work can be improved in future by utilizing a sizable dataset and integrating further factors, such as BC phase detection. The clinical use of BC treatment will benefit from this study. Its main goal is to fully cure or greatly prolong the lives of BC patients while guaranteeing a high quality of life.

KEYWORDS

- **artificial intelligence**
- **BC**
- **DT**
- **healthcare system**
- **KNN**

- **LR**
- **machine learning**
- **NB**
- **RF**
- **SVM**

REFERENCES

1. Ahmed S. Elkorany; Mohamed Marey; Khaled M. Almustafa; Zeinab F. Elsharkawy *"BC Diagnosis Using Support Vector Machines Optimized by Whale Optimization and Dragonfly Algorithms"* Published in: IEEE Access (Volume: 10); INSPEC Accession Number: 21863064 DOI: 10.1109/ACCESS.2022.3186021; Publisher: IEEE; Date of Publication: 24 June 2022.
2. Deepika S, KapilaaRamanathan, Devi N; *"Prediction of BC Using SVM Algorithm"*; International Journal of Applied Engineering Research ISSN 0973-4562 Volume 16, Number 4 (2021) pp. 316–320 © Research India Publications. http://www.ripublication.com
3. Epimack, Michael, He Ma, Hong, Li, and Shouliang Qi; *"An Optimized Framework for BC Classification Using Machine Learning"*; Academic Editor: Ernesto Roldan-Valadez; Volume 2022 | Article ID 8482022 | https://doi.org/10.1155/2022/8482022, ;Published18 Feb 2022.
4. Kristoffersen Edward Mayce R. Lomboy; Hernandez, Rowell M." *A comparative performance of BC classification using hyper-parameterized machine learning models"*; International Journal of Advanced Technology and Engineering Exploration; 10801101. DOI:10.19101/IJATEE.2021.874380Bhopal; Vol. 8, Iss. 82, (Sep 2021).
5. Muhammed Abd-Elnaby[a], Marco Alfonse[a], Mohamed Roushdy[b]. *"Classification of BC using microarray gene expression data: A survey"*; https://doi.org/10.1016/j.jbi.2021.103764; Journal of Biomedical Informatics; Volume 117, May 2021, 103764.
6. Onyinyechi Jessica Egwom, Mohammed Hassan, Jesse Jeremiah Tanimu, Mohammed Hamada, Oko Michael Ogar *"An LDA–SVM Machine Learning Model for BC Classification"* Publisher's Note: MDPI stays neutral with regard to jurisdictional claims in published maps and institutional affiliations; Biomedinformatics 2022, 2, 345–358. https://doi.org/ 10.3390/biomedinformatics2030022; Published: 26 June 2022.
7. M. Lakshmitha, A. Abdul Hayum.; *"Svm based approach of detection And classification of tumors in Mammography"*; International Journal of Engineering Applied Sciences and Technology, 2020; Vol. 5, Issue 1, ISSN No. 2455-2143, Pages 436–440 ; Published Online May 2020 in IJEAST.
8. Onyinyechi Jessica Egwom, Mohammed Hassan, Jesse Jeremiah Tanimu, Mohammed Hamada, Oko Michael Ogar; *"An LDA–SVM Machine Learning Model*

for BC Classification"; MDPI stays neutral with regard to jurisdictional claims in published maps and institutional affiliations.

9. Sharmin Ara, Annesha Das, Ashim Dey; *"Malignant and Benign BC Classification using Machine Learning Algorithms"*; Publisher: IEEE.
10. SrwaHasan Abdulla, Ali Makki Sagheer, Hadi Veisi; *"BC Classification Using Machine Learning Techniques: A Review*"; Basic Science Department, Sulaimani University, Iraq, Department of Computer Technical Engineering, Al-Qalam University College, Iraq. 3 Department of Computer Engineering, University of Tehran, Iran; Received: 12 June 2021; Revised: 09 July 2021; Accepted: 14 August 2021; Published online: 19 August 2021.
11. M. Ashok Kumar, I. Laurence Aroquiaraj, *"Adaptive Divergence Weight Firefly Algorithm (ADWFA) with Improved K-Means Algorithm and Adaptive Neuro Fuzzy Inference System (ANFIS) for type 2 diabetes mellitus prediction"*, Journal of Advanced Research in Dynamical and Control Systems, (ISSN: 1943-023X).
12. M. Ashok Kumar, I. Laurence Aroquiaraj, *"Improved K-Means with Adaptive Divergence Weight Bat Algorithm (IKM-ADWBA) and Feature Selection for Type 2 Diabetes Mellitus Prediction"* International Journal of Recent Technology and Engineering, (ISSN: 2277-3878), 2019.

CHAPTER 7

An Energy-Efficient Approach-Based Prediction in a Sensor Cloud System by Using Machine Learning Methods

ANBAR HUSSAIN[1], MUHAMMED IJAZ[1], YOUNAS AZIZ[1], NAEEM HUSSAIN[2], and BILAL AHMED[1]

[1]*School of Computer Science, Central South University, Hunan Province, PR China*

[2]*School of Computer Science, Shenzhen University, Guangdong Province, PR China*

ABSTRACT

The novel energy efficient-based prediction in the sensor cloud system (SCS) is used to monitor weather and climate variations such as temperature, humidity, wind, sea level pressure, dew point, cloud cover, precipitation, visibility, and events. This study proposed an advanced solution for SCS prediction via using long-short term memory (LSTM) and support vector machine (SVM) prediction models to minimize communication and accuracy prediction. In the traditional sensor cloud model, communication requests from users are constantly routed through the sensor cloud to the wireless sensor network, resulting in increased energy consumption in sensors. In our approach, sensors interface with the cloud once every 24 h, and the cloud system answers the majority of requests using prediction algorithms. The use of prediction models prevents sensors from consuming less energy. The LSTM model's inferences predicted the temperature and humidity for a certain day with a 96% accuracy rate, whereas the SVM

Real-Time Artificial Intelligence (AI): Key Motivations, Technologies, Platforms, and Use Cases. Pethuru Raj, B. Sundaravadivazhagan, V. Kavitha, B. Narendra Kumar Rao, & Hannah Vijaykumar (Eds.)

DOI: 10.1201/9781998511358-7

model's predictions were 90% average accuracy with highest AUC values. Our analysis demonstrates that the LSTM algorithm is appropriate since its accuracy is 96% with similar AUC values. Thus, testing data prediction in SCSs, our proposed sensor cloud model predicts future sensor data, allowing cloud-level responses to user requests while reducing power consumption by limiting sensor network transmission.

7.1 INTRODUCTION

The sensor cloud system (SCS) enables user access to sensors through a fusion of wireless sensor networks (WSNs) and cloud computing within a cloud environment [1]. To fulfill user requests for sensor access via the cloud, data needs to be transmitted to and from WSNs, which consumes additional energy. Therefore, it is crucial for the SCS to operate with high energy efficiency. Providing information to end users puts a significant strain on the sensor network, resulting in substantial energy loss and limitations on sensor battery life. Consequently, managing power consumption in a WSN is a fundamental requirement that has a direct impact on the maintenance of the sensor network. Each user sends requests to the responsive network every second by utilizing the sensor cloud.

The increased data transmission rate on this network results in higher energy consumption. As a result, the cloud system needs to deploy future sensor data prediction 24 h in advance, which would significantly reduce energy consumption. A model for energy-efficient integration of cloud and sensor networks, known as the autoregressive integrated moving average (ARIMA) model, is developed by combining an artificial neural network (ANN) prediction model with two activation functions and a prediction technique called ARIMA [2]. The ANN model is employed to analyze nonlinear components, and the output from the ANN, along with input data from ARIMA, is utilized to examine the linear aspects [3].

In another study [4], a clustering algorithm is proposed, which allows nodes to select cluster heads based on criteria such as communication costs and energy levels. As a result, cluster leaders are more likely to form around nodes that possess greater available energy as opposed to those with lesser residual energy. This promotes a more balanced and impartial distribution of energy throughout the network. Therefore, it becomes crucial to establish an energy usage prediction model for influential data-gathering nodes to curtail energy consumption during data transmission and enhance the network's longevity.

Furthermore, the author outlines a method to prevent sensor nodes from redundantly transmitting data that the sink node can already predict. The author introduces an energy-efficient data collection approach based on the ARIMA model for predictive purposes [5]. The study [6, 7] collects and analyzes data from a private cloud to demonstrate the existence of heat fluctuations. The authors introduced a hybrid prediction model that seeks to reduce data transmission by merging the ARIMA model, known for its linear prediction capabilities, with a neural network, which excels in nonlinear predictions. The author [8] proposed hybrid machine learning strategies, including the use of both an ANN and ARIMA models to minimize the quantity of data that requires transmission. The researcher utilized a long-short term memory (LSTM) neural network in the examination of forthcoming sea surface temperatures, as described in Ref. [9]. The accuracy of the predictions is improved by integrating temporal and geographical data into this LSTM model, which is utilized to forecast future SST values.

However, the authors [10–14] employed LSTM models for temperature prediction in different environments utilizing different datasets and produced valid prediction results. The lightweight thermal prediction system uses two pathways thermal predictors to regulate run-time management choices and research optimal lightweight in [15]. Likewise, a hybrid machine learning model, mentioned in Ref. [16], merges the ARIMA model with an LSTM model for predictive purposes. Machine learning techniques, including the utilization of a multilayer perceptron (MLP), were applied to estimate building energy consumption by utilizing data from a WSN. The findings revealed that it beats conventional models.

Furthermore, for agricultural forecasting, the authors [17, 18] presented system-based methodologies that integrate LSTM. According to the findings of Wang and Qiu [19], the optimal throughput can be attained by integrating the planned managing energy scheme with the increased profile-energy scheme; therefore, the accurate solar irradiance model (ASIM) method introduced a hybrid approach and an uncertain finite element method to predict temperature fields under particular conditions [20]. This approach provided a simple implementation and successfully resolved concerns related to heat conduction. Furthermore, the author suggested the DMM model in Ref. [21]. This model incorporates an LSTM powered by recurrent neural network (RNN) to decrease average latency and signaling overhead, while also augmenting overall throughput.

This chapter seeks to assess the precision of machine learning systems in forecasting humidity and temperature, while also reducing the deviation between the actual and predicted values of these variables. The neural network was used to obtain the smallest resolvable temperature discrepancy between the forecast and actual observations. On the basis of past data, multilayered perception was taught to anticipate minimum and maximum temperature and humidity. A machine learning model that utilizes gradient boosting is provided for the purpose of temperature prediction. A variety of experiments were performed, and information was collected from diverse sites. Following this, the efficacy of the temperature forecast was evaluated. The experimental results demonstrate the accurate prediction of both temperature and humidity by our model. We began with the LSTM model, which had a 96% accuracy rate in predicting the temperature and humidity for a certain day, and the support vector machine (SVM) model, which had a 90% accuracy rate. Our research shows that the LSTM algorithm is satisfied by 96% accuracy with similar AUC.

The research experiment employs machine learning algorithms to improve the accuracy of predictions. There are three primary contributions to this study. First, this study contributes to the machine learning method, the sensor cloud processes observed data to predict the temperature and humidity accuracy. Second, this study designed for two deep learning algorithms which are of the long short-term algorithm and SVM discovered various expected outputs. Furthermore, we employed machine learning techniques to demonstrate the feasibility of this methodology for prospective clients. We independently utilized LSTM and SVM in our experimental work to validate the predictive capabilities of each model, which resulted in the attainment of precise predictions. The present study improves the accuracy of predictions through the utilization of machine learning methodologies.

The structure of the subsequent sections of this chapter is as follows: A succinct summary of the extant papers is presented in Section 7.2. Section 7.3 outlines the experimental design, while Section 7.4 presents the study's results and accompanying discussion. Lastly, in Section 7.5, this chapter concludes by highlighting the main findings of the study, addressing limitations, and outlining potential directions for future research.

7.2 RELATED WORK

Sensor cloud prediction occurs by findings of machine learning algorithms. Article [22] provided a double-stack LSTM approach for estimating humidity, temperature, and wind speed for 24 and 72 h in advance and utilized metrological data from several towns in Morocco for 15 years, beginning in 2000 and ending in 2015. The study found that LSTM provides the most accurate weather forecasts and may be utilized to address several rain-related issues. The study employed the R-prop algorithm with a logistic activation function to construct a sensor cloud model for predicting energy consumption, ultimately aiming to enhance the energy efficiency of sensor networks. Sensors are constructed in a collection of independent clusters, with as little cluster rearrangement as possible to save energy usage.

Consequently, the study [23] employs a wide range of instruments for energy measuring in order to identify enhanced input and output processes for energy conservation. To anticipate the future energy consumption of virtual computers, a computation approach that conserves energy in cloud systems was employed. Using the algorithm minimum spanning tree protocol, the network's lifespan was diminished. Utilizing a heuristic strategy, backbone nodes were chosen to lower the power consumption of WSN. A minimal number of servers was chosen to complete a task. This minimal selection was made with the intention of performing load balancing across all the servers. Hence, the methods [24] indicate that variations in solar radiation have a significant influence on the efficiency of systems. Therefore, in order to facilitate timely measurements, an intelligent model utilizing least squares support vector machines (LS-SVM) is required to predict solar insolation for the following day. With an impressive 99.94% accuracy rate, the results indicate that LS-SVM is an exceptionally precise method for estimating solar insolation at a specific location.

The study [25] proposed a novel method for predicting solar radiation, including sun radiations, maximum and minimum temperatures. The study used ANNs and genetic algorithms. Numerous studies also utilized SVM and machine learning approach as a prediction method helped to cut costs and improve performance. Using Internet of Things (IoT) technologies, the study suggested a thermal comfort solution for small buildings. Sensors and IoT enable the dissemination of data and the monitoring of the distribution of all the parameters. Data from the past is most useful for prediction since it enables the collection of abundant data resources for predicting heating capacity using machine learning techniques. As the input is multidimensional and nonlinear, SVM was unable to reach the required prediction accuracy.

Deep learning enables more precise feature mining and model building, which opens the door to innovative prospects.

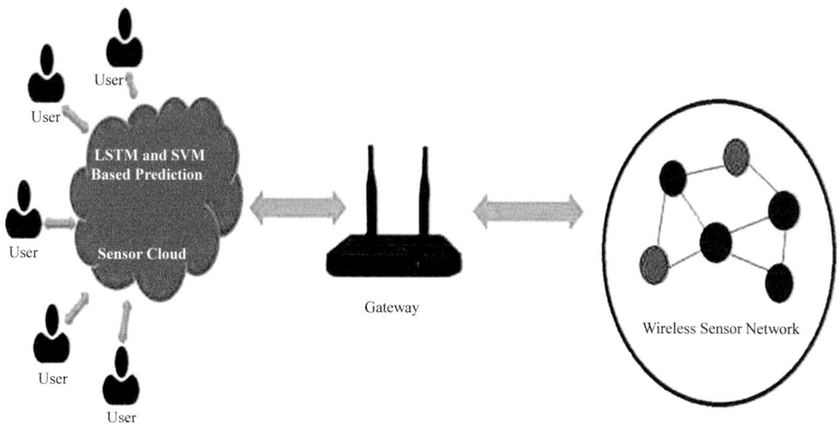

FIGURE 7.1 Proposed model based on LSTM and SVM.

Weather forecasting has been a hot issue for many years, and numerous efforts have been made to make it more mature. Massive amounts of research have been conducted in order to offer a potential system for weather forecasting [26]. Subsequent to its optimal performance on massive data sets, the LSTM deep learning model is employed to develop prediction models. For deep learning predictions of precipitation, the model varies from location to location owing to environmental differences; models created for one location would never function in other locations. A comparable case study was conducted in Ethiopia on the coffee production using the big dataset. This extensive inquiry makes a significant contribution to the development of a deep learning model, which relies on real dataset records of meteorological characteristics collected from the country's weather stations [27]. Several studies indicate that the conventional method for prediction does not accurately perform; hence, there was a need for a more accurate forecast. However as compared to others, the deep learning method is considered a great asset to the conventional method [28]. Prominent machine learning algorithms were utilized in this research endeavor to establish a relationship between a variety of pertinent climatology series of data as independent variables and local temperature as the dependent variable. Furthermore, expanding on the experiment described in the study [29], novel machine learning models were developed to predict air pollution levels using initial stage data, LSTM, and ARIMA models.

Furthermore, the authors [30–32] illustrate the development and evaluation of a novel algorithm designed to forecast outdoor air temperature by employing historical data in conjunction with validated finite results and pattern approximate matching. Traditionally, the SVM algorithm is used to predict weather, temperature, wind, and the main function of SVM is to classify binary data.

7.3 SYSTEM MODELS

The strategy for the suggested energy-efficient sensor-cloud model, deep learning algorithms LSTM and SVM explain how to employ prediction in the sensor-cloud to increase the energy efficiency of physical WSNs with enhanced methods to satisfy the accuracy prediction in SCS.

7.3.1 ENERGY EFFICIENT BASED SENSOR MODEL

The primary determinants of the use of energy in sensor networks with wireless connections are the transmission rate and the minimal separation distance among sensor nodes. As data sensing requires more energy, reducing data transmission can aid in energy conservation. Requests that originate from across the entire network are transmitted to sensor network through the clouds within an SCS. The frequency of user queries from the entire network results in increased energy consumption for sensor nodes that process these requests. This chapter offered cloud-based prediction of future sensor data using LSTM and SVM. Using LSTM and SVM, Figure 7.1 depicts the suggested energy-efficient sensor cloud model. To anticipate temperature and humidity, our model employs an LSTM and SVM-based prediction approach for accurate findings. Figure 7.2 illustrates how to make predictions using the LSTM and SVM models.

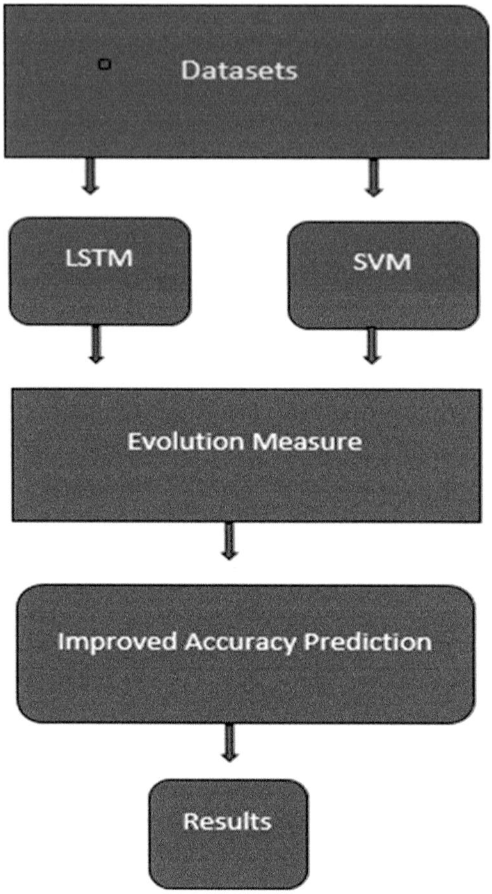

FIGURE 7.2 LSTM and SVM-based prediction model.

7.3.2 LONG-SHORT TERM MEMORY (LSTM)

The concept of LSTM was introduced by Hochreiter and Schmidhuber [33]. Additionally, LSTM was employed to tackle the challenge of vanishing gradients encountered in RNNs [34], achieved through the integration of various neural networks like artificial RNNs. LSTM neural networks are typically employed in deep learning architectures. Same as traditional feed-forward neural networks, LSTM networks include recurrent connections, allowing them to manage not only individual data points but also complete sequences of data. Thanks to its unique characteristics, LSTM has become more commercially successful innovations in the area of artificial intelligence. This success is attributed to its ability to effectively classify and make predictions using time series data, which is particularly valuable when there may be variable time intervals between critical events in a time series.

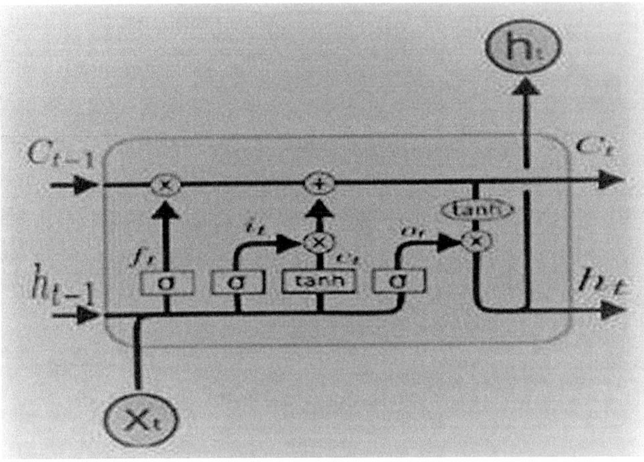

FIGURE 7.3 LSTM model.

Source: Reprinted from Ref. [36]. Copyright © 2020 Elsevier B.V.

The LSTM model is composed of a memory unit and three interacting gates: the input gate, forget gate, memory cell, and output gate. The memory cell is responsible for maintaining the previous state's status. As shown in Figure 7.3, the input gate regulates the amount of network input data retained in the unit state at the current time, denoted as "t." The forget gate determines at time "t - 1" whether information should be discarded or carried forward as reserved information to the input gate. Additionally, the output gate specifies which data should be produced as output. Equations 7.1–7.6 succinctly summarize the changes made within the LSTM layers.

These equations represent components of an LSTM model, a type of recurrent neural network. In Equation 7.1, C is the cell state, determined by input gate i, input data x at time t, previous cell state h at time $t - 1$, and bias bi

$$C = \sigma\,(Vi\ xt + Wi\ ht - 1) + bi \ldots \ldots\ldots \tag{7.1}$$

Equation 7.2 calculates the forget gate ft using its own weights and the same input parameters.

$$ft = \sigma\,(Vf\ xt + Wf\ ht - 1) + bf \tag{7.2}$$

$$ft = tanh\,(bf\ Vc\ xt + Wc\,ht - 1 + c) \tag{7.3}$$

$$\tilde{C}tt = ft \otimes - \quad + it \otimes Ct \tag{7.4}$$

$$ot = (xt + Wo\,ht - 1) + bo \tag{7.5}$$

$$ht = \otimes tanh\,(ct) \tag{7.6}$$

where t denotes time, it is input data, Wo and Vo express the weight matrices, ho is the hidden state, bias is denoted by bo, $tanh$ and σ are the activation function of $tanh$ and sigmoid, respectively. Ct, Ct ft, and ot stand for memory cell, input gate, forget gate, and output gate, respectively. The symbol ⊗ shows component-wise operations. At the end output ht is obtained by output gate and information in memory cell.

7.3.3 SUPPORT VECTOR MACHINE (SVM)

SVM is a machine learning method employed for the classification and prediction of data. SVMs work by identifying the optimal hyperplane that effectively separates a dataset into two distinct groups. In contrast to traditional machine learning algorithms, SVM stands out for its exceptional accuracy, achieved through the utilization of a relatively small amount of training data. We utilize the SVM algorithm to identify meteorological elements and forecast the weather. The fundamental notion of SVM is used to categorize binary data. The strategy that decomposes a multiclass problem into numerous separate binary classification work is used in the majority of previous approaches. These methods give results in inseparable cases, lowering classification accuracy. The classification or prediction system can be used on tasks after obtaining the input data and applying the preprocessing method [31]. In the context of linear classification, SVMs are designed to handle low-dimensional problems that are otherwise unsolvable. By leveraging a kernel function, these SVMs can transform data from lower dimensions to higher ones, enabling linear division. The linear reparability of data can be determined by whether a linear function completely divides a sample. If this is not the case, linear separation is not possible. Conceptually, this linear function corresponds to a three-dimensional point space, with an "s" line serving as a linear function within this space. The SVM model is based on gathering data points $x[i]$ from R^n, categorized into two groups. If $x[i]$ belongs to the first group, then it is assigned $y[i] = 1$, and for the second group, $y[i]$ is set to -1. In this learning scenario, we aim to find the right hyperplane described as $wx + b = 0$, where:

$y[i]\,(w \cdot x[i] + b) \geq 1$, and we aim to maximize $2/\|w\|$, which is equivalent to minimizing $\|w\|^2/2$ by applying the principle of duality; we can frame the problem as described in the following equation:

$$\text{Maximize } \sum a[j] - 2\sum a[j] \times a[i] \times y[j] \times y[i] \times x[j]$$
$$\text{subject to } a[i] \times C\sum[i] \times y[i] = 0. \tag{7.7}$$

This formulation helps us find the optimal hyperplane in the SVM model.

Furthermore, $(x[i] \times x[j])$ creates an internal sum of two vectors because in nonlinear items describes the internal result of the nuclear component $K(x[i], x[j])$ (the internal space of the subsequent vector products performance through mapping into a high dimensional kernel) instead of $x[i]$, $x[j]$.

7.4 RESULTS AND DISCUSSION

To conduct an experiment, it is necessary to collect data; we used Kaggle platform and collected data of first 5 years (2010–2015) of Madrid, Spain [35] for simulation. Following data cleaning and visualization, data was divided into testing and training data set to evaluate the accuracy of data on nontrained data. The LSTM and SVM algorithms are used individually on the same data set for weather prediction. Hence, unique methods adopt to improve the accuracy prediction to algorithm findings. The forecast values were obtained from a wide variety of sensors, and the determinants utilized for prediction included temperature (in Celsius), wind speed (in miles per hour), humidity level, dew points, sea pressure, and additional pertinent parameters. They were all crucial in predicting. It was a novel system in which all the queries were sent to the cloud rather than initially to sensors.

Data was cleaned after collection, and additional statistical approaches were used to check for skewness. Related characteristics were compared to their measurements, and a scale was evaluated to align all measurements for the best results. Various graphs were created to examine the data's extremes and minimum readings. Figures 7.4 and 7.5 are created to show the actual data as well as the expected mean temperature values via LSTM. However, generally display graphics will assist in the best visualization, so the data is large, several characteristics were chosen for effective forecasting. Hence, all features with strong correlation were excluded, and other unique features functioned well in producing efficient outcomes. These two machine learning algorithms individually perform prediction. Furthermore, we compare the accuracy predictions of both the algorithms. Consequently, LSTM finding accuracy predictions are much more satisfied with SVM. Sensor data is two-dimensional data in which time is plotted against other features; thus, this research will perform an experiment to determine exact accuracy, near or even more than the determined target. After applying algorithms it was seen that LSTM performs far better than SVM, because of time-series factor. SVM is one of the suitable techniques which usually gives high accuracy but as this is the data collected on massive scale, reading after every hour from sensors and plotted against time to for time series prediction LSTM played much satisfied.

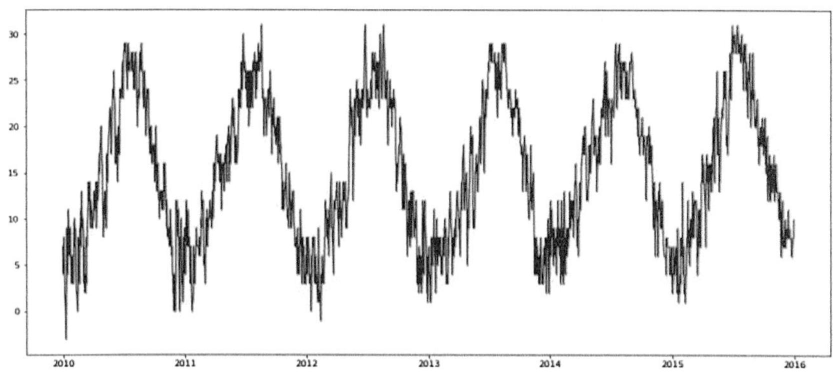

FIGURE 7.4 Date vs mean temperature (actual).

7.4.1 PREDICTION USING LSTM

Integration of a cloud-based predictive model utilizing LSTM is suggested. This study selects 3 years of evaluation data and 2 years of data for training from the specified dataset, respectively,, for the objectives of training and testing. Thirty thousand measurements, 5000 test samples, and 20,000 training samples comprise the dataset. The target parameters of our simulation are mean temperature and mean humidity. In addition to these, the following variables serve as predictors: (i) Mean Sea Level Pressure; (ii) Mean Dew Point; (iii) Cloud Cover; (iv) Mean Wind; (v) Speed; (vi) Precipitation; and (vii) Mean Speed. The state of visibility Mean (viii) Direction of Events I Wind. The mean humidity measurement and prognosis are illustrated in Figures 7.6 and 7.7, respectively. While the proposed date is plotted along the X-axis and the humidity is indicated along the Y-axis with accuracy of the forecast 96%.

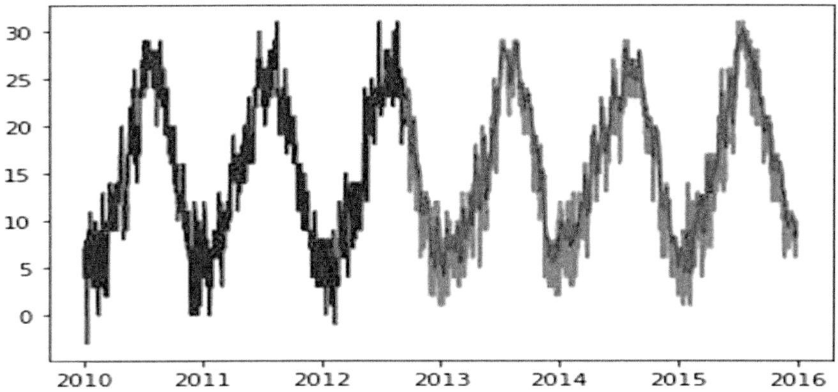

FIGURE 7.5 Date vs mean temperature (prediction).

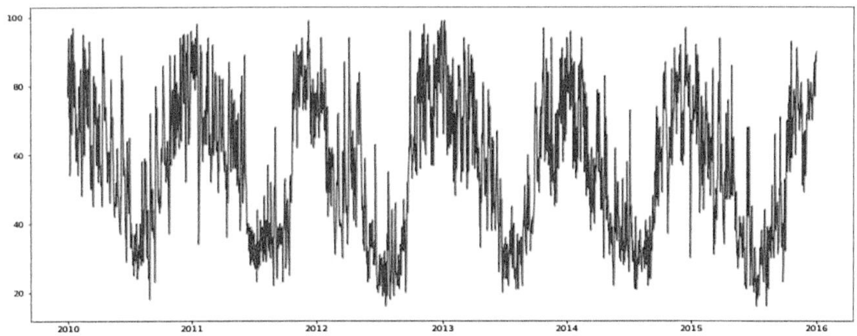

FIGURE 7.6 Date vs mean humidity (actual).

7.4.2 PREDICTION USING SUPPORT VECTOR MEMORY (SVM)

The dataset proposed for this study covers a 5-year period, specifically spanning from 2010 to 2015. This data is categorized into three distinct groups: training data, used for creating models for the aforementioned systems; validation data, employed to determine the most effective system parameters; and testing data, which plays no part in model construction or parameter selection. Upon inputting actual data into the SVM model and making predictions with a remarkable 90% accuracy, we were able to visualize the results in Figures 7.8 and 7.9. The X-axis indicates the years while Y-axis, scaled uniformly, represents temperature and humidity.

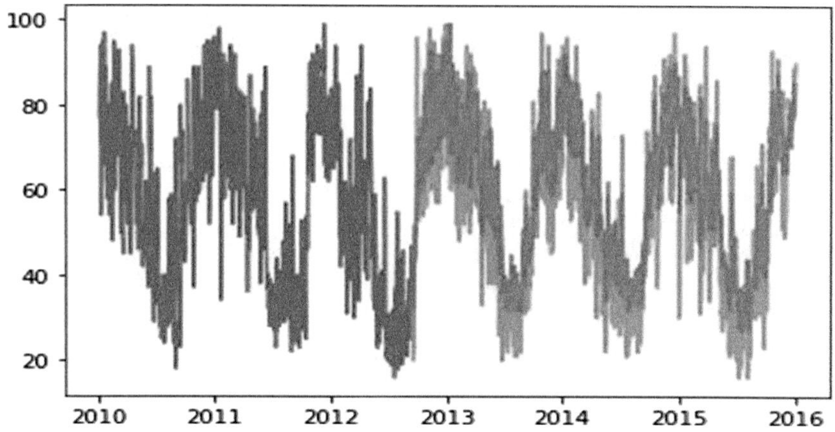

FIGURE 7.7 Date-mean-temperature prediction.

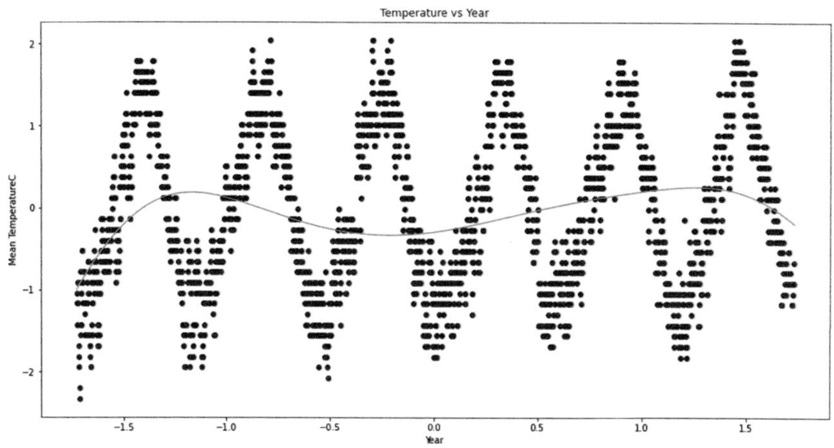

FIGURE 7.8 Date vs mean temperature.

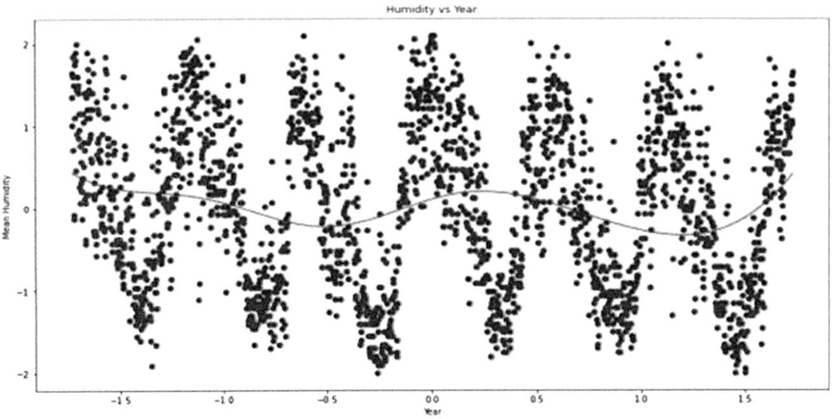

FIGURE 7.9 Date vs mean humidity.

7.4.3 *AVERAGE ACCURACY PREDICTION VIA SVM AND LSTM*

After determining the average accuracy values and AUC by contrasting the result tables of SVM and LSTM, it was found that SVM show highest AUC value while also having accuracy that is comparable to LSTM. However, LSTM also has comparable accuracy and a very similar AUC value. Hence, the given findings proved that long-short term LSTM outperform with SVM and traditional prediction algorithm for sensor cloud-based system with accurate accuracy.

TABLE 7.1 Average Accuracy Values and AUC of SVM.

Accuracy	Recall	Precision	AUC
0.8917	0.3187	0.8289	0.6486
0.9105	0.1216	0.3642	0.5975
0.8653	0.2098	0.6214	0.5863

7.5 CONCLUSION

The technique of LSTM for energy efficiency prediction in SCS which is compared with SVM peer to peer. Through simulation, we assessed our strategy's effectiveness and contrasted it to that of the LSTM and SVM models used independently. In the model proposed in our study, sensor communication with the cloud occurs on a daily basis, as opposed to continuous communication, with the cloud system responding to requests through the utilization of prediction algorithms. This approach of reduced communication intervals serves to extend the lifespan of the sensor batteries. Employing an LSTM-based prediction model with a singular activation function, the system achieved a notable 96% accuracy in forecasting both temperature and humidity for the upcoming day. Considering the performance of both the algorithms on the identical dataset, the configurations for each method were fine-tuned. SVM demonstrated a 90% accuracy rate and a maximum AUC value that was comparable to 96% accuracy rate of LSTM with similar accuracy rate. The highest value

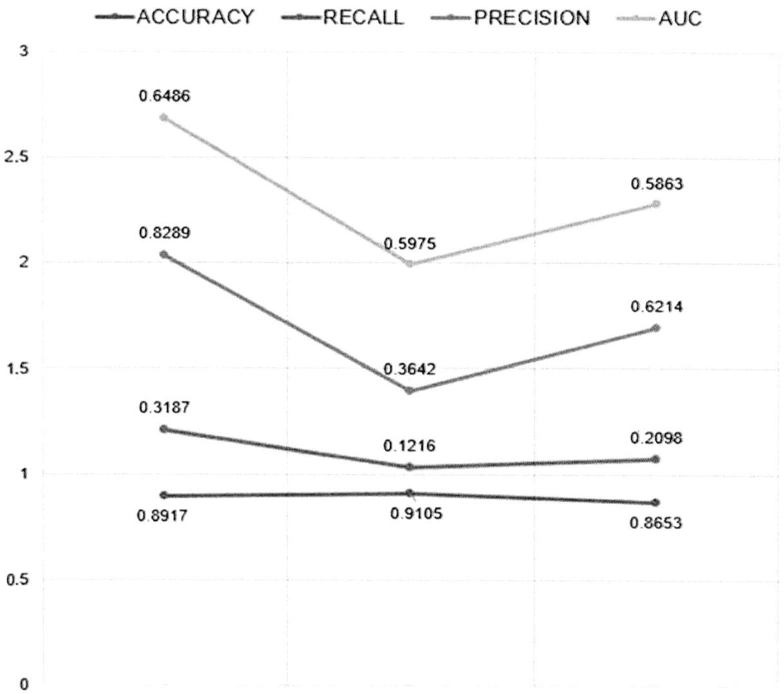

FIGURE 7.10 Average accuracy and AUC of SVM.

TABLE 7.2 Average Accuracy and AUC for LSTM Algorithm.

Accuracy	Recall	Precision	AUC
0.6710	0.9147	0.6118	0.6608
0.6489	0.8854	0.5709	0.6730
0.6216	0.8324	0.5809	0.6486

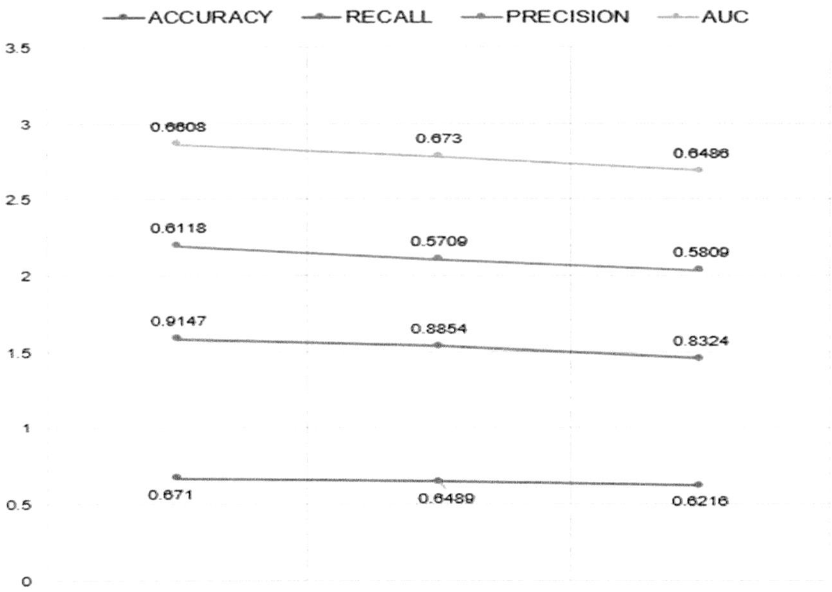

FIGURE 7.11 Average accuracy values and AUC of LSTM.

of AUC for SVM is 0.6486, 0.5975, and 0.5863 while 0.6608, 0.6730, and 0.6486 for LSTM. Long-term short memory is unquestionably the finest method for obtaining and delivering high-accuracy prediction outcomes when compared to the SVM and other traditional algorithms. As a matter of fact, following the test and simulations to increase accuracy prediction, LSTM surpass SVM for energy-efficient sensor clouds; however, our suggested model will serve as a guide for future research.

ETHICAL APPROVAL

This study adheres to all ethical guidelines during its preparation. We have verified that this manuscript has not been previously published and is not currently being considered by any other journal. Ethical approval and consent are not required for this study.

AVAILABILITY OF DATA AND MATERIALS

The dataset utilized in this study is available upon reasonable request from the corresponding author.

COMPETING INTEREST

There are no competing interests declared by the authors.

ACKNOWLEDGMENT

The authors are especially thankful to the Central South University, Hunan, China and Computer Department of Karakorum International University, Pakistan to support for research lab environment.

KEYWORDS

- **ANN**
- **cloud computing**
- **Internet of things**
- **LSTM**
- **sensor cloud**
- **wireless sensor network**

REFERENCES

1. Wang, T., Ke, H., Zheng, X., Wang, K., Kumar, A., & Liu, A. (2019). Big data cleaning based on mobile edge computing in industrial sensor- cloud. IEEE Transactions on Industrial Informatics, *PP*(99), 1–1.
2. Sharma, A., & Kakkar, A. (2020). A review on solar forecasting and power management approaches for energy- harvesting wireless sensor networks. *International Journal of Communication Systems, 33*(8), e4366.
3. Das, K., & Das, S. (2020). Data prediction and forecasting techniques-based energy efficient cloud-integrated sensor network. *Science & Technology Asia, 25*(4), 81–92.
4. Peng, J., Liu, T., Li, H., & Guo, B. (2013). Energy-efficient prediction clustering algorithm for multilevel heterogeneous wireless sensor networks. *International Journal of Distributed Sensor Networks*, *9*(2), 678214.
5. Liu, C., Wu, K., & Tsao, M. (2005). Energy efficient information collection with the arima model in wireless sensor networks. In *GLOBECOM'05. IEEE Global Telecommunications Conference.* IEEE.
6. Ilager, S., Ramamohanarao, K., & Buyya, R. (2020). Thermal prediction for efficient energy management of clouds using machine learning. *IEEE Transactions on Parallel and Distributed Systems, 32*(5), 1044–1056.
7. Askari Moghadam, R., & M. Keshmirpour. (2011). Hybrid ARIMA and neural network model for measurement estimation in energy-efficient wireless sensor

networks. In *International Conference on Informatics Engineering and Information Science* (pp. 35–48). Springer.

8. Das, K., Das, S., & Mohapatra, A. (2020). A novel energy-efficient sensor cloud model using data prediction and forecasting techniques. *Karbala International Journal of Modern Science, 6*(3), 1–10.
9. Yang, Y., Dong, J., Sun, X., Lima, E., Mu, Q., & Wang, X. (2017). A CFCC-LSTM model for sea surface temperature prediction. *IEEE Geoscience and Remote Sensing Letters, 15*(2), 207–211.
10. Mtibaa, F., Nguyen, K.-K., Azam, M., Papachristou, A., Venne, J.-S., & Cheriet, M. (2020). LSTM-based indoor air temperature prediction framework for HVAC systems in smart buildings. *Neural Computing and Applications, 32*(23), 17569–17585.
11. Liu, J., Zhang, T., Han, G., & Gou, Y. (2018). TD-LSTM: Temporal dependence-based LSTM networks for marine temperature prediction. *Sensors, 18*(11), 3797.
12. Fang, Z., Crimier, N., Scanu, L., Midelet, A., Alyafi, A., & Delinchant, B. (2021). Multi-zone indoor temperature prediction with LSTM-based sequence to sequence model. *Energy and Buildings, 245*, 111053.
13. Park, I., Kim, H. S., Lee, J., Kim, J. H., Song, C. H., & Kim, H. K. (2019). Temperature prediction using the missing data refinement model based on a long short-term memory neural network. *Atmosphere, 10*(11), 718.
14. Ayele, T. W., & Mehta, R. (2018). Real time temperature prediction using IoT. In *Second International Conference on Inventive Communication and Computational Technologies (ICICCT)* (pp. 1114–1117). IEEE.
15. Zhang, K., Guliani, A., Ogrenci-Memik, S., Memik, G., Yoshii, K., Sankaran, R., & Beckman, P. (2017). Machine learning-based temperature prediction for runtime thermal management across system components. *IEEE Transactions on Parallel and Distributed Systems, 29*(2), 405–419.
16. Xu, D., Zhang, Q., Ding, Y., & Zhang, D. (2022). Application of a hybrid ARIMA-LSTM model based on the spei for drought forecasting. *Environmental Science and Pollution Research, 29*(3), 4128–4144.
17. Chammas, M., Makhoul, A., & Demerjian, J. (2019). An efficient data model for energy prediction using wireless sensors. *Computers & Electrical Engineering, 76*, 249–257.
18. Guillén-Navarro, M. A., Martínez-España, R., Llanes, A., Bueno-Crespo, A., & Cecilia, J. M. (2020). A deep learning model to predict lower temperatures in agriculture. *Journal of Ambient Intelligence and Smart Environments, 12*(1), 21–34.
19. Wang, C., & Qiu, Z. (2015). Hybrid uncertain analysis for temperature field prediction with random, fuzzy and interval parameters. *International Journal of Thermal Sciences, 98*, 124–134.
20. Qureshi, H. K., Saleem, U., Saleem, M., Pitsillides, A., Lestas, M., et al. (2017). Harvested energy prediction schemes for wireless sensor networks: Performance evaluation and enhancements. *Wireless Communications and Mobile Computing, 2017*.
21. Mohanty, S. N., Lydia, E. L., Elhoseny, M., Al Otaibi, M. M. G., & Shankar, K. (2020). Deep learning with LSTM based distributed data mining model for energy efficient wireless sensor networks. *Physical Communication, 40*, 101097.
22. Endalie, D., Haile, G., & Taye, W. (2022). Deep learning model for daily rainfall prediction: Case study of Jimma, Ethiopia. *Water Supply, 22*(3), 3448–3461.

23. Smagulova, K., & James, A. P. (2019). A survey on LSTM memristive neural network architectures and applications. *The European Physical Journal Special Topics, 228*(10), 2313–2324.
24. Ekici, B. B. (2014). A least squares support vector machine model for prediction of the next day solar insolation for effective use of PV systems. *Measurement, 50*, 255–262.
25. Olatomiwa, L., Mekhilef, S., Shamshirband, S., Mohammadi, K., Petković, D., & Sudheer, C. (2015). A support vector machine–firefly algorithm-based model for global solar radiation prediction. *Solar Energy, 115*, 632–644.
26. Akhter, M. N., Mekhilef, S., Mokhlis, H., Almohaimeed, Z. M., Muhammad, M. A., Khairuddin, A. S. M., Akram, R., & Hussain, M. M. (2022). An hour-ahead pv power forecasting method based on an RNN-LSTM model for three different PV plants. *Energies, 15*(6), 2243.
27. Endalie, D., Haile, G., & Taye, W. (2022). Deep learning model for daily rainfall prediction: Case study of Jimma, Ethiopia. *Water Supply, 22*(3), 3448–3461.
28. Azari, B., Hassan, K., Pierce, J., & Ebrahimi, S. (2022). Evaluation of machine learning methods application in temperature prediction. *Environmental Engineering, 8*, 1–12.
29. Mani, G., & Volety, R. (2021). A comparative analysis of LSTM and ARIMA for enhanced real-time air pollutant levels forecasting using sensor fusion with ground station data. *Cogent Engineering, 8*(1), 1936886.
30. Wang, Y., Bai, Y., Yang, L., & Li, H. (2021). Short time air temperature prediction using pattern approximate matching. *Energy and Buildings, 244*, 111036.
31. Praba, M. B., Martin, A. J., Srivastava, S., & Rana, A. (2018). Weather monitoring system and rainfall prediction using SVM algorithm. *International Journal of Research in Engineering, Science and Management, 1*(10).
32. Mohandes, M. A., Halawani, T. O., Rehman, S.; & Hussain, A. A. (2004). Support vector machines for wind speed prediction. *Renewable Energy, 29*(6), 939–947.
33. Hochreiter, S.; & Schmidhuber, J. (1997). Long short-term memory. *Neural Computation, 9*(8), 1735–1780.
34. Biehl, M. (1999). Supervised sequence labelling with recurrent neural networks. *Neural Network, 1*.
35. de Castro, J. S. (2017). Meteorological data of Madrid, Spain, Kaggle Data, V1. https://www.kaggle.com/ Julian simon/ weather_madrid_lemd_1997_2015.csv. (Accessed January 10, 2019).
36. Nguyen, H. P., Liu, J., & Zio, E. (2020). A long-term prediction approach based on long short-term memory neural networks with automatic parameter optimization by Tree-structured Parzen Estimator and applied to time-series data of NPP steam generators. Applied Soft Computing, 89, 106116.

CHAPTER 8

Using Online Predictions and Machine Learning for Fraud Detection in Online Transactions

KASSIM KALINAKI[1,2], WASSWA SHAFIK[3,4], and KHAIRUL EAHSUN FAHIM[5]

[1]*Department of Computer Science, Islamic University in Uganda, Mbale, Uganda*

[2]*Borderline Research Laboratory, Kampala, Uganda*

[3]*School of Digital Science, Universiti Brunei Darussalam, Gadong, Brunei Darussalam*

[4]*Dig Connectivity Research Laboratory (DCRLab), Kampala, Uganda*

[5]*Information Technology Management, ZNRF University of Management Sciences, Dhaka, Bangladesh*

ABSTRACT

Effective and efficient fraud detection techniques are critical in the current surge in online transactions to ensure sustainable communities and cities. A promising approach to real-time fraud detection is the utilization of online predictions, which leverage machine learning (ML) algorithms to assess the likelihood of fraudulent activities. This chapter investigates the application of online predictions within a financial institution, specifically for detecting fraudulent transactions. Moreover, it delves into the data preprocessing techniques to adequately prepare the data for analysis, the

Real-Time Artificial Intelligence (AI): Key Motivations, Technologies, Platforms, and Use Cases. Pethuru Raj, B. Sundaravadivazhagan, V. Kavitha, B. Narendra Kumar Rao, & Hannah Vijaykumar (Eds.)

DOI: 10.1201/9781998511358-8

ML algorithms utilized for prediction purposes, and the evaluation metrics adopted to measure the prediction model's performance. Furthermore, this chapter thoroughly examines the practical implications of using online predictions for fraud detection, including the potential benefits, challenges, and prospects. By providing comprehensive insights into the use of online predictions for fraud detection, this chapter effectively underscores the critical role played by ML techniques in combatting fraud within online transactions and avails future perspectives and trends in fraud detection. Finally, the future of fraud detection using online predictions is poised for significant advancements. These trends will enable organizations to stay ahead of evolving fraud techniques, protect their assets and customers, and maintain trust and acceptance in an increasingly digital and interconnected world.

8.1 INTRODUCTION

In recent years, the world has witnessed an exponential rise in online transactions due to the digital revolution which has transformed how we conduct business and interact with one another, offering unparalleled convenience and accessibility. However, this rapid expansion of online transactions has also introduced new challenges, chief among them being the alarming increase in online fraud [20]. As technology advances, so do the tactics of fraudsters, making it imperative to develop sophisticated fraud detection systems to protect businesses and consumers. Conventional strategies for countering fraudulent activities, including rule-based frameworks and manual inspection procedures, face a formidable challenge in keeping pace with the continuously evolving tactics adopted by fraudsters [42].

The cost of online fraud, both in terms of financial losses and reputational damage, is now a big challenge for businesses and financial institutions globally [47]. Moreover, online financial transactions' sheer magnitude makes manual scrutiny of irregularities infeasible [18]. This predicament engenders challenges in formulating methodologies capable of discerning deviant data patterns. Developing advanced fraud detection systems capable of analyzing vast amounts of transactional data in real time is imperative to address this pressing issue. Emerging techniques, including online predictions and machine learning (ML) algorithms, offer powerful tools to tackle online fraud's complex and dynamic nature.

Essential in detecting fraud in online transactions, online predictions, also known as real-time predictions or streaming predictions, enable businesses to swiftly identify and respond to potentially fraudulent activities by analyzing incoming data streams in real time [9, 36]. An approach commonly applied in online prediction is anomaly detection, in which algorithms analyze transactional data to identify patterns and behaviors that significantly deviate from the norm. For instance, if a customer typically makes small purchases but suddenly initiates a significant transaction, it could raise suspicion and trigger an anomaly alert, indicating potential fraud [22]. Hence, businesses can promptly detect anomalies and take immediate action to mitigate risks and minimize losses by establishing behavior models using historical data. Behavioral analysis is another powerful technique that leverages customer behavior patterns and historical transactional data to identify deviations in typical online user activities [70].

ML algorithms learn from past fraudulent and nonfraudulent behavior instances, enabling real-time predictions based on ongoing customer interactions. For example, studies have indicated that if a customer's purchasing behavior suddenly changes, such as making transactions in unusual locations or at irregular times, it may signify fraudulent activity [70]. The behavioral analysis allows businesses to continuously monitor customer behavior and promptly detect suspicious activities, enhancing fraud prevention measures and safeguarding against financial losses. Network analysis is another popular real-time technique that utilizes transactions' interconnectedness to uncover fraudulent behavior patterns by examining relationships between entities [41]. For instance, network analysis can detect the pattern and raise alerts if multiple customers sharing the same IP address are involved in fraudulent activities across different accounts and merchants. Real-time analysis of complex networks enables businesses to proactively identify and disrupt organized fraudulent networks, mitigating potential risks and protecting against substantial financial losses.

Consisting of supervised, unsupervised, or deep learning (DL) techniques, ML algorithms have revolutionized fraud prediction by enabling the development of accurate and robust models capable of detecting complex and previously unseen fraud patterns [52, 65]. In supervised learning, algorithms trained on labeled data can identify specific transaction attributes associated with fraudulent behavior. This approach allows the algorithm to identify patterns and features indicative of fraudulent behavior. For instance, a supervised learning algorithm can recognize specific transaction attributes, such as transaction amount, location, or time of day, often associated with fraudulent activities [1, 14].

Leveraging a labeled dataset allows businesses to develop highly accurate models that continuously improve their fraud detection capabilities. Unsupervised learning algorithms excel in detecting anomalies and clusters within unlabeled data, allowing them to identify emerging or previously unknown fraud patterns. In contrast to supervised learning paradigms, unsupervised learning algorithms meticulously scrutinize unannotated datasets, proficiently discerning distinct clusters or aberrations residing within the dataset. Transactions exhibiting conspicuous deviations from the established norms are vigilantly earmarked as plausible instances of fraudulent activity [56]. These algorithms are particularly useful in detecting emerging or previously unknown fraud patterns. For example, consider a situation where a fraudster devises a new method to exploit vulnerabilities in a payment system. An unsupervised learning algorithm can detect unusual transaction patterns associated with this new form of fraud, even without prior labeled examples. Unsupervised learning techniques allow businesses to adapt to evolving fraud schemes and detect fraud that may go undetected by traditional rule-based systems.

DL algorithms, such as neural networks, process large amounts of transactional data and capture intricate features and nonlinear relationships, enhancing fraud detection accuracy [21]. This enables the detection of subtle fraud patterns that might be challenging to identify using traditional methods. Accordingly, DL models can learn hierarchical representations of transaction features and identify patterns indicative of fraudulent behavior, even if they are not explicitly defined by human experts [35]. Moreover, DL algorithms excel in learning from raw data and automatically discovering complex fraud indicators, contributing to enhanced fraud detection accuracy. By leveraging ML capabilities, businesses can enhance their fraud detection systems, improving accuracy and staying ahead of fraudsters in the ever-evolving landscape of online transactions. In response to the pressing issue of addressing the escalating sophistication exhibited by malevolent actors, this chapter unfolds an exhaustive exploration of the utilization of online predictions in financial institutions, with a specific focus on identifying fraudulent transactions. It accentuates diverse methodologies for data preprocessing, the deployment of ML algorithms for prediction objectives, the adoption of evaluation metrics to gauge model performance in prediction, and the tangible implications of employing online predictions for fraud detection. In addition, this chapter encompasses the inherent advantages, encountered challenges, and prospects associated with such endeavors.

8.1.1 CONTRIBUTIONS OF THE CHAPTER

We highlight the main contributions of the chapter, as shown below.

1. A thorough examination of the application of online predictions within financial institutions for the purpose of detecting fraudulent transactions.
2. A highlight of different data preprocessing techniques employed in the context of fraud detection using online predictions.
3. A discussion of ML algorithms utilized for prediction purposes in fraud detection.
4. An examination of the evaluation metrics adopted for measuring the prediction performance of the models used in fraud detection.
5. An exploration of the practical implications of employing online predictions for fraud detection in financial institutions.
6. A comprehensive exploration of the potential benefits of using online predictions for fraud detection, such as enhanced efficiency, proactive detection, and improved customer trust.
7. A discussion of the future prospects of utilizing online predictions for fraud detection.

8.1.2 CHAPTER ORGANIZATION

The rest of this chapter is organized as follows; Section 8.2 depicts the data preprocessing techniques for fraud detection. Section 8.3 details the evaluation metrics for assessing fraud detection model performance. Section 8.4 discusses the different case studies of fraud detection in financial institutions. Section 8.5 highlights the various practical implications of online predictions for fraud detection. The challenges of using ML techniques for fraud detection are depicted in Section 8.6. Future perspectives and trends in fraud detection using online predictions are detailed in Section 8.7. The conclusion is as shown in Section 8.8.

8.2 DATA PREPROCESSING TECHNIQUES FOR FRAUD DETECTION

Techniques for preprocessing data are critical for enhancing fraud detection's overall quality and performance. These techniques include data cleansing, feature selection, feature scaling, dealing with imbalanced data, temporal analysis, data transformation, and categorical variable handling [69]. Data preprocessing techniques play a pivotal role in enhancing the effectiveness of fraud detection systems. These techniques encompass several critical steps, including data cleansing to eliminate errors and inconsistencies, feature selection to identify relevant attributes, feature scaling for equal contribution of features, addressing imbalanced data, considering temporal patterns, data transformation for distribution suitability, and handling categorical variables. In the context of fraud detection, these techniques collectively enhance data quality, minimize noise, deal with missing values, and outliers, ensuring equitable treatment of imbalanced data, capturing temporal trends, and converting categorical data into numerical representations. Consequently, these preprocessing strategies lead to the creation of more accurate and robust fraud detection models, benefiting not only fraud detection but also domains such as finance, healthcare, and marketing, where data reliability is paramount for informed decision-making. Once data is preprocessed, organizations can proceed to evaluate various metrics to assess the performance of their fraud detection models. Organizations can enhance data quality, minimize noise and inconsistencies, deal with missing values and outliers, normalize data distributions, capture temporal connections, and translate categorical information into understandable numerical representations by implementing these strategies [31]. Beyond fraud detection, these comprehensive data preprocessing techniques have extensive utility, benefiting domains such as finance, healthcare, marketing, and many others, where precise and trustworthy data analysis is critical for decision-making and problem-solving. After data is preprocessed, the next step is to evaluate the different metrics to correctly assess the fraud detection model. Some key metrics for this assessment are described in the section below.

8.3 EVALUATION METRICS FOR ASSESSING FRAUD DETECTION MODEL PERFORMANCE

To make sure that fraud detection models are successful at detecting and preventing fraudulent activity, it is crucial to evaluate their performance.

It is essential to use proper evaluation criteria that accurately reflect the model's performance as fraud tendencies continue to change. Evaluation metrics are crucial for assessing the performance of fraud detection models, particularly in the face of evolving fraud patterns. These metrics include accuracy, which measures the proportion of correctly classified instances but can be misleading in imbalanced datasets. Precision and recall assess the model's ability to identify fraud while minimizing false positives, with the F1-Score offering a balanced evaluation, especially valuable for imbalanced datasets. The receiver operating characteristic (ROC) curve and area under the curve (AUC) illustrate performance across different thresholds, while specificity and the false positive rate gauge the model's ability to correctly classify legitimate transactions and quantify false positives' impact. The precision-recall curve and average precision help manage unbalanced datasets, emphasizing fraud detection. Collectively, these metrics provide a comprehensive understanding of a fraud detection model's effectiveness, enabling organizations to make informed decisions and continually improve their systems in response to evolving fraud tendencies, all while maintaining high accuracy and managing false positives. Some of these essential assessment metrics commonly deployed to evaluate the proficiency of fraud detection models are described below. These processes help organizations judge the efficiency and dependability of their fraud detection systems [43].

8.3.1 ACCURACY

Accuracy is a crucial evaluation statistic that gauges the accuracy of the model's predictions [19]. It computes the ratio of correctly classified instances (fraudulent and valid) to the summation of instances in the dataset. While accuracy is an intuitive performance metric, it can be misleading when working with imbalanced datasets in which the majority class dominates. As a result, accuracy should be understood with other metrics to provide a thorough knowledge of model performance.

8.3.2 PRECISION AND RECALL

Precision and recall are often used metrics that reveal information about the model's capacity to identify fraudulent instances while minimizing false positives correctly. Of all cases labeled as fraudulent, precision represents the percentage of accurately recognized fraudulent instances [57]. Recall (also known as positive rate or sensitivity) computes the percentage of fraudulent incidents previously identified out of all real fraudulent cases. Precision and recall must be balanced to successfully achieve a balance between precisely recognizing fraud and avoiding false positives.

8.3.3 F1-SCORE

The F1-score is a weighted mean of precision and recall and offers a single result that combines precision and recall, thereby providing a balanced evaluation of model performance [58]. The F1-score is very beneficial when dealing with imbalanced datasets since it considers both false positives and false negatives. It weighs precision and recall equally and provides an overall assessment of the model's efficacy.

8.3.4 RECEIVER OPERATING CHARACTERISTIC (ROC) CURVE AND AREA UNDER THE CURVE (AUC)

This is a graphical depiction of how well the model performs at various categorization levels [19]. As the threshold for identifying occurrences is modified, it shows the true positive rate (recall) versus the false positive rate (1-specificity). The AUC represents the area under the ROC curve, which provides a single scalar value representing the model's capacity to differentiate between fraudulent and authentic instances. AUC greater than one indicates superior overall performance.

8.3.5 SPECIFICITY AND FALSE-POSITIVE RATE

Specificity assesses the model's ability to classify valid cases as nonfraudulent. It computes the proportion of legitimate occurrences that were correctly detected out of all legitimate instances. The false positive rate, in conjunction with specificity, gauges the percentage of authentic occurrences erroneously classified as spurious [16]. These measurements are vital for determining the effect of false positives on the efficiency of fraud detection systems.

8.3.6 PRECISION-RECALL CURVE AND AVERAGE PRECISION

The precision-recall curve elegantly portrays the delicate balance between recall and precision across diverse classification thresholds. Its profound utility becomes especially pronounced in the context of imbalanced datasets, where the primary focus lies in detecting fraudulent cases, prioritizing sensitivity over the broader spectrum of accuracy. By calculating the average precision overall recall levels, the average precision provides a single metric that sums up the model's performance [58].

Evaluating the performance of fraud detection algorithms necessitates carefully examining various measures that capture various aspects of model efficacy. ROC curve, Accuracy, F1-score, precision, recall, AUC, specificity, false-positive rate, precision-recall curve, and average precision are critical evaluation metrics for fraud detection algorithms. Organizations can acquire a comprehensive knowledge of the model's ability to detect fraudulent actions while limiting false positives by examining these metrics, allowing them to make educated decisions and continuously improve their fraud detection systems. All these metrics play a critical role in solving real-world studies, described elaborately with relevant recent literature reviews in the following subsection.

8.4 CASE STUDIES OF FRAUD DETECTION IN FINANCIAL INSTITUTIONS

The rise of digital transactions has heightened the growing need for robust fraud detection systems in financial institutions. These case studies delve into adopting online predictions for fraud detection in a financial institution, highlighting the problems encountered, techniques used, and results obtained. Several research works have focused on various strategies for detecting fraud in various sectors. Rasmus et al. [26] focus on automating the mapping of financial transfers to matching accounts in accounting by employing character-level word embeddings and ML. When considering a staggering 473 companies individually, the proposed system triumphs over traditional methods, showcasing its superiority with a remarkable average top-1 accuracy of 80.50%. When trained on 44 firms from 28 sectors, it also displays generalization across various organizations and industries, reaching a test accuracy of over 80% [26].

Lebichot et al. [34] presented deep transfer learning techniques to detect fraudulent activities using credit cards, explicitly moving models from online to in-person transactions. Their solutions showed encouraging results on more than 80 million transactions, surpassing cutting-edge benchmarks. In their thought-provoking discourse, Sivanantham et al. [59] delved into the fraudulent activities using the credit cards and shed light on the scarcity of studies exploring the intricacies of analyzing genuine credit card data. They pitted ML algorithms against credit card datasets, unveiling the varying degrees of efficacy in their wake. They improved recall, accuracy, precision, and F1 score in identifying fraud using a hybrid algorithm model with a voting classifier. The study also addressed how different classifiers perform in identifying fraudulent transactions and identified significant dataset variables contributing to higher accuracy [59].

Ghosh and Prabu [17] addressed the growing issue of robocalls and their impact on telecommunication users. They proposed a methodology utilizing two ensemble models, Random Forest and XGBoost, to classify robocalls using a dataset acquired from the Federal Trade Commission. The study aimed to analyze call records and identify patterns to differentiate robocalls from legitimate calls. Through meticulous analysis of precision, responsiveness, and temporal efficiency, the evaluation of the models unveiled a resounding testament to the potency of ML methodologies in discerning and discernibly differentiating robocalls [17]. The issue of fraud, misuse, and waste in health insurance claims that jeopardizes the provision of high-quality healthcare was addressed by Nortey et al. [46].

They suggested using a Bayesian mixture model and Bayesian quantile regression techniques to examine health insurance claim data and spot possibly dubious claims. The models were used on a dataset of 115 people with cardiovascular illnesses. With an overall accuracy of 92%, the analysis revealed that 25 out of 115 claims (21.7%) were perhaps questionable. The usefulness of the Bayesian quantile regression model for identifying abnormalities in health insurance claims is demonstrated by this study. Dong et al. [15] discusses the difficulty of identifying corporate fraud, emphasizing how time-consuming this effort is. Researchers have typically relied on financial information and wording from financial statements. However, the authors provide an analytical approach that deploys unstructured datasets from financial social media platforms to evaluate the likelihood of corporate fraud, led by systemic functional linguistics. They put together a special dataset that includes both fraudulent and nonfraudulent businesses and social media data from before the suspected fraud infraction. The system automatically extracts various signals, including sentiment, emotion, subject, lexical, and social network aspects. Then, to detect fraud, these features are fed into ML classifiers. Their algorithm's effectiveness is assessed and contrasted with industry standards. Dong et al. [15] discuss the difficulty of identifying corporate fraud, emphasizing how time-consuming this effort is. Knuth and Ahrholdt [32] addresses the issue of online consumer fraud and the lack of information about risk indicators in transaction data.

To evaluate data from one of the world's top online stores, they use a decision tree as a data mining tool. They discover many fraud patterns, as a result, improving the distinction between fraudulent and legitimate online buying transactions. The research outcomes offer valuable insights into risk indicators, influencing variable selection, and the innovation of strategies for fraud prevention and systems within practical and theoretical contexts [25]. Addressing the demand for automated techniques to detect fraudulent healthcare providers, the authors propose a data-driven technique the categorization of healthcare fraud using Medicare claims data.

To meticulously assemble richly annotated datasets, the openly available datasets from the Centers for Medicare & Medicaid Services (CMS) have been adeptly harnessed by the authors and have judiciously fortified them with supplementary provider summary attributes. Moreover, they have introduced an adapted cross-validation methodology to ensure the robustness of the evaluation outcomes. The datasets have been subjected to rigorous scrutiny employing extreme gradient boosting and random forest learners, a testament to their outperforming prowess when contrasted with the initial datasets employed in analogous endeavors. This investigation underscores the paramount importance of comprehending and adeptly preparing data in the context of ML applications for the detection of healthcare fraud. In so doing, it lays the cornerstone for future research endeavors in this distinctive realm. In their exploration of fraud detection within the auto insurance sector through predictive modeling, Aslam et al. [10] have adroitly leveraged a publicly accessible dataset pertaining to car insurance, employing the Boruta algorithm for the identification of influential features. The authors have proficiently deployed three predictive models, specifically naïve Bayes (NB), support vector machine (SVM), and logistic regression (LR) in order to fashion a mechanism for the detection of fraudulent activities. The comprehensive assessment of these models incorporates an ensemble of six metrics that derive their essence from the confusion matrix. The findings conspicuously highlight the exceptional accuracy of the SVM and underscore the pinnacle achievement in terms of the f-measure score realized by LR. The research uniquely identifies fault, base policy, and the age of the policyholder as the most consequential attributes in this context. These discernments not only augment the arsenal of tools available for fraud detection in the auto insurance domain but also cast illumination on the conceptualization of efficacious strategies for the prevention of fraudulent activities. Furthermore, the envisaged framework exemplifies its potential for addressing real-time challenges within the auto insurance sector. Table 8.1 summarizes the case studies above.

8.5 PRACTICAL IMPLICATIONS OF ONLINE PREDICTIONS FOR FRAUD DETECTION

Online predictions for fraud detection have significant practical implications for various industries, particularly cybersecurity and financial services. By leveraging advanced ML algorithms, online prediction models can analyze vast amounts of data in real time and detect fraudulent activities, thereby mitigating risks and protecting individuals, businesses, and institutions.

8.5.1 REAL-TIME FRAUD DETECTION

Online prediction models enable organizations to detect fraudulent activities in real time [61]. Traditional rule-based systems often rely on predefined rules and thresholds, which may not capture emerging fraud patterns [2]. With online predictions, anomalies and suspicious patterns can be identified promptly, allowing for immediate action and reducing the potential impact of fraudulent activities.

8.5.2 ENHANCED ACCURACY

Online prediction models leverage ML algorithms that continuously learn and adapt to new data [68]. This iterative learning process enables the models to improve their accuracy over time. By analyzing large volumes of historical and real-time data, these models can identify complex fraud patterns and evolve to detect new and previously unknown types of fraud [29].

TABLE 8.1 Summary of the Case Studies on Fraud Detection Using ML.

Case Study Reference	Techniques	Dataset	Findings
[26]	ML, Word Embeddings	Financial transaction texts	Achieved an impressive 80.50% mastery in accurately linking financial transfers to accounting accounts, surpassing the baseline models that neglected transaction texts or relied on mundane word-counting methods. This exceptional performance extended across diverse companies and sectors, demonstrating remarkable test accuracy exceeding 80% after training on data from 44 companies across 28 sectors.
[34]		Credit card transactions (e-commerce and face-to-face)	Investigated deep transfer learning methods for detecting credit card fraud, focusing on transferring classification models from online to in-person transactions. Expanded a technique from Ganin et al. [76] and proposed a unique domain adaption strategy. Compared to cutting-edge benchmarks, the methodologies were evaluated on a dataset of more than 80 million transactions, and they produced promising results.
[59]	Hybrid Machine Learning Approach (Voting Classifier)	Credit card datasets	Applied ML algorithms to detect credit card fraud. Used a hybrid algorithm model involving a voting classifier. Achieved good accuracy, precision, recall, and F1 score in fraud detection. Identified important variables in the dataset for improved accuracy and discussed various classifiers' performance in detecting fraud.
[17]	Ensemble Methods (Random Forest, XGBoost)	Robocall dataset from the Federal Trade Commission	XGBoost and Random Forest were suggested as two ensemble models for classifying robocalls. Analyzed the models' accuracies, sensitivity, and processing times while evaluating the dataset. Robocalls could be identified and distinguished from actual calls using ML algorithms.
[46]	Bayesian Quantile Regression, Bayesian Mixture Model	Healthcare insurance claims	Bayesian mixture and quantile regression models were used to identify anomalies in health insurance claims. When the models were applied, 115 people with cardiovascular illnesses were included in a dataset and identified 25 claims (21.7% of them) as possibly dubious, resulting in a 92% overall accuracy. Showed how well the Bayesian quantile regression model works to find health insurance data anomalies.

TABLE 8.1 *(Continued)*

Case Study Reference	Techniques	Dataset	Findings
[15]	Information from financial, social media, and machine learning classifiers	Data from financial, social media platforms	Proposed a framework for analytics that would use unstructured data from financial social media sites to detect corporate fraud. Built a particular dataset, including fraudulent and nonfraudulent businesses and social media data from before the claimed fraud violation. Machine learning classifiers were fed extracted signals like sentiment, emotion, subject, lexical, and social network properties. Using financial ratios and language-based features, they evaluated the performance of their algorithm in comparison to standard methodologies. They have shown how financial social media data can be used to enhance established fraud detection techniques.
[32]	Decision Tree (Data Mining)	Online shopping transaction data	A decision tree was utilized as a data mining tool to discover consumer fraud risk indicators and combinations in online shopping transaction data. Data from one of the world's top online shops was analyzed, and fraud trends were detected, improving the separation of fraudulent and legitimate online buying transactions. The findings give useful insights for future practical and theoretical work on the design of fraud prevention measures and systems.
[25]	Data-Centric AI, Supervised Learning (Extreme Gradient Boosting, Random Forest)	Medicare claims data	A data-centric approach to enhance healthcare fraud categorization with the help of Medicare claims data was presented. Deploying publicly accessible data from the CMS, we created nine large-scale labeled datasets for supervised learning. We enriched the datasets with new provider summary features to reduce target leakage and suggested an improved cross-validation technique. The datasets were evaluated using high gradient boosting and random forest learners, which regularly outperformed the original datasets. The significance of data interpretation and preparation in ML applications for healthcare fraud detection was emphasized.

TABLE 8.1 *(Continued)*

Case Study Reference	Techniques	Dataset	Findings
[10]	LR, SVM, NB	Car insurance dataset	Innovatively, a cutting-edge system was developed for proactively detecting fraud in the realm of vehicle insurance. Employing the ingenious Boruta algorithm, a meticulous process of feature selection unearthed the utmost vital aspects. Leveraging the power of naive Bayes, SVMs, and LR, a triumvirate of prediction models were employed. To assess performance, a comprehensive evaluation involved six metrics derived from the labyrinthine confusion matrix. Remarkably, the SVM exhibited unparalleled accuracy, while LR shone brightest in terms of the f-measure score. Notably, fault, the base policy, and the policyholder's age emerged as the crux factors in this discerning analysis. Armed with real-time capabilities, this platform holds immense potential in addressing pertinent challenges within the auto insurance industry.

8.5.3 REDUCED FALSE POSITIVES

False positives occur when legitimate transactions or activities are incorrectly flagged as fraudulent. Online prediction models can help minimize false positives by considering multiple data points and applying sophisticated algorithms [75]. By accurately identifying fraudulent behavior and reducing false positives, organizations can avoid unnecessary disruptions for legitimate users, improve customer experience, and save resources on manual investigation of false alarms [72].

8.5.4 PREVENTION OF FINANCIAL LOSSES

Fraudulent activities can lead to significant financial losses for individuals and organizations. Online prediction models offer proactive fraud prevention by identifying suspicious behavior before financial harm occurs [5, 30]. By swiftly detecting and blocking fraudulent transactions, organizations can minimize financial losses, protect customer assets, and safeguard financial stability [71].

8.5.5 IMPROVED CUSTOMER EXPERIENCE

Effective fraud detection mechanisms are crucial for maintaining customer trust and satisfaction. Online prediction models enable organizations to detect and prevent fraud without inconveniencing genuine customers [71]. By accurately distinguishing between legitimate and fraudulent activities, organizations can provide a seamless user experience, reduce transaction friction, and build long-term customer loyalty [4].

8.5.6 ADAPTIVE AND SCALABLE SOLUTIONS

Online prediction models can be designed to adapt and scale with evolving fraud patterns and increasing data volumes. As fraudsters employ new techniques and technologies, these models can learn and detect emerging patterns, ensuring ongoing protection [3]. Moreover, online predictions can handle large-scale data processing, enabling organizations to analyze vast information efficiently and identify fraudulent activities across diverse channels and platforms.

8.5.7 COMPLIANCE AND REGULATORY REQUIREMENTS

Many industries, such as finance, healthcare, and telecommunications, have strict compliance and regulatory requirements for fraud detection and prevention. Online prediction models can aid organizations in meeting these obligations by providing robust fraud detection capabilities [49]. By implementing advanced fraud detection mechanisms, businesses can demonstrate their commitment to compliance, mitigate legal risks, and protect sensitive customer information [1].

8.5.8 ENHANCED RISK MANAGEMENT

Online predictions for fraud detection contribute to comprehensive risk management strategies. Organizations can reduce operational, financial, and reputational risks by identifying and preventing fraudulent activities. Fraud detection models can be integrated into broader risk management frameworks to provide a holistic approach to security and mitigate potential vulnerabilities across various business processes [53]. These implications enable organizations to stay ahead of fraudsters, protect their customers and assets, and maintain trust in an increasingly digital world. The following section presents the challenges of ML for fraud detection.

8.6 CHALLENGES OF USING ML TECHNIQUES FOR FRAUD DETECTION

While ML techniques offer momentous advantages for fraud detection, they also come with several challenges that organizations must address. Within this section, some notable challenges with ML techniques are demonstrated below.

8.6.1 IMBALANCED DATASETS

Fraudulent activities are often rare compared to legitimate transactions, resulting in imbalanced datasets where the number of positive (fraudulent) instances is much smaller than that of negative (legitimate) instances. This class imbalance can lead to biased models that have a higher tendency to misclassify fraudulent activities as fair [48]. Balancing class distributions necessitates meticulous data sampling strategies, including oversampling or undersampling, to optimize the model's capacity for effective learning amid the constraints of a sparse presence of positive instances [62].

8.6.2 EVOLVING FRAUD PATTERNS

Fraudsters constantly adapt and evolve their techniques to bypass detection systems. Fraud patterns continually change, and historical data alone may not capture the latest fraud indicators [63]. To stay effective, ML models must be regularly updated and trained on fresh data. Continuous monitoring and feedback loops are necessary to identify emerging fraud patterns and update the models accordingly [66].

8.6.3 FEATURE ENGINEERING AND DATA QUALITY

The quality and relevance of the data used for training ML models significantly impact their performance. Identifying the most informative features and extracting meaningful insights from complex and high-dimensional data can be challenging [12]. Additionally, data quality issues like missing values, outliers, and inconsistent data can affect the performance of the models [27, 28]. Careful preprocessing, feature selection, and data cleansing techniques are essential to ensure accurate and reliable fraud detection [50].

8.6.4 ADVERSARIAL ATTACKS

Fraudsters are increasingly sophisticated and may attempt to manipulate or deceive ML models. Adversarial attacks involve intentionally modifying or injecting data to bypass fraud detection systems [38]. For example, attackers may attempt to disguise fraudulent activities as routine transactions. Developing robust models resistant to adversarial attacks requires advanced techniques, such as anomaly detection, model validation, and adversarial training, to ensure that the models can identify and mitigate such attacks [27, 28, 54, 60].

8.6.5 INTERPRETABILITY AND EXPLAINABILITY

ML models, especially complex ones like deep neural networks, are often considered black boxes, making understanding how they arrive at their predictions challenging. Interpretability and explainability are critical for fraud detection systems to gain trust and acceptance from stakeholders [8]. Organizations must balance model performance and interpretability by adopting techniques like feature importance analysis, model explanations, and rule-based postprocessing to provide transparency and insights into decision-making [23].

8.6.6 INTEGRATION AND DEPLOYMENT

Integrating ML models into the existing fraud detection systems can be complex. Organizations must ensure seamless infrastructure integration, including data pipelines, real-time processing systems, and operational workflows [64]. Deploying and monitoring models in a production environment requires robust infrastructure, version control, and continuous monitoring to ensure optimal performance and adaptability [44].

8.6.7 REGULATORY COMPLIANCE AND PRIVACY CONCERNS

Fraud detection involves processing and analyzing sensitive customer data, which raises privacy concerns and compliance requirements, for example, under general and other data protection regulations [37]. Organizations must handle data securely, anonymize or encrypt sensitive information, and ensure compliance with relevant rules while maintaining the accuracy and efficacy of their fraud detection systems [13]. Addressing these challenges requires technical expertise, domain knowledge, and ongoing investment in research and development.

8.7 FUTURE PERSPECTIVES AND TRENDS IN FRAUD DETECTION USING ONLINE PREDICTIONS

The future of fraud detection using online predictions holds several exciting perspectives and trends, as demonstrated below.

8.7.1 ADVANCED ANOMALY DETECTION TECHNIQUES

Traditional rule-based systems are often limited in detecting complex and evolving fraud patterns. Future fraud detection systems will rely on advanced anomaly detection techniques, such as unsupervised ML algorithms and DL models [51]. These techniques can identify subtle anomalies and patterns that may indicate fraudulent activities, even in large and complex datasets.

8.7.2 CONTINUAL LEARNING AND ADAPTIVE MODELS

Fraudsters continually adapt their tactics, requiring fraud detection systems to be agile and adaptable. Future systems will employ continual learning approaches, where models can learn and update themselves in real time as new data becomes available [55]. This allows for better detection of emerging fraud patterns and quicker adaptation to changing attack techniques.

8.7.3 GRAPH ANALYTICS AND NETWORK ANALYSIS

Fraud often involves complex networks and interconnected activities. Future fraud detection systems will leverage graph analytics and network analysis techniques to uncover hidden connections and relationships between fraudulent entities [11]. By examining the relationships and interactions within networks, these systems can identify coordinated fraud schemes and detect suspicious patterns that may go unnoticed with traditional approaches.

8.7.4 FUSION OF MULTIPLE DATA SOURCES

Future systems will integrate and analyze diverse data sources to enhance fraud detection accuracy. This includes traditional structured data, such as transaction records and user profiles, and unstructured data from several sources, for example, social media, text messages, and call recordings [40]. By leveraging NPLs and sentiment analysis techniques, fraud detection systems can gain deeper insights and context to detect fraudulent activities.

8.7.5 REAL-TIME PREDICTIVE ANALYTICS

Timely detection is critical to fraud prevention. Future systems will emphasize real-time predictive analytics, enabling instant decision-making and intervention [45]. By utilizing high-speed data processing and real-time streaming analytics, fraud detection models can quickly analyze data as it flows in, immediately identifying and preventing fraudulent activities [33].

8.7.6 EXPLAINABLE AI FOR TRANSPARENCY

As ML models become more complex, the need for explainability and transparency becomes increasingly important. Future fraud detection systems will focus on developing explainable AI techniques that provide insights into model predictions and decision-making processes [73]. This helps build trust with stakeholders and enables investigators to understand the rationale behind fraud alerts.

8.7.7 COLLABORATION AND DATA SHARING

Fraud detection is a global challenge that requires collaboration and information sharing among organizations. Future trends will involve increased collaboration between industries, government agencies, and cybersecurity organizations to share data, insights, and best practices [7]. Such collaborations can help build comprehensive fraud detection systems and provide a broader perspective on emerging fraud trends.

8.7.8 ENHANCED USER AUTHENTICATION METHODS

As fraudsters exploit vulnerabilities in traditional authentication methods, future fraud detection systems will explore advanced user authentication techniques. This includes biometric authentication, behavioral analysis, and multifactor authentication systems that provide more robust security measures and reduce the risk of unauthorized access and identity theft [6, 24].

8.7.9 INTEGRATION OF BLOCKCHAIN TECHNOLOGY

Blockchain technology offers immutable and transparent transaction records, making it a promising tool for fraud detection. Future systems may integrate blockchain technology to enhance fraud prevention, authentication, and auditability [74]. By leveraging blockchain's decentralized nature, fraud detection systems can provide secure and tamper-proof records of transactions, ensuring transparency and reducing the risk of fraud [39].

8.7.10 ETHICAL CONSIDERATIONS AND PRIVACY PROTECTION

With the increasing use of personal data in fraud detection, future systems will prioritize ethical considerations and privacy protection. Organizations must ensure compliance with data protection regulations and implement privacy-enhancing techniques, for instance, differential privacy and secure data-sharing protocols, to safeguard sensitive information while maintaining effective fraud detection capabilities [67]. Therefore, the future of fraud detection using online predictions is poised for significant advancements. These trends will enable organizations to stay ahead of evolving fraud techniques, protect their assets and customers, and maintain trust in an increasingly digital and interconnected world.

8.8 CONCLUSION

In conclusion, this chapter presents a comprehensive investigation into applying online predictions to detect fraudulent transactions within financial institutions. This chapter offers valuable contributions to researchers, practitioners, and students in this field by addressing the urgent need to combat the increasingly sophisticated techniques employed by fraudsters. The chapter's exploration of data preprocessing techniques provides researchers and practitioners with practical insights into effectively preparing and cleansing data for accurate fraud detection. This knowledge enhances the reliability and efficiency of prediction models, enabling proactive identification and prevention of fraudulent activities. Discussing various ML algorithms empowers researchers and practitioners to select appropriate models tailored to their specific fraud detection needs. By understanding the strengths and limitations of different algorithms, they can make informed decisions when designing prediction systems, optimizing performance, and adapting to evolving fraud patterns. Adopting rigorous evaluation metrics allows researchers and practitioners to assess the performance of prediction models effectively. This enables them to measure accuracy, precision, recall, and other relevant indicators, facilitating fraud detection systems' continuous improvement and refinement. This chapter's focus on the practical implications of using online predictions for fraud detection provides tangible benefits to researchers, practitioners, and students. They understand the potential advantages, including improved detection rates, reduced financial losses, enhanced efficiency, and increased customer trust. This knowledge empowers them to implement and utilize online prediction systems effectively, bolstering the security and integrity of financial institutions. Furthermore, by addressing the challenges and prospects of employing online predictions, this chapter equips researchers, practitioners, and students to anticipate and overcome potential obstacles. This knowledge enables them to tackle data privacy concerns, enhance model interpretability, and develop robust defenses against adversarial attacks, ensuring fraud detection systems' continued success and resilience. In summary, this chapter significantly contributes to fraud detection within financial institutions by offering a comprehensive exploration of the application of online predictions. Its insights into data preprocessing techniques, ML algorithms, evaluation metrics, and practical implications empower researchers, practitioners, and students to detect and prevent fraudulent transactions effectively. Leveraging online predictions enables financial institutions to fortify their defenses against fraudsters, safeguarding their operations and maintaining the trust of their stakeholders.

KEYWORDS

- **blockchain technology**
- **DL algorithms**
- **fraudulent transactions**
- **machine learning**
- **online predictions**
- **real-time fraud detection**

REFERENCES

1. Afriyie, J. K., Tawiah, K., Pels, W. A., Addai-Henne, S., Dwamena, H. A., Owiredu, E. O., Ayeh, S. A., & Eshun, J. (2023a). A supervised machine learning algorithm for detecting and predicting fraud in credit card transactions. *Decision Analytics Journal*, *6*, 100163. https://doi.org/10.1016/J.DAJOUR.2023.100163
2. Ahmad, H., Kasasbeh, B., Aldabaybah, B., & Rawashdeh, E. (2023). Class balancing framework for credit card fraud detection based on clustering and similarity-based selection (SBS). *International Journal of Information Technology (Singapore)*, *15*(4). https://doi.org/10.1007/s41870-022-00987-w
3. Akinbowale, O. E., Mashigo, P., & Zerihun, M. F. (2023). The integration of forensic accounting and big data technology frameworks for internal fraud mitigation in the banking industry. *Cogent Business and Management*, *10*(1), 2163560. https://doi.org/10.1080/23311975.2022.2163560.
4. Al Balawi, S., & Aljohani, N. (2023). Credit-card fraud detection system using neural networks. *International Arab Journal of Information Technology*, *20*(2). https://doi.org/10.34028/iajit/20/2/10
5. Ali, A. A., Khedr, A. M., El-Bannany, M., & Kanakkayil, S. (2023). A powerful predicting model for financial statement fraud based on optimized XGBoost ensemble learning technique. *Applied Sciences, 13*(4), 2272. https://doi.org/10.3390/app13042272.
6. Alli, A. A., Kassim, K., Mutwalibi, N., Hamid, H., & Ibrahim, L. (2021). Secure fog-cloud of things: Architectures, opportunities and challenges. In M. Ahmed, & P. Haskell-Dowland (Eds.), *Secure Edge Computing* (1st Ed., pp. 3–20). CRC Press. https://doi.org/10.1201/9781003028635-2.
7. Alsadhan, N. (2023). A multi-module machine learning approach to detect tax fraud. *Computer Systems Science and Engineering*, *46*(1), 241–253. https://doi.org/10.32604/csse.2023.033375
8. Arockiam, J. M., & Pushpanathan, A. C. S. (2023). Mapreduce-iterative support vector machine classifier: Novel fraud detection systems in healthcare insurance industry. *International Journal of Electrical and Computer Engineering*, *13*(1), 756. https://doi.org/10.11591/ijece.v13i1.pp756-769
9. Ashwin, V., Menon, V., Devagopal, A. M., Nived, P. A., & Udayan Divya, J. (2023). Detection of fraudulent credit card transactions in real time using SparkML and Kafka. *Lecture Notes in Networks and Systems*, *540*, 285–295. https://doi.org/10.1007/978-981-19-6088-8_26
10. Aslam, F., Hunjra, A. I., Ftiti, Z., Louhichi, W., & Shams, T. (2022). Insurance fraud detection: Evidence from artificial intelligence and machine learning. *Research in International Business and Finance*, *62*, 101744. https://doi.org/10.1016/J.RIBAF.2022.101744
11. Chen, A. Q., Wu, H. L., Wang, T., Wang, X. Z., Sun, H. B., & Yu, R. Q. (2023). Intelligent analysis of excitation-emission matrix fluorescence fingerprint to identify and quantify adulteration in camellia oil based on machine learning. *Talanta, 251,* 123733. https://doi.org/10.1016/j.talanta.2022.123733
12. Chhabra Roy, N., & Prabhakaran, S. (2023). Sustainable response system building against insider-led cyber frauds in banking sector: A machine learning approach. *Journal of Financial Crime*, *30*(1). https://doi.org/10.1108/JFC-12-2021-0274

13. De Graeve, M., Birse, N., Hong, Y., Elliott, C. T., Hemeryck, L. Y., & Vanhaecke, L. (2023). Multivariate versus machine learning-based classification of rapid evaporative Ionization mass spectrometry spectra towards industry based large-scale fish speciation. *Food Chemistry*, *404*(1), 134632. https://doi.org/10.1016/j.foodchem.2022.134632
14. Dhankhad, S., Mohammed, E. A., & Far, B. (2018). Supervised machine learning algorithms for credit card fraudulent transaction detection: A comparative study. In *Proceedings - 2018 IEEE 19th International Conference on Information Reuse and Integration for Data Science, IRI 2018* (pp. 122–125). https://doi.org/10.1109/IRI.2018.00025
15. Dong, W., Liao, S., & Zhang, Z. (2018). Leveraging financial social media data for corporate fraud detection. *Journal of Management Information Systems*, *35*(2), 461–487. https://doi.org/10.1080/07421222.2018.1451954
16. Georgieva, S., Markova, M., & Pavlov, V. (2019). Using neural network for credit card fraud detection. In *AIP Conference Proceedings* (Vol. *2159*(1), p. 030013). https://doi.org/10.1063/1.5127478
17. Ghosh, M., & Prabu, P. (2019). Empirical analysis of ensemble methods for the classification of robocalls in telecommunications. *Academia.Edu*, *9*(4), 3108–3114. https://doi.org/10.11591/ijece.v9i4.pp3108-3114
18. Gianini, G., Ghemmogne Fossi, L., Mio, C., Caelen, O., Brunie, L., & Damiani, E. (2020). Managing a pool of rules for credit card fraud detection by a Game Theory based approach. *Future Generation Computer Systems*, *102*, 549–561. https://doi.org/10.1016/J.FUTURE.2019.08.028
19. Gupta, A., Lohani, M. C., & Manchanda, M. (2021). Financial fraud detection using naive Bayes algorithm in highly imbalance data set. *Journal of Discrete Mathematical Sciences and Cryptography, 24*(5), 1559–1572. https://doi.org/10.1080/09720529.2021.1969733
20. Hasan, I., & Rizvi, S. (2022). AI-Driven fraud detection and mitigation in e-Commerce transactions. *Lecture Notes on Data Engineering and Communications Technologies*, *90*, 403–414. https://doi.org/10.1007/978-981-16-6289-8_34
21. Hemdan, E. E. D., & Manjaiah, D. H. (2022). Anomaly credit card fraud detection using deep learning. *Studies in Big Data*, *91*, 207–217. https://doi.org/10.1007/978-3-030-75855-4_12
22. Hilal, W., Gadsden, S. A., & Yawney, J. (2022). Financial fraud: A review of anomaly detection techniques and recent advances. *Expert Systems with Applications*, *193*, 116429. https://doi.org/10.1016/J.ESWA.2021.116429
23. Jáñez-Martino, F., Alaiz-Rodríguez, R., González-Castro, V., Fidalgo, E., & Alegre, E. (2023). A review of spam email detection: Analysis of spammer strategies and the dataset shift problem. *Artificial Intelligence Review*, *56*(2), 1145–1173. https://doi.org/10.1007/s10462-022-10195-4
24. Jayanthi, E., Ramesh, T., Kharat, R. S., Veeramanickam, M. R. M., Bharathiraja, N., Venkatesan, R., & Marappan, R. (2023). Cybersecurity enhancement to detect credit card frauds in health care using new machine learning strategies. *Soft Computing*, *27*, 7555–7565. https://doi.org/10.1007/s00500-023-07954-y
25. Johnson, J. M., & Khoshgoftaar, T. M. (2023). Data-centric AI for healthcare fraud detection. *SN Computer Science*, *4*(4), 389. https://doi.org/10.1007/s42979-023-01809-x

26. Jørgensen, R. K., & Igel, C. (2021). Machine learning for financial transaction classification across companies using character-level word embeddings of text fields. *Intelligent Systems in Accounting, Finance and Management*, *28*(3), 159–172. https://doi.org/10.1002/isaf.1500.
27. Kalinaki, K., Namuwaya, S., Mwamini, A., & Namuwaya, S. (2023a). Scaling up customer support using artificial intelligence and machine learning techniques. In *Contemporary Approaches of Digital Marketing and the Role of Machine Intelligence* (pp. 23–45). IGI Global. https://doi.org/10.4018/978-1-6684-7735-9.ch002
28. Kalinaki, K., Thilakarathne, N. N., Mubarak, H. R., Malik, O. A., & Abdullatif, M. (2023b). Cybersafe capabilities and utilities for smart cities. In *Cybersecurity for Smart Cities* (pp. 71–86). Springer, Cham. https://doi.org/10.1007/978-3-031-24946-4_6
29. Karunachandra, B., Putera, N., Wijaya, S. R., Suryani, D., Wesley, J., & Purnama, Y. (2023). On the benefits of machine learning classification in cashback fraud detection. *Procedia Computer Science*, *216*(9), 364–369. https://doi.org/10.1016/j.procs.2022.12.147
30. Khan, M. Z., Shaikh, S. A., Shaikh, M. A., Khatri, K. K., Mahira Abdul Rauf, Kalhoro, A., & Muhammad Adnan. (2023). The performance analysis of machine learning algorithms for credit card fraud detection. *International Journal of Online and Biomedical Engineering (IJOE)*, *19*(03), 82–98. https://doi.org/10.3991/ijoe.v19i03.35331
31. Kim, H., & Kim, H. (2022). Contextual anomaly detection for high-dimensional data using Dirichlet process variational autoencoder. *IISE Transactions, 55*(5), 433–444. https://doi.org/10.1080/24725854.2021.2024925
32. Knuth, T., & Ahrholdt, D. C. (2022). Consumer fraud in online shopping: Detecting risk indicators through data mining. *International Journal of Electronic Commerce*, *26*(3), 388–411. https://doi.org/10.1080/10864415.2022.2076199
33. La Morgia, M., Mei, A., Sassi, F., & Stefa, J. (2023). The doge of wall street: Analysis and detection of pump and dump cryptocurrency manipulations. *ACM Transactions on Internet Technology*, *23*(1). https://doi.org/10.1145/3561300
34. Lebichot, B., Le Borgne, Y.-A., He-Guelton, L., Oblé, F., & Bontempi, G. (2020). Deep-learning domain adaptation techniques for credit cards fraud detection. In *Recent Advances in Big Data and Deep Learning* (pp. 78–88). https://doi.org/10.1007/978-3-030-16841-4_8
35. Li, R., Liu, Z., Ma, Y., Yang, D., & Sun, S. (2023). Internet financial fraud detection based on graph learning. *IEEE Transactions on Computational Social Systems*, *10*(3), 1394–1401. https://doi.org/10.1109/TCSS.2022.3189368
36. Lubna, A., H.V, A. S., & Manjunath, K. N. (2023). ATM weapon, fraud detection and cybercrime prediction. In *International Conference on Advances in Electronics, Communication, Computing and Intelligent Information Systems (ICAECIS)* (pp. 528–533). https://doi.org/10.1109/ICAECIS58353.2023.10170287
37. M. Younis, R. (2023). A survey on using machine learning and deep learning based Iris recognition. *Academic Journal of Nawroz University*, *12*(1), 47–54. https://doi.org/10.25007/ajnu.v12n1a1677
38. Masood, M., Nawaz, M., Malik, K. M., Javed, A., Irtaza, A., & Malik, H. (2023). Deepfakes generation and detection: State-of-the-art, open challenges, countermeasures, and way forward. *Applied Intelligence*, *53*(4), 3974–4026. https://doi.org/10.1007/s10489-022-03766-z

39. Md, A. Q., Narayanan, S. M. S. S., Sabireen, H., Sivaraman, A. K., & Tee, K. F. (2023). A novel approach to detect fraud in Ethereum transactions using stacking. *Expert Systems, 40*(7), e13255. https://doi.org/10.1111/exsy.13255
40. Momeny, M., Neshat, A. A., Jahanbakhshi, A., Mahmoudi, M., Ampatzidis, Y., & Radeva, P. (2023). Grading and fraud detection of saffron via learning-to-augment incorporated Inception-v4 CNN. *Food Control*, *147*(2), 109554. https://doi.org/10.1016/j.foodcont.2022.109554
41. Nanduri, J., Liu, Y. W., Yang, K., & Jia, Y. (2020). Ecommerce fraud detection through fraud islands and multi-layer machine learning model. *Advances in Intelligent Systems and Computing*, *1130 AISC*, 556–570. https://doi.org/10.1007/978-3-030-39442-4_41
42. Narayan, A., Madhu Kumar, S. D., & Chacko, A. M. (2023). A review of financial fraud detection in e-commerce using machine learning. *Smart Innovation, Systems and Technologies*, *327*, 237–248. https://doi.org/10.1007/978-981-19-7524-0_21
43. Nassif, A. B., Talib, M. A., Nasir, Q., & Dakalbab, F. M. (2021). Machine learning for anomaly detection: A systematic review. *IEEE Access*, *9*, 78658–78700. https://doi.org/10.1109/ACCESS.2021.3083060
44. Ni, L., Li, J., Xu, H., Wang, X., & Zhang, J. (2023). Fraud feature boosting mechanism and spiral oversampling balancing technique for credit card fraud detection. *IEEE Transactions on Computational Social Systems, PP(99), 1–16.*. https://doi.org/10.1109/TCSS.2023.3242149
45. Ning, W., Chen, S., Qiang, F., Tang, H., & Jie, S. (2023). A credit card fraud model prediction method based on penalty factor optimization AWTadaboost. *Computers, Materials and Continua*, *74*(3), 5951–5965. https://doi.org/10.32604/cmc.2023.035558
46. Nortey, E. N. N., Pometsey, R., Asiedu, L., Iddi, S., & Mettle, F. O. (2021). Anomaly detection in health insurance claims using Bayesian quantile regression. *International Journal of Mathematics and Mathematical Sciences*, 1–11. https://doi.org/10.1155/2021/6667671
47. Pandey, A., Jaiswal, H., Vij, A., & Mehrotra, T. (2022). Case study on online fraud detection using machine learning. In *2nd International Conference on Advance Computing and Innovative Technologies in Engineering* (pp. 48–52). https://doi.org/10.1109/ICACITE53722.2022.9823538
48. Prabhakaran, N., & Nedunchelian, R. (2023). Oppositional cat swarm optimization-based feature selection approach for credit card fraud detection. *Computational Intelligence and Neuroscience*, *2023,* 2693022. https://doi.org/10.1155/2023/2693022
49. Rahman, M. J., & Zhu, H. (2023). Predicting accounting fraud using imbalanced ensemble learning classifiers – evidence from China. *Accounting and Finance, 2023*, 2693022. https://doi.org/10.1111/acfi.13044
50. Rtayli, N., & Enneya, N. (2023). Credit card fraud detection using predictive features and machine learning algorithms. *International Journal of Internet Technology and Secured Transactions*, *13*(2), 159. https://doi.org/10.1504/ijitst.2023.129578
51. Selvam, K., & Berkmans, J. (2023). A widespread survey on machine learning techniques and user substantiation methods for credit card fraud detection. *International Journal of Business Intelligence and Data Mining*, *22*(1), 223. https://doi.org/10.1504/ijbidm.2023.10047750

52. Sadgali, I., Sael, N., & Benabbou, F. (2019). Performance of machine learning techniques in the detection of financial frauds. *Procedia Computer Science, 148*, 45–54. https://doi.org/10.1016/J.PROCS.2019.01.007
53. Safi, A., & Singh, S. (2023). A systematic literature review on phishing website detection techniques. *Journal of King Saud University - Computer and Information Sciences, 35*(April). https://doi.org/10.1016/j.jksuci.2023.01.004
54. Salekshahrezaee, Z., Leevy, J. L., & Khoshgoftaar, T. M. (2023). The effect of feature extraction and data sampling on credit card fraud detection. *Journal of Big Data, 10*(1). https://doi.org/10.1186/s40537-023-00684-w
55. Settipalli, L., & Gangadharan, G. R. (2023). WMTDBC: An unsupervised multivariate analysis model for fraud detection in health insurance claims. *Expert Systems with Applications, 215*(15), 119259. https://doi.org/10.1016/j.eswa.2022.119259
56. Sharma, N., & Ranjan, V. (2023). Credit card fraud detection: A hybrid of PSO and K-means clustering unsupervised approach. In *Proceedings of the 13th International Conference on Cloud Computing, Data Science and Engineering, Confluence* (pp. 445–450). https://doi.org/10.1109/CONFLUENCE56041.2023.10048876
57. Singh, A., Ranjan, R. K., & Tiwari, A. (2022). Credit card fraud detection under extreme imbalanced data: A comparative study of data-level algorithms. *Journal of Experimental & Theoretical Artificial Intelligence, 34*(4), 571–598. https://doi.org/10.1080/0952813X.2021.1907795
58. Sisodia, D., & Sisodia, D. S. (2022). Data sampling strategies for click fraud detection using imbalanced user click data of online advertising: An empirical review. *IETE Technical Review, 39*(4), 789–798. https://doi.org/10.1080/02564602.2021.1915892
59. Sivanantham, S., Dhinagar, S. R., Kawin, P., & Amarnath, J. (2021). Hybrid approach using machine learning techniques in credit card fraud detection. In *Advances in Intelligent Systems and Computing* (pp. 243–251). https://doi.org/10.1007/978-981-15-5029-4_19.
60. Shafik, W., & Kalinaki, K. (2023). Smart city ecosystem: An exploration of requirements, architecture, applications, security, and emerging motivations. In *Handbook of Research on Network-Enabled IoT Applications for Smart City Services* (pp. 75–98). IGI Global. https://doi.org/10.4018/979-8-3693-0744-1.CH005
61. Sofy, M. A., Khafagy, M. H., & Badry, R. M. (2023). An intelligent Arabic model for recruitment fraud detection using machine learning. *Journal of Advances in Information Technology*. https://doi.org/10.12720/jait.14.1.102-111
62. Srinivasan, S., & P, D. (2023). Enhancing the security in cyber-world by detecting the botnets using ensemble classification based machine learning. *Measurement Sensors, 25*(5), 100624. https://doi.org/10.1016/j.measen.2022.100624
63. Strelcenia, E., & Prakoonwit, S. (2023). Improving classification performance in credit card fraud detection by using new data augmentation. *AI*, 2023, *4*(1), 172–198. https://doi.org/10.3390/ai4010008.
64. Bc, A. (2023). A Study of Machine Learning Effectiveness in Detecting Credit Card Fraud: KNIME. *International Journal for Research in Applied Science and Engineering Technology, 11*(2), 4. https://doi.org/10.22214/ijraset.2023.49213
65. Thennakoon, A., Bhagyani, C., Premadasa, S., Mihiranga, S., & Kuruwitaarachchi, N. (2019). Real-time credit card fraud detection using machine learning. In *Proceedings of the 9th International Conference On Cloud Computing, Data Science and*

Engineering, Confluence (pp. 488–493). https://doi.org/10.1109/CONFLUENCE.2019.8776942

66. Tiwari, R. (2023). The application of AI and machine learning in the financial industry and its effects on risk management and fraud detection. *International Journal of Scientific Research in Engineering and Management, 07*(01). https://doi.org/10.55041/ijsrem17595
67. Ullah, Z., & Jamjoom, M. (2023). A smart secured framework for detecting and averting online recruitment fraud using ensemble machine learning techniques. *PeerJ Computer Science, 9,* e1234. https://doi.org/10.7717/peerj-cs.1234
68. Vanini, P., Rossi, S., Zvizdic, E., & Domenig, T. (2023). Online payment fraud: From anomaly detection to risk management. *Financial Innovation, 9*(1). https://doi.org/10.1186/s40854-023-00470-w
69. Wahidahwati, W., & Asyik, N. F. (2022). Determinants of auditors ability in fraud detection. *Cogent Business & Management, 9*(1). https://doi.org/10.1080/23311975.2022.2130165
70. Wang, S., Liu, Y., Zheng, C., & Lin, R. (2022). Purchase pattern based anti-fraud framework in online e-commerce platform using graph neural network. *Lecture Notes in Computer Science (Including Subseries Lecture Notes in Artificial Intelligence and Lecture Notes in Bioinformatics, 13605 LNAI*, 112–123. https://doi.org/10.1007/978-3-031-20500-2_9
71. Xia, R., & Alshameri, F. (2023). Credit card fraud detection: An evaluation of SMOTE resampling and machine learning model performance. *International Journal of Business Intelligence and Data Mining, 23*(1). https://doi.org/10.1504/ijbidm.2023.10044811
72. Xie, S., Liu, L., Sun, G., Pan, B., Lang, L., & Guo, P. (2023). Enhanced e-commerce fraud prediction based on a convolutional neural network model. *Computers, Materials and Continua, 75*(1), 1107–1117. https://doi.org/10.32604/cmc.2023.034917
73. Yang, J., Chen, K., Ding, K., Na, C., & Wang, M. (2023). Auto insurance fraud detection with multimodal learning. *Data Intelligence, 5*(2), 388–412. https://doi.org/10.1162/dint_a_00191
74. Yang, Z., Jianjun, L., Faqiri, H., Shafik, W., Talal Abdulrahman, A., Yusuf, M., & Sharawy, A. M. (2021). Green internet of things and big data application in smart cities development. *Complexity, 2021*(1). https://doi.org/10.1155/2021/4922697
75. Yu, W., Wang, Y., Liu, L., An, Y., Yuan, B., & Panneerselvam, J. (2023). A multi-perspective fraud detection method for multiparticipant e-commerce transactions. *IEEE Transactions on Computational Social Systems, 11*(2), 1564–1576. https://doi.org/10.1109/TCSS.2022.3232619
76. Y. Ganin et al., Domain-adversarial training of neural networks. *Journal of Machine Learning Research, 2016*(17), 1–35. Accessed: Mar. 29, 2025. [Online]. Available: http://jmlr.org/papers/v17/15-239.html

CHAPTER 9

Incremental Learning in Real-Time Artificial Intelligence

Y. V. R. NAGA PAWAN[1] and BHANU PRAKASH KOLLA[2]

[1]*Department of Computer Science and Engineering, Anurag Engineering College, Ananthagiri (V&M), Suryapet (Dist.), Telangana, India*

[2]*Department of Computer Science and Engineering, Koneru Lakshmaiah Education Foundation, Vaddeswaram, Andhra Pradesh, India*

ABSTRACT

Incremental learning (IL) allows an intelligent machine to continue receiving new knowledge while maintaining what it is learned based on a constantly advancing approach to real-time artificial intelligence (AI). This chapter examines the IL of real-time AI systems in detail. This chapter examines the IL of real-time AI systems in detail. The chapter starts with an address on inherent learning and how valuable it is for real-time systems. After that, it evaluates the drawbacks of traditional machine learning techniques, which find it extremely difficult to handle new data and dynamic circumstances because they require large amounts of online training datasets. The following section of this chapter explores numerous approaches to IL, including transfer learning, active learning, and online learning. It highlights the benefits and drawbacks of each strategy and demonstrates how to apply them to get around problems like data imbalance, idea drift, and catastrophic forgetting. This chapter also includes the case examples of how IL has been used successfully in real-time AI systems for speech, picture, and anomaly detection.

Real-Time Artificial Intelligence (AI): Key Motivations, Technologies, Platforms, and Use Cases.
Pethuru Raj, B. Sundaravadivazhagan, V. Kavitha, B. Narendra Kumar Rao, & Hannah Vijaykumar (Eds.)

DOI: 10.1201/9781998511358-9

These case studies demonstrate how IL improves system efficiency, reduces processing needs, and produces more precise and trustworthy predictions. A talk about the difficulties and advantages of incorporating IL into real-time AI systems closes this chapter. To ensure the consistency and correctness of the system, it accentuates the necessity for ongoing data streams, practical algorithms, and efficient validation techniques. Finally, this chapter presents a comprehensive analysis of IL in real-time AI systems, emphasizing the benefits, techniques, and challenges. This chapter shows that IL increases a system's performance and permits real-time information to be transmitted in changed environments.

9.1 INTRODUCTION

Today, many enterprises are holding data in large databases. It increases the database sizes exponentially without a limit [1]. Modern businesses, large retailers, online traders, telecommunications, share market traders, and production units with traditional and modern devices collect terabytes of data globally [2].

Traditional education was the starting point for developing new algorithms and techniques, like supervised, untrained, or reinforcement learning. A thorough examination of these techniques and their use in solving practical issues led to significant advancements in machine learning (ML). These methods are applied in various areas, for example, recommending systems, audio and image identification, fraud detection or natural language processing (NLP). They have provided perceptive justifications and valuable solutions to a range of problems. Traditional education has certain disadvantages. In dynamic or changing circumstances, training and test data distribution varies. Traditional models could have difficulty adjusting to idea drift or changes in the data distribution. Constantly retraining models from scratch whenever new data becomes available may also be expensive in terms of time and computation.

This chapter throws light on conventional machine-learning techniques and their disadvantages. This chapter also explores the disadvantages of traditional ML techniques that are difficult to adjust to new data and changing circumstances due to large offline training datasets, as well as their benefits in real-time systems by discussing incremental learning (IL) and its advantages. This chapter's next section examines various IL strategies, such as transfer, active, and online learning (OL). It outlines the advantages and disadvantages of each tactic and shows how to use it to avoid issues like data imbalance, concept drift, and catastrophic forgetting [3].

Additionally, this chapter provides real-world case studies of how IL has been used effectively in real-time artificial intelligence (AI) systems for anomaly, voice, and image recognition. These case studies show how IL increases system effectiveness, decreases processing requirements, and generates more accurate and reliable predictions. This chapter discusses the challenges and benefits of introducing IL into real-time AI systems. It emphasizes the need for continual data streams, practical algorithms, and effective validation mechanisms to ensure the system's consistency and accuracy.

This chapter concludes with a comprehensive review of IL in real-time AI systems, underlining its advantages, strategies, and difficulties. It also shows the system's adaptability in real time to changing environmental conditions and enhancing system performance with IL.

9.1.1 CONVENTIONAL ML TECHNIQUES

The early days of ML techniques were rule-based and expert systems. These systems primarily rely on manually designed rules and heuristics to make decisions or perform tasks based on specific domain knowledge. These approaches required human experts to encode knowledge explicitly into the systems. The systems are limiting in scalability and adaptability to new scenarios [4–6].

Later, with the advent of conventional learning, the emphasis shifted toward building and training the models. These models find patterns and relationships in the data.

Traditional learning has been instrumental in advancing ML through developing and applying supervised, unsupervised, and reinforcement learning algorithms. The said approaches have undergone extensive research. In many different real-world situations, these algorithms have been used extensively. They have significantly contributed to the progress and achievements in ML.

In ML, traditional learning refers to the conventional approach of training models on a fixed dataset to optimize their performance on that specific dataset. It assumes that the data used for training and testing represents the target problem and assumes a static data distribution. In traditional learning, the process typically involves the steps as shown in Figure 9.1 [7–9].

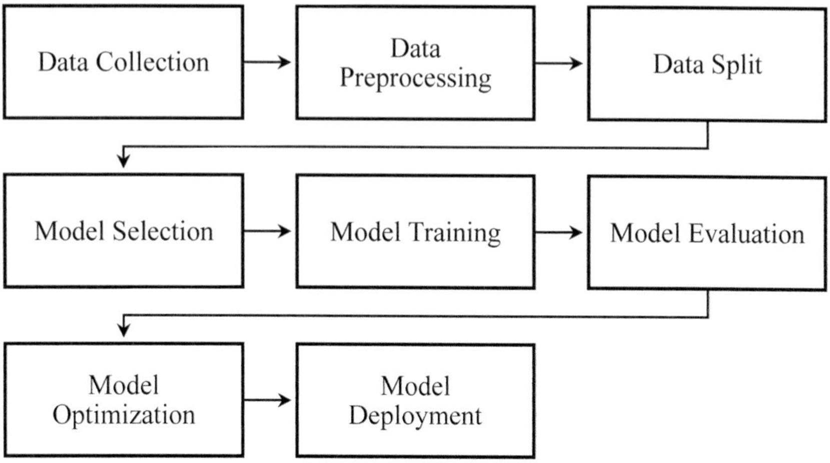

FIGURE 9.1 Steps involved in traditional learning.

1. Data Collection: Gathering a representative dataset that includes input features and corresponding target outputs (labels) for a given problem. The dataset should be diverse, comprehensive, and properly annotated.
2. Data Preprocessing: Cleaning and preparing the collected data for further analysis. This step involves removing duplicates, handling missing values, normalizing or scaling features, and potentially performing feature selection or extraction.
3. Dataset Split: The splitting of a dataset into training and testing parts. The selected or newly designed model is trained with a training set. The test set acts as an independent evaluation to assess the model's performance on untested data.
4. Model Selection: Choosing an appropriate learning algorithm or model architecture relies on the problem type and characteristics of the dataset. This selection involves considering factors such as the data's nature, the problem's complexity, and the available computational resources.
5. Model Training: The chosen model is trained to discover the underlying patterns and relationships between the input characteristics and target outputs using the training data. This method entails altering the model's parameters or weights based on an optimization technique to minimize a specified loss or error function.

6. Model Evaluation: Assessing the performance of the trained model using the testing data. Various evaluation metrics are used, depending on the problem, such as accuracy, precision, recall, F1 score, or mean squared error. This step helps gauge how well the model generalizes to unseen data and provides insights into its effectiveness.
7. Model Optimization: The trained model is iteratively fine-tuned, tuning hyperparameters to improve performance. This process may involve cross-validation, grid search, or Bayesian optimization to find the optimal configuration.
8. Model Deployment: Once satisfied with the model's performance, it can be deployed for making predictions or decisions on new, unseen data in real-world applications.

This process may involve integrating the model into a software system or creating an API for online inference.

These steps provide an overview of the traditional learning process. The specifications may vary depending on the problem domain, the complexity of the data, and the chosen learning algorithms.

Traditional learning approaches encompass a range of techniques used to train ML models on static datasets. Some frequently used traditional learning models [7] are listed below and shown in Figure 9.2.

1. Unsupervised Learning: Unsupervised learning aims to find patterns or structures in unlabeled data without explicit target labels. K-means clustering, density-depending spatial clustering of applications with noise (DBSCAN), and hierarchical clustering are clustering algorithms that combine related data points depending on their features' similarity. Well-known techniques like principal component analysis (PCA) and t-distributed stochastic neighbor embedding (t-SNE) reduce the dimensionality of the data, preserving the dataset structure. Association rule mining, anomaly detection, and generative models are other unsupervised learning techniques [10–22].
2. Supervised Learning: Supervised learning is a widely used approach where models are trained on labeled examples to learn the mapping between input features and corresponding target labels. It involves minimizing an objective function, such as mean squared error or cross-entropy loss, to optimize model parameters. Techniques like

linear regression, logistic regression, decision trees, and support vector machines (SVM) [23–31] are widely used.

3. Semi-Supervised Learning: The amalgamation of unsupervised and supervised learning is semi-supervised learning. It leverages a limited amount of labeled data and a larger pool of unlabeled data to train models. The idea is that the unlabeled data can provide additional information or improve generalization. Self-training, cotraining, and generative models (e.g., generative adversarial networks [GANs]) are commonly applied in semi-supervised learning [32–35].
4. Reinforcement Learning: Training an agent to learn optimum practices through feedback from the environment is part of reinforcement learning. An agent shall engage in a relationship with the environment, obtain an award or penalty based on his activities, and strive to maximize its total reward over time. Q-learning and deep Q-networks (DQNs) are popular algorithms used in reinforcement learning [36, 37].
5. Ensemble Learning: Several models are combined in ensemble learning to provide predictions or choices. The predictions made by each model in the ensemble are combined using methods like weighted average and majority voting. By utilizing a variety of models, ensemble methods—such as random forests and gradient boosting—improve prediction accuracy, robustness, and generalization [38–41].
6. Neural Networks: Neural networks are generally called artificial neural networks (ANN). ANNs learn complex patterns from large-scale datasets. Convolutional neural networks (CNNs) are used for analyzing images and videos. Recurrent neural networks (RNNs) are suitable for sequential data analysis. Transformers are effective in NLP. Conventional training methods, for example, stochastic descent gradients SGD and back propagation, are frequently applied to optimize neural network parameters. GANs, autoencoders, transformers, reinforcement learning networks (RLN), and neural tuning machines (NTM) are gaining success. Neural networks are rapidly evolving [22, 42–48].

These traditional learning approaches have been instrumental in various ML applications, forming the foundation for many advanced techniques. However, they typically assume a static dataset and may not adapt well to dynamic data distributions or evolving scenarios. IL approaches have emerged to address the limitations of traditional learning and enable continuous adaptation and improvement with new data.

Traditional learning techniques are effectively used in various fields, including fraud detection, NLP, audio and picture recognition, and recommendation systems. They have provided valuable insights and solutions to many practical problems.

This data-driven traditional learning approach offered the following advantages that include [41, 49–51]:

1. Automatic Feature Extraction: Traditional learning algorithms do not require human feature engineering since they can automatically extract pertinent features from raw data. The models can learn representations more suited for the task at hand.
2. Adaptability to Different Domains: Traditional learning models are domain-agnostic and can be applied to many problem domains. They do not rely on handcrafted rules specific to a particular domain, making them more flexible and applicable to diverse applications.
3. Scalability with Big Data: Traditional training methods are used to handle large data sets efficiently, enabling them to build models that will take advantage of the potential of big data.
4. Generalization to Unseen Data: Traditional learning models aim to generalize patterns learned from the training set to predict unseen or new data accurately. They strive to capture underlying relationships that hold across the dataset, allowing them to perform well on previously unseen instances.
5. Automation of Decision-Making: Traditional learning models can simplify decision-making processes by learning from data. They make predictions, classify objects, detect anomalies, and perform other tasks in a way that does not require any human intervention after training on the relevant data.

9.1.2 ISSUES WITH TRADITIONAL LEARNING

Traditional learning methods have certain drawbacks and need help adapting to different situations. The following are some challenges associated with traditional learning [52–55]:

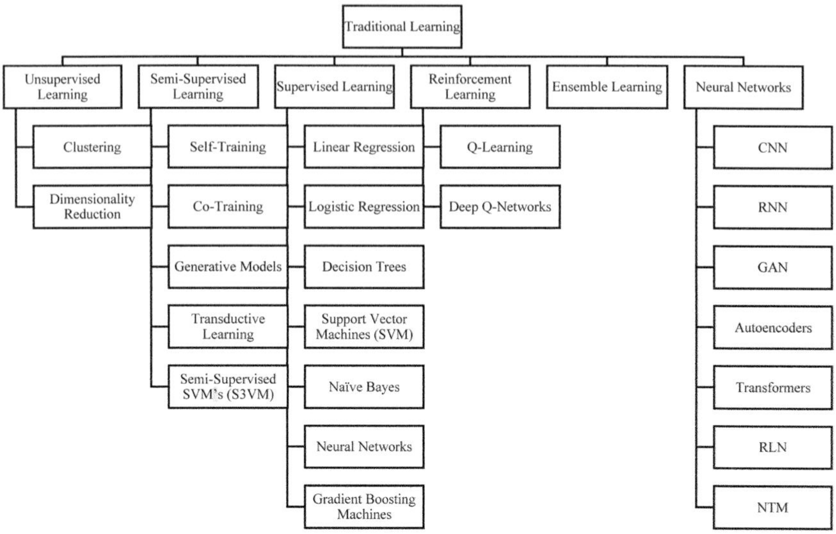

FIGURE 9.2 Traditional learning approaches.

1. Lack of Adaptability: Traditional learning models are typically trained on a fixed dataset and assume a static data distribution. They are not designed to handle dynamic or evolving data. It is time-consuming and computationally expensive; the model must be retrained from scratch when new data is augmented.
2. Memory Requirements: Batch learning typically requires storing the entire dataset in memory for training the model. It is almost impractical or infeasible for large-scale datasets. This method restricts the scalability of traditional learning approaches, especially when dealing with streaming or online data.
3. Limited Efficiency: When new data is constantly received or just a portion of the dataset is relevant, batch learning for model training might be ineffective as it involves analyzing the whole data set. There may be excessive computing costs if the training procedure is repeated after each update or data addition.
4. Inability to Handle Concept Drift: Concept drift is where the underlying data distribution changes over time. Traditional learning models struggle to adapt to concept drift, as they assume a stationary data distribution during training. When the data distribution changes, the model's performance may degrade significantly.

5. Loss of Previous Knowledge: When traditional models are trained on new data, they often need to remember or overwrite the knowledge acquired from previous training sessions. It is known as catastrophic forgetting. It limits the model's ability to retain previously learned patterns, leading to a lack of continuity in knowledge accumulation.
6. Suboptimal Resource Utilization: Traditional learning approaches must efficiently utilize the resources available during training. For instance, even though just a small percentage of the dataset may have changed considerably, the whole dataset is frequently utilized for retraining when new data is received. This results in redundant computations and inefficient use of computational resources.

Addressing these issues and adapting to dynamic environments requires adopting IL techniques. IL enables models to learn from new data while preserving past knowledge, adapt to concept drift, handle large-scale datasets efficiently, and continuously improve without retraining from scratch.

It assumes that the training and testing data follow the same distribution, which may not hold in dynamic or evolving scenarios. Traditional models need help adapting to changes in the data distribution or handling concept drift effectively. Furthermore, retraining models from scratch every time as the data is evolving can be time-consuming and computationally expensive.

IL or lifetime learning strategies have been developed to overcome these drawbacks, allowing models to gradually learn from new data and adapt to shifting surroundings, improving their adaptability and performance over time.

9.2 APPROACHES IN IL

9.2.1 ACTIVE LEARNING

Active learning (AL) is an ML approach where the algorithm actively selects and queries the most informative or uncertain instances from an unlabeled dataset for annotation by an oracle (human expert). The goal is to iteratively label the most relevant instances to improve the model's performance while minimizing the labeling effort [56].

9.2.1.1 POOL-BASED AND STREAM-BASED AL

AL can be categorized into pool-based and stream-based approaches. In pool-based AL, a fixed unlabeled pool of instances is available, and the model selects instances for labeling from this pool [57]. In stream-based AL, instances arrive in a streaming fashion, and the model must make labeling decisions on the fly. Stream-based AL requires efficient and adaptive query strategies to handle the continuous data stream [58].

9.2.1.2 AL PROCESS

The process involved in AL is shown below [56–59]:

1. Initialize the Model: Start with a small labeled dataset.
2. Select Query Strategy: Determine a query technique to decide which instances should be tagged next. The model's uncertainty, information gain, or other parameters serve as the foundation for the query approach. Uncertainty, margin, and entropy-based sampling are some of the most well-liked query techniques. The details of these strategies are discussed later.
3. Query Instance Selection: Apply a chosen query strategy to the unlabeled dataset and select the instances expected to provide the most helpful information for model improvement. These instances are then sent for annotation.
4. Annotation by Oracle: Present the selected instances to an oracle, a human expert, for annotation or labeling.
5. Model Update: Incorporate the newly labeled instances into the labeled dataset and retrain or update the model using the combined labeled dataset.
6. Iterate: Repeat steps 2–5 until a sufficient number of instances have been labeled or a desired level of performance is achieved.

The key idea behind AL is that the model actively selects the instances expected to be the most beneficial for improving its performance rather than randomly selecting instances for labeling. By carefully choosing the instances to label, AL can reduce the labeling effort required and achieve better model performance with a smaller labeled dataset.

AL has applications in various domains where labeled data is scarce or expensive, such as text classification, image recognition, and medical diagnosis. It allows the model to focus on the most informative instances and adaptively acquire knowledge to improve its predictions.

9.2.1.3 QUERY STRATEGIES

In AL, many query techniques are employed to choose cases for annotation. Several typical methods include:

9.2.1.3.1 AL Using Uncertainty Sampling

Uncertainty sampling is a widely used query strategy in AL, where instances with high uncertainty are selected for labeling by an oracle. The idea behind uncertainty sampling is that the model's uncertainty in predicting an instance's label can indicate its informativeness, and labeling such instances can potentially improve the model's performance [58, 60–62].

1. Train a Classifier: Train a classifier on an initially labeled dataset.
2. Obtain Predictions: Apply the trained classifier to the unlabeled instances and obtain each class's predicted probabilities or confidence scores.
3. Compute Uncertainty: Calculate the uncertainty of each instance using a measure such as entropy or margin. Standard uncertainty measures include:
    - Entropy: Compute the entropy of the predicted probabilities across the classes. Higher entropy indicates higher uncertainty.
    - Margin: The difference between expected probability with the highest and second-highest values is calculated. Reduced margin values signify increased uncertainty.
4. Select Instances with High Uncertainty: Sort the instances in descending order based on their uncertainty values. Select the instances with the highest uncertainty, as these are the instances where the model is least detailed about their labels.
5. Label the Selected Instances: Present the selected instances to an oracle (human expert) for annotation or labeling.
6. Update the Model: Incorporate the newly labeled instances into the labeled dataset and retrain or update the model using the combined labeled dataset.

7. Iterate: Repeat steps 2–6 until a sufficient number of instances have been labeled or a desired level of performance is achieved.

By focusing on instances with high uncertainty, uncertainty sampling aims to target challenging or ambiguous instances for the model. Labeling these instances can provide valuable information for the model to refine its decision boundaries and improve performance. It is a practical and intuitive approach to AL, allowing the model to actively seek instances that contribute the most to its learning process.

9.2.1.3.2 AL Using Margin Sampling

Margin sampling is a query strategy commonly used in AL to select instances for labeling based on the margin of uncertainty in a classifier's predictions. It focuses on selecting cases where the classifier is least confident or has the smallest margin between the predicted probabilities of the top two candidate classes [56, 61, 63, 64].

1. Train a Classifier: Train a classifier on an initially labeled dataset.
2. Obtain Predictions: Apply the trained classifier to the unlabeled instances and obtain predicted probabilities for each class.
3. Compute Margins: Compute the margins for each instance by taking the difference between the highest and second-highest predicted probabilities. A smaller margin indicates a higher level of uncertainty.
4. Select Instances with Small Margins: Sort the instances based on the margin values in an ascending order. Select the instances with the smallest margins, as these are the instances where the classifier is most uncertain or ambiguous.
5. Label the Selected Instances: Present the selected instances to an oracle (human expert) for annotation or labeling.
6. Update the Model: Incorporate the newly labeled instances into the labeled dataset and retrain or update the model using the combined labeled dataset.
7. Iterate: Repeat steps 2–6 until a sufficient number of instances have been labeled or a desired level of performance is achieved.

The margin sampling technique concentrates on cases challenging the classifier to correctly categorize by choosing examples with small margins.

In areas of the feature space where the model is ambiguous or error-prone, it helps to enhance performance.

Margin sampling is a popular query strategy in AL, along with other strategies, like uncertainty sampling, diversity sampling, and query-by-committee. The choice of query strategy depends on the specific problem, dataset, and the characteristics of the classifier being used. Evaluation and experimentation must determine the best approach for a given task.

9.2.1.3.3 AL Using Entropy-Based Sampling

Entropy-based sampling is a query strategy commonly used in AL, where instances with high entropy in their predicted class probabilities are selected for labeling. Entropy measures a probability distribution's level of information or uncertainty. AL chooses instances with high entropy because it believes they are more informative [57, 65–67], to reduce the uncertainty in the model's prediction.

1. Train a Classifier: Train a classifier on an initially labeled dataset.
2. Obtain Predictions: Apply the trained classifier to the unlabeled instances and obtain the predicted probabilities for each class.
3. Compute Entropy: Compute the entropy of the predicted probabilities for each instance. The entropy is calculated using the formula:

   entropy, H = -£ p log p

   where "p" represents the predicted probability of each class for a given instance.
4. Select Instances with High Entropy: Sort the instances in a descending order based on their entropy values. Choose the instances with the highest entropy because these are the ones where the model is least confident in its predictions or is least definite about them.
5. Label the Selected Instances: Present the selected instances to an oracle (human expert) for annotation or labeling.
6. Update the Model: Incorporate the newly labeled instances into the labeled dataset and retrain or update the model using the combined labeled dataset.
7. Iterate: Repeat steps 2–6 until a sufficient number of instances have been labeled or a desired level of performance is achieved.

Entropy-based sampling seeks to target situations when the model's predictions are unclear or uncertain by choosing cases with high entropy. The model performs better because the decision boundaries are refined, and the model's uncertainty is reduced by labeling these occurrences. Experimentation and evaluation with different threshold values or variations of entropy-based sampling can help determine the most effective approach for a specific task or dataset.

9.2.1.3.4 Various Other Sampling Methods

1. Diversity Sampling: Select instances that are diverse or dissimilar to the labeled cases already in the training set. The model explores different parts of the feature space and captures a more comprehensive data representation [68].
2. Query-by-Committee: Train multiple models or ensemble classifiers on the labeled data and select instances where the models disagree the most. This method focuses on instances more challenging for the ensemble to classify consistently. Labeling these instances can help resolve uncertainties and improve the model's performance [24].
3. Expected Model Change: Expected Model Change is a powerful query strategy in AL, as it allows for the selection of instances that can bring about the most substantial improvement in the model's behavior and generalization capabilities. It helps in efficiently utilizing the labeling resources by prioritizing cases that have the potential to make a significant impact on the model's performance [69].
4. Bayesian AL: It allows for explicit uncertainty modeling and provides a principled way to make informed decisions about querying instances for labeling. By incorporating Bayesian inference, it can handle small labeled datasets effectively, take model parameter uncertainty, and make reliable predictions in data-scarce scenarios [70, 71].

9.2.1.4 LABELING STRATEGIES

The effectiveness of AL depends on the labeling process. It is crucial to carefully design the process, ensure high-quality annotations, and address potential labeling biases or errors.

Labeling strategies in AL refer to the methods and criteria used to select cases from an unlabeled dataset for manual annotation or labeling. The goal is strategically choosing instances that will maximize the learning and performance of a machine-learning model [72–74]. The common labeling strategies [57, 61, 67, 70] in AL are given below:

1. Uncertainty Sampling: Select instances for labeling the current model find most uncertain or ambiguous. This strategy can include the following:
   - Least Confidence: Select instances with the lowest predicted class probability.
   - Margin Sampling: Select instances with the slightest difference between the top two predicted class probabilities.
2. Entropy: Select instances with the highest entropy of predicted class probabilities.
3. Diversity Sampling: Select various or dissimilar cases from the labeled instances already in the training set. This strategy aims to cover different regions of the feature space and capture a broader range of patterns.
4. Query by Committee: Use an ensemble of models or a committee of classifiers. Select instances where the models disagree the most in their predictions. This strategy assumes that instances where the models have high disagreement are more uncertain and informative.
5. Expected Model Change: Estimate the potential change or improvement in the model's performance upon labeling an instance. This strategy considers the expected impact of adding a new case to the training set based on factors such as the model's uncertainty and the distribution of unlabeled data.
6. AL with Human-in-the-Loop: Incorporate human expertise and preferences into the labeling process. This strategy involves active collaboration between the model and human annotators, where the annotators guide the selection of instances based on their domain knowledge or specific criteria.
7. Representative Sampling: Select instances representing different clusters or subgroups in the data. This strategy aims to ensure a balanced representation of the data distribution and prevent biased or skewed model learning.

8. Hybrid Strategies: Combine multiple labeling strategies to leverage their strengths and improve the efficiency and effectiveness of AL. For example, they combine uncertainty sampling with diversity sampling or incorporate human feedback with other strategies.

The strategic selection of the labeling strategy is based upon a specific problem, available resources, and data characteristics, extending the possibility to experiment with different strategies to find an apt strategy.

By intelligently selecting instances for labeling, AL reduces the annotation effort, accelerates the learning process, and improves model performance compared to traditional learning approaches that rely on large, fully labeled datasets.

9.2.1.5 AL WITH DIFFERENT MODEL TYPES

AL applies to a wide range of models, including deep learning models (e.g., neural networks) and conventional ML models (e.g., SVM, decision trees). The paradigm and the accessibility of AL libraries and frameworks may impact the specific implementation [75, 76].

9.2.1.6 AL IN THE PRESENCE OF CLASS IMBALANCE

AL can be instrumental in addressing class imbalance issues. By focusing on the instances from the minority class or instances with high uncertainty, AL helps to balance the training set and improve the model's performance on underrepresented classes [77, 78].

9.2.1.7 COMBINING AL WITH OTHER TECHNIQUES

AL is a dynamic and evolving area of research with numerous variations and extensions. Researchers continuously explore new query strategies, explore AL in different settings, and adapt the concept to various applications and problem domains. By intelligently selecting instances for labeling, AL provides a practical approach to improving the efficiency and effectiveness of the ML process [79].

AL can be combined with other techniques, such as transfer learning (TL), ensemble methods, and semi-supervised learning, to enhance model performance further and reduce the labeling effort [80].

9.2.2 TRANSFER LEARNING

TL can also be applied in IL, where the goal is continuously learning and adapting to new data while retaining knowledge from previous tasks or domains. IL poses challenges such as catastrophic forgetting, where the model needs to remember previously learned knowledge when exposed to new data. TL can help mitigate this issue by leveraging knowledge from previous tasks to facilitate learning on new tasks or data [81].

In IL scenarios, the pretrained model is a starting point, capturing general knowledge and representations from a large initial dataset. As new data is augmented, the model is fine-tuned on the latest data while retaining its previously learned knowledge. It allows the model to adapt to new tasks or domains without completely discarding its existing knowledge [82].

9.2.2.1 TL PROCESS

The process of incorporating TL into IL involves the following steps [83]:

1. Pretraining: Train a base model on a large, diverse dataset using traditional supervised or unsupervised learning techniques. The base model learns general features and representations from the initial dataset, capturing valuable knowledge.
2. IL Model: As new data becomes available, update the model to adapt to the new tasks or domains while retaining knowledge from previous tasks. Several strategies can be employed for IL:
3. Fine-tuning: Freeze the earlier layers of the pretrained model to preserve the learned representations. Update the later layers or add new layers to accommodate the new data. Train the model on the latest data, allowing it to adapt while leveraging the knowledge captured in the earlier layers.
4. Knowledge Distillation: Train a separate student model on the latest data while leveraging the pretrained model's predictions as a guide. The pretrained model acts as a teacher, transferring its

knowledge to the student model. The student model learns from both the latest data and the wisdom distilled from the teacher model.

5. Rehearsal: Periodically sample a portion of the previously seen data alongside the new data. Train the model on this combined dataset, including old and new data. Rehearsal helps retain knowledge from previous tasks, reducing catastrophic forgetting.
6. Network Expansion: Expand the model's architecture to accommodate new tasks or data. Add new layers or branches to capture further task-specific information. Freeze or update the existing layers minimally, focusing on training the new layers on the latest data.
7. Evaluation and Iteration: Assess the performance of the updated model on the new tasks or data. Evaluate the model's retention of knowledge from previous tasks. Iterate the IL process as new data continues to arrive.

The characteristics of a given task, the available resources, and the necessary trade-off between performing new tasks and retaining prior knowledge are critical when selecting IL strategies like fine tuning, information clarification, rehearsal or network expansion.

Models may use their prior knowledge to adapt to new information and reduce the requirement for significant retraining to maintain proficiency on previously acquired tasks by including TL in IL. In a dynamic environment, it provides ongoing learning and adaptation.

9.2.2.2 TYPES OF TL

There are several approaches to TL, each addressing different scenarios and objectives. Some commonly recognized types of TL [84–87]:

1. Inductive TL: Inductive TL aims to transfer knowledge from a source domain with labeled data to a target domain with different but related data. The trained models are applied to enhance performance on a target task once the model has been trained in the source domain.
2. Transductive TL: Transductive TL involves transferring knowledge from a labeled subset of the target domain to the remaining

unlabeled data in the same domain. This type of TL is proper when labeled data is limited and expensive to obtain for the target domain.

3. Domain Adaptation: Transferring knowledge from one domain (the source) to another (the target), where data distributions may be different, is the basic goal of domain adaptation. The goal is to adapt the model's representations and predictions to perform well on the target domain, even though the training data is primarily from the source domain.
4. Multitask Learning: Multitask learning involves training a model to learn multiple related tasks simultaneously. The idea is that shared knowledge across tasks can improve overall performance. This type of TL is applicable when numerous tasks have common underlying patterns or features.
5. One-shot Learning: One-shot learning deals with scenarios where only a single or a few examples per class are available for training. TL is used to generalize knowledge from a large, labeled dataset to learn to recognize new classes with limited labeled samples.
6. Zero-shot Learning: Zero-shot learning refers to training a model to recognize classes or concepts not seen during training. Instead, the model learns to associate seen and unseen classes using auxiliary information, such as class attributes or semantic embeddings.

These TL approaches provide strategies to leverage prior knowledge from related tasks, domains, or auxiliary information to enhance learning and performance on new tasks or data. The choice of TL type depends on the specific problem, available data, and the desired learning objectives.

9.2.2.3 IMPLEMENTATION OF TL

TL implementation using Python involves utilizing pretrained models and leveraging their learned representations to improve performance on new tasks or datasets. Implementing TL involves the following steps:

1. Import Dependencies: Begin by importing the necessary libraries and modules.
2. Load Pretrained Model: Choose a pretrained model suitable for the task. Examples include VGG16, ResNet, Inception, or BERT. Load the pretrained model using the corresponding library.

3. Modify the Model: Depending on the specific task, one may need to modify the pretrained model. Typically, it involves removing or freezing some layers and adding new layers specific to the task. The new layers are usually initialized randomly or with predefined weights.
4. Data Preparation: Prepare the dataset for training. In this stage, preprocessing procedures related to the data are carried out, including image scaling, input normalization, and data splitting into test, validation, and training sets.
5. Feature Extraction: The pretrained model extracts features from the given dataset. Pass the data through the pretrained model to a specific layer and extract the output as features.
6. Train the Model: A pretrained model is modified by adding new layers, and it will be trained on a specific task. Compile the model with an appropriate loss function, optimizer, and metrics. Train the model using labeled data, a smaller dataset particular to the task.
7. Fine-tuning (Optional): This step is optional. If desired, the model is fine-tuned by unfreezing some of the earlier layers of the pre-trained model. This step allows the layers to be updated during training, adapting them to the specific task.
8. Evaluate and Test: A separate test set evaluates the model's performance. Calculate relevant metrics to assess its accuracy, precision, recall, or any other suitable metrics for the task.
9. Predictions: Use the trained model to predict new, unseen data and analyze the results.

9.2.3 ONLINE LEARNING

OL is also an IL or lifelong learning. Models are continually trained and changed in this machine-learning paradigm in response to new available data. Unlike traditional batch learning, where models are trained on fixed datasets in a one-time process, OL allows for incremental updates to the model based on new observations.

In OL, the model learns from individual or small batches of data samples in sequential order. As new samples arrive, the model updates its parameters to incorporate the latest information and improve performance.

This iterative process enables the model to adapt to changes in the data distribution, handle concept drift, and learn in dynamic environments.

OL is beneficial when data is generated continuously or when it is not feasible or practical to store and retrain models on large, fixed datasets. It is commonly employed in real-time applications, streaming data analysis, online recommendation systems, fraud detection, and other domains where new data arrives incrementally [88].

Key features and considerations in OL include [89]:

1. Efficiency: OL enables models to learn and update in real-time or near real-time, allowing fast adaptation to changing data.
2. Scalability: OL allows models to scale with increasing data volumes, as they can update the model parameters incrementally without requiring complete retraining on the entire dataset.
3. Memory Usage: OL processes the data incrementally; it avoids storing large datasets in memory, making it memory-efficient.
4. Concept Drift: OL models can adapt to changes in the underlying data distribution, making them suitable for handling concept drift, where the statistical properties of the data change over time.
5. A trade-off between New and Old Data: OL models must balance leveraging new data to adapt the model and preserving knowledge learned from previous data to avoid catastrophic forgetting.

Standard algorithms and techniques used in OL include online gradient descent, stochastic gradient descent, adaptive learning rates, and methods for handling imbalanced data streams.

OL provides a flexible and adaptive approach to ML, allowing models to learn and improve as new data arrives continuously. It offers advantages in dynamic and evolving environments where models must adapt quickly and efficiently to changing data patterns.

9.2.3.1 TYPES OF OL

In OL, several types or categories describe different aspects of the learning process. These categories can help classify and understand the specific characteristics and requirements of OL approaches. The following are some common types or categories of OL:

1. Batch OL: In this type, the model is updated periodically using mini-batches of data. The model parameters are updated in batches rather than after each data point. This approach improves efficiency by reducing the frequency of parameter updates.
2. Online Gradient Descent: Online gradient descent is a popular algorithm for OL. It updates the model parameters based on the gradient of the loss function computed for each data point. This method changes the parameters after every observation, allowing the model to learn progressively.
3. Online Sequential Learning: In this type, the model is trained and updated sequentially as new data arrives. The model learns from each new data point and updates its parameters accordingly. This approach is suitable for scenarios where data arrive one at a time or in a streaming fashion.
4. Online Passive Learning: In passive OL, the model is trained using the incoming data but without actively selecting or querying specific instances for training. The model learns passively from the available data stream without external guidance.
5. Active OL: Active OL involves selecting specific instances from the incoming data stream for annotation or labeling. The model identifies samples with high uncertainty or potential learning value and queries for their labels. This approach reduces labeling efforts by selectively choosing informative instances for training.
6. Online Reinforcement Learning: Online reinforcement learning is an OL where an agent interacts with an environment and learns to take actions to maximize rewards. The agent learns by trial and error, continuously updating its policy based on the observed rewards and feedback from the environment.
7. Online Clustering: Online clustering algorithms aim to cluster data points online. They incrementally update the clusters as new data arrives, allowing for dynamic clustering in evolving data streams.

These categories are not mutually exclusive; different OL methods can combine multiple approaches or techniques. The choice of the OL category depends on the specific requirements, constraints, and characteristics of the learning problem and the available data stream.

9.2.3.2 OL PROCESS

In the context of TL, OL refers to a method where a pretrained model is continuously updated or adapted with new data as it becomes available. It allows the model to incorporate new information and adjust its knowledge over time without requiring complete retraining from scratch.

In traditional training transfer, a pretrained model is usually set in place, and only task-specific layers on the top of an already trained model are trained using one particular dataset. However, the entire model, including the pretrained layers, can be updated in OL using new incoming data.

OL in TL can be beneficial in scenarios where the data distribution or the task itself may change over time. It can adapt to emerging patterns and maintain its performance on the current task as it updates pretrained models with new data.

The process of OL in IL typically involves the following steps:

1. Data Collection: In OL, data is collected incrementally over time. This task can be done through various sources such as streaming data, real-time user interactions, or dynamic datasets.
2. Model Initialization: Initially, a model is initialized with some default parameters or weights. The initial model selection relies on the specific learning algorithm and the nature of the problem.
3. Online Training: As new data instances arrive, they are presented individually or in small batches to the learning algorithm. The model is updated incrementally using the latest data to adapt to the changing patterns and characteristics of the data.
4. Model Evaluation: The model's performance may be evaluated after it has been updated using online evaluation metrics or a different validation set. This step helps to monitor the model's progress and make decisions about further updates or adjustments.
5. Feedback and Adaptation: The model receives input, and modifications are made to enhance its performance in light of the assessment results.
6. This step can involve fine-tuning the model parameters, updating feature representations, or incorporating new knowledge into the model.
7. Iteration: The above steps are repeated iteratively as new data arrives. The model continuously learns and adjusts to enhance its predictions and generalizations, utilizing the latest data.

9.2.3.3 IMPLEMENTATION OF OL

Numerous phases involve implementing OL to analyze data progressively and update the model with new observations. The stages needed to implement OL are broken out in further depth as follows:

1. OL Algorithm Selection: Select an algorithm suitable for OL, such as online gradient descent, online passive learning, or online reinforcement learning. The choice depends on the specific problem and data characteristics.
2. Initialize the Model: Create an initial model using the chosen algorithm and set the initial parameters. This step includes defining the model architecture, setting hyperparameters, and initializing weights.
3. Obtain New Observations: Retrieve new data points or observations as they become available from a data stream, database, or any other data source.
4. Preprocess the New Observations: Preprocess the new observations to ensure they are in a suitable format for the OL algorithm. This step may involve data cleaning, normalization, feature scaling, or any other necessary data transformations.
5. Update the Model: Feed the preprocessed data into the OL algorithm to update the model parameters based on the new observations. The specific update process depends on the chosen algorithm.
    - For online gradient descent, update the model weights using the gradient computed from the new observation's loss function.
    - Incorporate the new observations into the model for online passive learning without actively selecting or querying for labels.
    - For online reinforcement learning, update the model's policy based on the observed rewards and feedback from the environment.
6. Evaluate the Model: Evaluate the improved model's performance with appropriate assessment measures. This step can involve using a validation set or comparing the model's predictions with ground truth labels.
7. Repeat Steps 3–6: Continuously repeat steps 3–6 as new observations arrive. As a result, the model can rapidly adapt and adjust to evolving data patterns.

8. Optional: Handle Concept Drift: If concept drift is expected in the data, consider monitoring and adapting the model to detect and handle changes in the underlying data distribution. This step may involve techniques like drift detection algorithms, model retraining, or concept drift adaptation methods.
9. Model Persistence: Optionally, save the trained model parameters at regular intervals or checkpoints to maintain a record of the learned knowledge. This step allows you to restore the model and continue learning from where you left off in case of system interruptions or for future use.

9.3 COMPARISON OF AL, TL, AND OL

AL, TL, and OL are three distinct approaches to ML, each with its characteristics and goals. A systematic comparison of these three learning approaches is tabulated in Table 9.1:

For instance, the individual problem, data collection, or setting might alter the applicability and efficacy of specific techniques. When choosing the most suitable learning method, careful consideration of the particular requirements and limitations is necessary.

9.4 MORE IL USE CASES

Use cases in respect to AL, TL, and OL are as follows:

1. AL use cases:
   - Object Detection: AL can be used to select informative image regions for annotation in object detection tasks, reducing the need for extensive manual labeling and improving the efficiency of training object detection models.
   - Text Classification: AL can be applied in text classification tasks, where it helps select representative and uncertain text samples for annotation, leading to better classification performance with fewer labeled examples.
   - Drug Discovery: AL can assist in discovering new drugs by selecting compounds with the highest potential for activity or

TABLE 9.1 Comparison of Active, Transfer, and OL.

	Goal
AL	AL aims to iteratively select the most informative instances from a large pool of unlabeled data and query their labels from an oracle (human or expert) to improve the model's performance.
TL	Through information from one task or domain (the source task/domain), TL seeks to enhance learning and performance on a separate but related task or domain (the target task/domain).
OL	As fresh or new data becomes available, OL seeks to continually update the model's parameters, enabling it to generate predictions in real-time and adjust to shifting data distributions.
	Data Usage
AL	AL focuses on selecting instances for which the model is uncertain or has low confidence, actively querying labels for those instances, and using them to update the model. It aims to minimize the amount of labeled data required for training.
TL	TL leverages knowledge from a pretrained model or a source domain and applies it to a target domain. The pretrained model's parameters or representations are fine-tuned using less labeled data specific to the target task or domain.
OL	OL processes data instances one at a time or in small batches, updating the model's parameters incrementally as new data arrives. It does not require access to the entire dataset simultaneously and can adapt to evolving data distributions.
	Model Adaptation
AL	AL focuses on improving the model's generalization performance by selecting informative instances for labeling. The model itself is not explicitly adapted or updated during AL.
TL	TL adapts a pretrained model to a new task or domain by fine-tuning the model's parameters using a smaller labeled dataset specific to the target task or domain. The pretrained model's knowledge is transferred and used as a starting point for learning in the new context.
OL	The model can adjust to data arrival and produce predictions in real-time because OL constantly modifies the model's parameters as new data come in. Every new batch or instance of data is used to update the model.
	Data Efficiency
AL	AL aims to reduce the annotation effort by selecting the most informative instances for labeling. It focuses on labeling only the cases expected to impact the model's performance most, leading to improved data efficiency.

TABLE 9.1 *(Continued)*

TL	TL can improve data efficiency by reusing knowledge from a pretrained model. It leverages the labeled data from the source domain or task and requires a smaller amount of labeled data specific to the target domain or task.
OL	OL can be data-efficient by updating the model incrementally using new data instances. It does not require retraining the model on the entire dataset, thus reducing computational resources and time.
	Use Cases
AL	AL is beneficial when labeled data is limited and expensive. It is commonly used when manual annotation is time-consuming or costly, such as in medical diagnosis, text classification, or image recognition.
TL	TL is sound when there is a shortage of labeled data in the target domain or task. A TL model to improve learning and performance in a new domain, such as image classification, NLP, or recommendation systems, is used when knowledge acquired in a related area or task can enhance learning and performance.
OL	OL is useful when real-time predictions or adaptability to changing data distributions are required.
	Scalability
AL	The scalability of AL depends on the selection strategy used to query labels for informative instances. Some selection strategies may require computations over the entire unlabeled dataset, which can be computationally expensive for large datasets.
TL	TL can be scalable if the pretrained model's parameters are fixed, and only the task-specific layers are fine-tuned using a smaller labeled dataset. Conversely, fine-tuning the entire model on a large dataset may require significant computational resources.
OL	OL can be highly scalable as it processes data instances incrementally, allowing for real-time predictions and efficient updates of model parameters. It can handle large datasets and adapt to streaming data.
	Generalization and Robustness
AL	AL focuses on selecting informative instances for labeling, which can help improve the model's generalization performance and robustness by actively addressing regions of uncertainty or difficulty.
TL	TL can improve generalization and robustness by leveraging knowledge from a pretrained model. The pretrained model has learned from a large dataset or a related task, leading to better feature representations and generalization capabilities on the target task or domain.
OL	The generalization and robustness of OL depend on the model's adaptability to changing data distributions. OL algorithms must handle concept drift and update the model effectively to maintain performance.

TABLE 9.1 *(Continued)*

	Interpretability
AL	AL does not directly impact model interpretability. However, the labeled instances selected for labeling during AL can help build interpretable models by focusing on cases that are representative or challenging for the model.
TL	The interpretability of TL depends on the pretrained model used as the starting point. Some pretrained models, such as linear or decision trees, are inherently more interpretable than complex deep-learning models.
OL	The interpretability of OL depends on the specific learning algorithm used. Some OL algorithms, such as online linear models, are more interpretable due to their simplicity and incremental update mechanisms.

toxicity testing, thus reducing the cost and time required for experimental validation.

2. TL use cases:
   - Image Recognition: TL is used widely for image recognition tasks, where pretrained models trained in large datasets, for example, Image Net, have been tuned to specific image classification tasks that improve performance and reduce the need for large labeled data sets.
   - Natural Language Processing: TL has demonstrated effectiveness in various NLP tasks, like sentiment analysis, named entity identification, and machine translation, by utilizing pretrained language models like BERT or GPT to transfer knowledge to particular tasks.
   - Computer Vision in Healthcare: TL has been employed in medical image analysis tasks, enabling models to learn from large publicly available datasets and transferring that knowledge to diagnose specific diseases or conditions with limited labeled medical data.
3. OL use cases:
   - Recommender Systems: OL is commonly used in recommender systems to provide personalized recommendations in real-time as user preferences and item interactions are observed. The model continuously learns from user feedback and adapts to changing user preferences.
   - Fraud Detection: OL is effective in fraud detection applications, where the model needs to learn from streaming transaction data and adapt to new fraud patterns in real-time, enabling timely identification and prevention of fraudulent activities.
   - Adaptive Control Systems: OL is used in adaptive control systems, where the model has to evolve and modify its behavior based on real-time sensory inputs to ensure safe and optimal performance in dynamic contexts. The examples of these systems include robots and autonomous cars.

These use cases highlight the adaptability and usefulness of learning strategies in various contexts and tasks. The needs, constraints, and available resources must all be considered when deciding which learning technique is ideal for a particular use case.

9.5 BENEFITS AND DRAWBACKS ON IL

9.5.1 BENEFITS

IL offers several benefits compared to traditional batch learning approaches. The critical benefits of IL [90–92] are specified below:

1. Adaptability: IL enables models to update and react in real-time when new data becomes available. It enables flexible model updates in a dynamic setting with constantly evolving dynamics and data.
2. Efficiency: IL is more efficient regarding memory usage and computational resources. Instead of retraining the model from scratch with the entire dataset, only the new data or relevant subsets must be processed, reducing the computational burden and enabling scalability.
3. Continuous Learning: IL facilitates continuous learning scenarios where the model can learn and improve without discarding previously known knowledge. It allows the model to accumulate knowledge gradually, making it well-suited for applications where data arrives in a streaming fashion or in small batches.
4. Cost Reduction: IL reduces the need for extensive data annotation or labeling efforts. Selectively choosing the most informative instances to update the model minimizes the dependency on fully labeled datasets, resulting in cost savings for acquiring labeled data.
5. Faster Deployment: IL enables faster deployment of ML models in real-world applications. It allows models to learn and update slowly, thus shortening the time needed for model development and training to be easier to use and adapt to new scenarios.
6. Handling Concept Drift: IL is effective in handling concept drift, which refers to the changes in the underlying data distribution over time. The model can adapt to concept drift by continuously updating its parameters and adapting to new patterns or changes in the data distribution.
7. Personalization and Customization: IL allows for personalized and customized models. As new data is added, the model can adapt

specifically to the preferences and characteristics of individual users or target-specific subgroups within the data, enabling personalized recommendations, adaptive systems, or targeted interventions.

These benefits make IL valuable in various domains, such as real-time prediction, anomaly detection, recommendation systems, adaptive control systems, and many others, where continuous learning and adaptability to changing environments are crucial.

9.5.2 DRAWBACKS

While IL offers several benefits, certain drawbacks to consider are [92]:

1. Catastrophic Forgetting: IL models are prone to catastrophic forgetting, which is when the model forgets previously learned information when it learns new data. As the model adapts to recent observations, it may overwrite or lose the knowledge it had previously acquired. It may be problematic if the latest data represents only some of the distribution or if the model needs to maintain an understanding of multiple tasks.
2. Complexity: IL can be more complex than traditional batch learning. It requires careful handling of data streams, updating model parameters, and managing memory usage. Ensuring efficient and effective learning from new observations while preserving previous knowledge can be challenging and may require sophisticated algorithms and techniques.
3. Need for Data Ordering: In IL, the order in which data arrives or is processed can impact the model's performance. A nonrandom or biased arrangement for the incoming data might result in better-than-ideal results in learning. Selecting an appropriate data ordering strategy or handling data imbalance becomes essential to maintain model performance.
4. Lack of Global Optimization: IL typically focuses on local optimization, updating the model based on recent observations. It can limit the model's ability to achieve a globally optimal solution since it cannot access the entire dataset. The lack of

global optimization can sometimes lead to suboptimal model performance.

5. Sensitivity to Noisy Data: IL models can be sensitive to noisy or erroneous data points. Suppose the new observations contain significant noise or outliers; the model's performance can be negatively affected, especially if the noise is incorporated into the model during the incremental update process. Robust preprocessing techniques or outlier detection methods may be required to mitigate this issue.
6. Limited Exploration: IL may limit the model's ability to explore the entire input space thoroughly. Since the model focuses on updating based on the most recent observations, it may need to explore more frequent or underrepresented regions of the input space. It can impact the model's generalization ability and may require additional techniques to encourage exploration.

When adopting IL, it is critical to consider these disadvantages and develop suitable mitigation techniques. These issues may be addressed, and regularization, replay mechanisms, ensembling, and task-specific techniques can enhance the overall performance of IL models.

9.6 APPLICATIONS IN IL IN REAL-TIME CASE STUDIES

IL is applied in the industries and sectors such as:

1. Healthcare
2. Finance and banking
3. Manufacturing and supply chain
4. Retail and e-Commerce
5. Telecommunications
6. Energy utilities
7. Transportation and logistics
8. Media and entertainment
9. Weather
10. Agriculture
11. Environment sustainability

9.6.1 CASE STUDIES

Area	Domain	Case Study Name	Description
Agriculture and Weather	Crop Disease Detection	Early Detection of Plant Diseases in Indian Agriculture [93]	A crop disease detection model that continually learns from photos of plants, meteorological information, and disease patterns unique to Indian agriculture is created using IL approaches. The technology can reliably diagnose and categorize plant illnesses by adjusting the models, enabling prompt treatments, and reducing crop losses for Indian farmers.
	Crop Yield Prediction	Crop Yield Prediction Model for Indian Farmers [94]	A crop yield forecast model that constantly learns from past meteorological data, soil conditions, and crop-specific factors is created using IL approaches. The model can forecast crop yields for various Indian areas by modifying the models and giving farmers helpful information for improved planning, resource allocation, and decision-making.
	Rainfall Prediction	Rainfall Prediction Model for Indian Regions [95]	A rainfall prediction model that constantly learns from historical rainfall data, satellite imaging, and atmospheric variables unique to India is created using IL. The technology can accurately anticipate rainfall by modifying the models, enabling farmers to design irrigation systems and plant crops and harvests with more knowledge.
	Crop Recommendation	Crop Recommendation System for Indian Farmers [96]	A crop recommendation system that continually learns from soil properties, climatic data, and market movements in the Indian agricultural environment is created using IL approaches. The system may recommend viable crop options to farmers by modifying the models, considering soil fertility, water accessibility, and market demand.
	Drought Monitoring and Mitigation	Global Drought Monitoring and Mitigation [97]	A global drought monitoring and mitigation system that continually learns from several data sources, such as soil moisture levels, temperature, and historical drought trends, is developed using IL. By modifying the models, the system can identify and track drought situations, enabling preventive actions like resource allocation, crop diversification, and water conservation tactics to lessen the effects of drought on world agriculture.

TABLE *(Continued)*

Area	Domain	Case Study Name	Description
	Irrigation Optimization	Smart Irrigation System for Indian Farms [98]	An intelligent irrigation system that continually learns from information on soil moisture, weather patterns, and crop water needs is developed using IL. The technology can optimize irrigation schedules, save water, and increase crop output for Indian farmers, particularly in water-scarce areas.
	Weather Forecasting	Weather Forecasting Model for Indian Regions [99]	A weather forecasting model that continually learns from real-time meteorological data, satellite images, and historical weather trends in India is created using IL approaches. The system may deliver precise and localized weather forecasts by modifying the models, assisting farmers and agricultural stakeholders in making knowledgeable decisions about irrigation, crop management, and harvest planning.
	Climate Modelling	Climate Change Impact Assessment [100]	IL approaches involve developing climate models that constantly learn from climatic data, greenhouse gas emissions, and past climate trends. By transforming the IL models, researchers and scientists can assess climate change impacts at domestic and international levels, help shape policy choices and develop plans for adaptation and mitigation.
	Storm Tracking and Early Warning Systems	Real-Time Storm Tracking and Early Warning [101]	The development of storm monitoring and early warning systems that continually learn from radar data, storm traits, and past storm trends is accomplished through IL. The system can track storms precisely in real-time, forecast their direction and strength, and send out early alerts to citizens and first responders, all of which lessen the effect of severe weather occurrences. It is accomplished by adjusting the models.
	Improved Weather Prediction	High-Resolution Weather Forecasting Model [102]	Weather forecasting systems are improved via IL approaches by updating the model based on current weather information, satellite images, and atmospheric conditions. The improvements in the integrated local models will make them more trustworthy and accurate, helping decision-makers in a number of areas, such as crisis management, road transport, or agriculture, to be better prepared for possible catastrophic weather events.

TABLE *(Continued)*

Area	Domain	Case Study Name	Description
Healthcare	Disease Diagnosis and Prediction	DeepMind's AI for Diabetic Retinopathy Detection [103]	A primary cause of blindness, diabetic retinopathy, may be detected and predicted using AI models developed using IL approaches. A significant number of retinal scans, medical information, and patient outcomes are regularly analyzed by the system to improve its performance. By adjusting these models, doctors could quickly identify patients with diabetes retinopathy and be able to predict the critical condition that evolves, giving them an opportunity for timely intervention and individualized treatment.
	Disease Diagnosis and Prediction	IBM Watson for Oncology [104]	IBM Watson for Oncology uses IL to support doctors in deciding on cancer therapy. The most recent medical research, patient information, and treatment results are constantly used to improve the system. It offers individualized therapy suggestions and continuously updates its knowledge base to enable more precise and current cancer diagnosis and treatment regimens.
	Medical Imaging Analysis	Google's AI for Breast Cancer Detection in Mammograms [105]	IL is employed to train AI models that can analyze mammograms and detect early signs of breast cancer. The system continuously learns from a vast dataset of mammograms, biopsy results, and patient data. By adopting the IL models, radiologists can convalesce the prediction accuracy and efficiency of breast cancer screening, leading to early detection, better patient treatment, and early and faster recovery.
	Medical Imaging Analysis	DeepMind's AI for Retinal Disease Detection [106]	DeepMind's AI models for retinal disease detection employ IL. The models are trained on large-scale datasets of retinal scans and continually updated with new data. IL makes it possible to identify and diagnose conditions like diabetic retinopathy early on, which benefits patients by empowering medical practitioners to make better judgements.
	Drug Discovery and Development	Atomwise's AI for Drug Discovery [107]	IL techniques are used to develop AI drug discovery and virtual screening models. The system continuously learns from molecular databases, chemical structures, and biological activity data. Researchers may find promising

TABLE *(Continued)*

Area	Domain	Case Study Name	Description
			drug candidates, improve lead compounds, and speed up drug discovery by modifying the models. It will lead to the development of new medications and therapies that are specific to the diagnosed illness.
	Predictive Analytics for Patient Outcomes	Predictive Analytics in Intensive Care Units (ICUs) [108]	In ICUs, IL is used to create models that predict patient outcomes. Real-time patient data, vital signs, test findings, and medical records train the system constantly. Healthcare professionals may improve patient care and lower death rates by modifying the models to forecast problems, identify high-risk patients, and make wise treatment decisions.
Finance and Banking	Fraud Detection	PayPal's Fraud Detection System [109]	A fraud detection system that continually learns from transaction data, user behavior, and previous fraud tendencies is created using IL approaches. By modifying the models, PayPal can recognize and stop fraudulent behaviors in real time, resulting in safe transactions and client protection.
		Citibank's Fraud Detection System [110]	Citibank uses IL to detect fraud in real time during financial transactions. Customer transaction patterns, past fraud data, and new fraud tendencies are constantly used to improve the system. It protects consumer accounts by adjusting its models and rules in real time to spot questionable transactions and thwart fraud.
	Credit Risk Assessment	Capital One's Credit Scoring Models [111]	The credit scoring models Capital One uses to determine credit risk are improved via IL approaches. New consumer information, repayment histories, and market trends regularly update the models. Therefore, the bank can make wise lending decisions and efficiently manage risk thanks to the more precise and dynamic credit risk assessments that have been made possible.
	Credit Scoring	LendingClub's Credit Scoring Model [112]	A credit scoring model that continually learns from borrower data, loan performance, and credit history is built using IL. By modifying the models, LendingClub evaluates loan applicants' credit worthiness, makes precise lending decisions, and efficiently manages credit risk.

TABLE *(Continued)*

Area	Domain	Case Study Name	Description
	Risk Assessment	JPMorgan Chase's Risk Management System [113]	IL techniques are utilized to develop a risk assessment system that continuously learns from market data, economic indicators, and historical risk events. JPMorgan Chase can evaluate and control several risks, such as credit, operational, and market risks, by modifying and tuning the models.
	Algorithmic Trading	Citadel's Trading Algorithms [114]	Algorithmic trading models that continually learn from market data, price changes, and trading signals are developed using IL. By modifying the models, Citadel can make data-driven trading decisions, improve trading tactics, and seize market opportunities.
Economics	Financial Market Prediction	Stock Market Forecasting [115]	Stock market trends and fluctuations are predicted using IL techniques. The algorithms learn from previous stock prices, real-time financial data, and news sentiment analysis. Economic researchers and traders can enhance their trading techniques by modifying the models to make better investment judgements.
	Demand Forecasting	Retail Sales Forecasting [116]	In the retail sector, IL is used to predict consumer demand. The algorithms learn from sales data, marketing initiatives, and outside variables like seasonality and economic indices. By modifying the models, retailers may optimize inventory management, supply chain operations, and pricing strategies based on precise demand projections.
	Economic Policy Analysis	Macroeconomic Forecasting [117]	Forecasting methods based on IL are used for macroeconomic variables, including GDP growth, inflation, and unemployment rates. The models constantly improve based on policy changes, global economic indices, and real-time economic data. Policymakers and economists may examine the effects of various policy scenarios and create more accurate forecasts by changing the models.
Manufacturing and Supply chain	Quality Control and Defect Detection	Bosch's Quality Control System [118]	A quality control system that continually learns from sensor data, manufacturing parameters, and historical defect records is created using IL approaches. Bosch can discover and categorize problems in real time using modified models, resulting in excellent product quality and less waste.

TABLE *(Continued)*

Area	Domain	Case Study Name	Description
	Predictive Maintenance	Siemens' Predictive Maintenance Solution [119]	A predictive maintenance system that continually learns from sensor data, equipment performance, and previous maintenance records is created using IL. Siemens can minimize downtime and increase operational efficiency by modifying the models to forecast equipment failures, plan maintenance proactively, and optimize maintenance tasks.
	Supply Chain Optimization	Walmart's Supply Chain Optimization [120]	Walmart's supply chain processes are optimized using IL approaches. Real-time data on inventory levels, consumer demand, and transportation logistics train the system constantly. Walmart may save costs and increase customer happiness by adjusting the algorithms to optimize inventory levels, demand predictions, and shipping routes.
		General Electric's Predictive Maintenance Solution [121, 122]	In manufacturing facilities, equipment failures are predicted using IL approaches. The system learns from real-time sensor data, previous maintenance logs, and equipment performance measurements. Manufacturers may reduce downtime, plan maintenance proactively, and discover probable faults early on by modifying the models, which boosts productivity and reduces costs.
	Demand Forecasting	Amazon's Demand Forecasting Model [123]	IL is applied to develop a demand forecasting model that learns from historical sales data, market trends, and customer behavior. By adapting the models, Amazon can forecast demand accurately, optimize inventory management, and improve product availability, increasing customer satisfaction and reducing stockouts.
Retail and E-Commerce	Personalized Recommendations	Amazon's Recommendation System [124]	A personalized recommendation system that continually learns from consumer browsing habits, purchase histories, and product qualities is developed using IL approaches. By modifying the models, Amazon can provide clients with tailored product recommendations, enhancing their shopping experience and boosting revenue.

TABLE *(Continued)*

Area	Domain	Case Study Name	Description
	Customer Segmentation	Sephora's Customer Segmentation Model [125]	A customer segmentation model is created using IL, which constantly learns from consumer demographic information, transaction history, and browsing habits. By modifying the models, Sephora may divide its clientele into several groups according to their tastes, enabling customized advertising campaigns and one-on-one service.
	Dynamic Pricing	Uber's Surge Pricing Algorithm [126]	IL techniques are utilized to develop a dynamic pricing algorithm that continuously learns from real-time data on supply and demand, traffic conditions, and user behavior. By adapting the models, Uber can dynamically adjust prices to match demand, maximize revenue, and optimize driver availability.
	Inventory Management	Zara's Inventory Optimization System [127]	An inventory optimization system that continually learns from sales data, seasonal trends, and market demand is developed using IL. By modifying the models, Zara may reduce stockouts and surplus inventory, increase overall inventory management effectiveness, and optimize inventory levels.
	Pricing Optimization	Dynamic Pricing in Online Retail [128]	Pricing methods in e-commerce are optimized via IL. The system learns from market dynamics, competitive pricing, and client demand patterns. By modifying the models, E-commerce systems may dynamically alter pricing, maximize revenue, and improve profit margins.
	Customer Churn Prediction	Customer Retention in Subscription-based E-commerce [129]	In subscription-based e-commerce enterprises, consumer attrition is predicted using IL approaches. The system learns from user behavior, engagement indicators, and customer data. E-commerce platforms may detect clients who are likely to leave, take proactive steps to keep them and increase customer loyalty by changing the models.
Customer service and experience	Chatbot Assistance	Chatbot Customer Support [130]	Customer service chatbots are trained using IL techniques. The technology continually picks up new information from consumer interactions, comments, and previous chat logs. Chatbots better comprehend client inquiries, deliver detailed replies, and give individualized help by adjusting the models and improving the customer service experience.

TABLE *(Continued)*

Area	Domain	Case Study Name	Description
	Sentiment Analysis	Social Media Monitoring [131]	The client sentiment on social media networks is assessed using IL. Through real-time social media data, client evaluations, and feedback, the system constantly improves. Companies may enhance customer happiness by adjusting the models to acquire insights into consumer sentiment, spot developing patterns, and respond to customer issues proactively.
	Personalized Marketing	Targeted Marketing Campaigns [132]	Marketing strategies are tailored using IL approaches based on consumer preferences and behaviors. The system learns from consumer information, purchasing history, and interaction patterns. By modifying the models, businesses may send personalized offers, promote pertinent items, and customize marketing messaging, increasing client engagement and conversion rates.
	Customer Lifetime Value Prediction	Customer Retention and Loyalty [133]	IL is applied to predict lifetime value, enabling companies to identify high-value customers and design retention strategies. The system continuously learns from customer data, transaction history, and engagement metrics. By adapting the models, companies can identify potential churn risks, offer personalized incentives, and enhance customer loyalty.
	Customer Segmentation	Market Segmentation for Marketing Campaigns [134]	The customers are segmented Using IL based on their interests, habits, and demographics. The client information, transaction histories, and reactions to marketing initiatives are all constantly used to improve the models. By modifying the models, marketers may better target specific client categories and increase the efficiency of their advertising campaigns.
Telecommunications	Customer Churn Prediction	AT&T's Customer Churn Prediction Model [135]	A customer churn prediction model that continually learns from customer use data, billing data, and historical churn records is created using IL approaches. By modifying the algorithms, AT&T may identify at-risk consumers, forecast customer churn probabilities, and conduct targeted retention initiatives, boosting customer retention rates.

TABLE *(Continued)*

Area	Domain	Case Study Name	Description
	Customer Churn Prediction	Telco Customer Churn Prediction [136]	To predict client attrition, telecommunications corporations frequently use IL. The model is regularly updated to recognize clients in danger of leaving a company by tracking customer behavior, service consumption, and feedback. As a result, proactive client retention techniques, such as tailored offers or individualized interventions, may be used to reduce turnover rates.
	Network Traffic Analysis	Verizon's Network Traffic Analysis System [137]	A network traffic analysis system that continually learns from real-time network data, device information, and user behavior is developed using IL. Verizon can detect network abnormalities, recognize possible security risks, and optimize network performance by modifying the models, resulting in dependable and secure communication services.
	Service Quality Monitoring	T-Mobile's Service Quality Monitoring Solution [138]	A service quality monitoring system that continually learns from customer input, network performance measurements, and service problems is developed using IL approaches. T-Mobile can monitor service quality in real time, proactively detect and address issues, and provide a more excellent customer experience by modifying the models.
	Predictive Maintenance for Telecom Infrastructure	Vodafone's Predictive Maintenance Solution [139]	A predictive maintenance solution for telecom infrastructure, such as cell towers and network hardware, is created using IL. The system learns from sensor data, maintenance logs, and past failure patterns. Vodafone can foresee equipment failures, improve maintenance plans, and reduce downtime by modifying the models, resulting in uninterrupted telecom services.
Energy and Utilities	Energy Load Forecasting	E.ON's Energy Load Forecasting Model [140]	An energy load forecasting model that constantly learns from past energy consumption data, weather trends, and consumer behavior is created using IL approaches. By adapting the models, E.ON can accurately estimate energy demand, optimize energy production and distribution, and guarantee resource utilization.

TABLE *(Continued)*

Area	Domain	Case Study Name	Description
	Asset Condition Monitoring	National Grid's Asset Condition Monitoring System [141]	An asset condition monitoring system that continually learns from sensor data, maintenance logs, and past failure trends is developed using IL. By modifying the models, National Grid can predict impending asset failures, plan maintenance in advance, and maximize asset performance, resulting in dependable and economical operations.
	Smart Grid Optimization	Pacific Gas and Electric Company's Smart Grid Optimization [142]	IL approaches optimize the smart grid infrastructure for effective load balancing, demand response, and energy distribution. By modifying the models, Pacific Gas and Electric Company may optimize grid operations, incorporate renewable energy sources, and dynamically alter energy distribution, decreasing energy waste and increasing system dependability.
	Water Management	Singapore's Smart Water Management System [143]	An intelligent water management system that continually learns from sensor data, weather conditions, and use patterns is developed using IL. Singapore can ensure sustainable and effective water management by modifying the models to optimize water distribution, find leaks, and apply water conservation measures.
	Energy Optimization in Power Plants	Siemens' Energy Optimization Solution [144]	Energy use and efficiency in power plants are optimized using IL approaches. The system continually picks up new skills from meteorological data, previous plant performance, and real-time energy data. By modifying the models, power plant operators may modify operating parameters, maximize energy production, and lessen environmental effects, resulting in energy savings and sustainability.
	Energy Management	Smart Grid Systems [145]	Innovative grid systems employ IL to enhance energy management and usage. The algorithms continually pick up new information from real-time data on energy use, the availability of renewable energy sources, and weather predictions. These systems may dynamically decide on energy distribution, load balancing, and demand response by modifying the models, which results in more effective and sustainable energy use.

TABLE *(Continued)*

Area	Domain	Case Study Name	Description
	Renewable Energy Forecasting	Solar and Wind Energy Forecasting [146, 147]	Forecasting models for renewable energy are created using IL, which constantly learns from meteorological information, solar radiation levels, wind speeds, and historical energy generation trends. Energy suppliers and grid operators can adequately forecast the production of solar and wind energy, better the distribution of energy, and enhance the integration of renewable energy sources into the system by modifying the models.
Environment and Sustainability	Waste Management	Intelligent Waste Sorting [148]	Waste sorting procedures use IL to increase the effectiveness of recycling. The system learns from real-time sensor data and user feedback to recognize various trash kinds and categorize them for the proper recycling techniques. Waste management systems may improve recycling rates and lessen environmental impact by routinely updating the models.
	Water Resource Management	Water Consumption Monitoring [149]	Systems for managing water resources use IL to track and manage water usage. Real-time data, including consumption patterns, weather, and water quality characteristics, train the system constantly. These systems can detect water waste, forecast water demand, and enable effective water allocation by modifying the models, which helps to manage water resources sustainably.
	Ecosystem Monitoring:	Biodiversity Monitoring [150]	Systems for monitoring biodiversity use IL approaches to evaluate and assess the state of the ecosystem. Real-time data from environmental sensors, satellite photography, and citizen science observations are constantly used to improve the systems. These systems can track changes in biodiversity, spot endangered species, and aid conservation efforts by modifying the models.
Transportation and Logistics	Traffic Prediction	Traffic Flow Prediction [151]	Traffic flow and congestion patterns are predicted using approaches for IL. The algorithm regularly picks up new information from past traffic patterns, real-time traffic data, and outside variables like the weather and current events. Transportation authorities may enhance overall traffic management by modifying the models to predict traffic congestion better, adjust the timing of traffic signals, and offer real-time traffic data.

TABLE *(Continued)*

Area	Domain	Case Study Name	Description
	Traffic Management	Intelligent Traffic Systems [152]	Intelligent traffic systems use IL to improve traffic flow and lessen congestion. The algorithms use real-time traffic data, such as vehicle movement, volume, and historical patterns, to learn continually. These systems can offer dynamic traffic control methods, such as route suggestions and signal timing modifications, by adjusting the models to increase traffic efficiency.
	Demand Forecasting	Transportation Demand Forecasting [153]	The demand for public transport services is forecasted using IL. The system regularly picks up new information from customer information, ticket sales, and past usage trends. By modifying the models, transportation organizations may enhance the overall passenger experience, optimize their service schedules, and allocate resources more effectively.
	Route Optimization	Delivery Route Optimization [154]	IL techniques are utilized to optimize delivery routes for logistics and transportation companies. The system continuously learns from real-time data on road conditions, traffic congestion, and delivery constraints. By adapting the models, companies can generate optimal routes, minimize delivery time and costs, and improve overall logistics efficiency.
	Vehicle Performance Monitoring	Fleet Management and Maintenance [155, 156]	The performance of a fleet of cars is tracked and optimized via IL. Real-time sensor data, vehicle diagnostics, and historical performance measurements are all constantly used to improve the system. Fleet managers may save money and enhance operational efficiency by modifying the models to identify possible maintenance issues better, maximize fuel economy, and decrease vehicle downtime.
	Supply Chain Management in Logistics	UPS's Package Routing Optimization [157]	In logistics operations, IL improves package routing and delivery. From real-time information on package volume, traffic patterns, and consumer preferences, the system continually learns. By modifying the models, logistics organizations may increase customer satisfaction and cut operating costs by optimizing delivery routes, lowering transportation costs, and improving delivery efficiency.

TABLE *(Continued)*

Area	Domain	Case Study Name	Description
Media and Entertainment	Content Recommendation	Netflix's Content Recommendation System [158]	A content recommendation system that continually learns from user-watching history, ratings, and preferences is created using IL approaches. Netflix may improve user engagement and boost customer satisfaction by modifying the models by personalizing content recommendations for each user.
	Sentiment Analysis	Twitter's Sentiment Analysis for TV Shows [159]	A sentiment analysis system that continually learns from Twitter data and user responses to TV broadcasts is created using IL. Twitter can analyze audience sentiment in real time by modifying the models and giving TV programme producers, marketers, and content providers helpful information.
	Ad Targeting	Google's Ad Targeting Model [160]	An ad targeting model that continually learns from user behavior, demographics, and online activities is developed using IL approaches. By adjusting the models, Google can offer tailored advertisements to users, enhancing ad relevancy and boosting advertising performance.
	Content Moderation	Facebook's Content Moderation System [161]	A content moderation system that constantly learns from user-reported material, automatic flagging, and moderation judgements is developed using IL. Facebook can detect and eliminate harmful information in real time by changing the models, delivering a safer and more exciting platform for its users.
	Personalized Music Recommendations	Spotify's Music Recommendation System [162]	A personalized music recommendation system that continually learns from user listening behaviors, playlists, and preferences is created using IL approaches. Spotify can provide users with customized music suggestions by modifying the models, improving their music discovery experience, and raising user engagement.
	Video Content Analysis	YouTube's Video Content Analysis System [163]	A video content analysis system that continually learns from user interactions, video metadata, and content qualities is created using IL. By modifying the models, YouTube may categorize and suggest videos depending on their content, optimizing user video viewing experience and search results.

TABLE *(Continued)*

Area	Domain	Case Study Name	Description
	News Personalization	The New York Times News Personalization System [164]	A new customization system that continually learns from user reading preferences, article preferences, and subject interests is created using IL approaches. The New York Times may give tailored news suggestions to readers by modifying the algorithms, boosting user interaction, and enhancing the relevancy of news content.
	Movie Genre Classification	Netflix's Movie Genre Classification Model [165]	A movie genre categorization model is created using IL, which constantly learns from user ratings, movie information, and watching habits. Netflix can effectively categorize movies into categories by modifying the models, which enables accurate suggestions and improves the user's movie-choosing process.

9.7 ADVANTAGES AND LIMITATIONS IN IMPLEMENTATION OF IL IN RT

9.7.1 ADVANTAGES

The implementation of IL in real-time scenarios offers several advantages, including [166–169]:

1. Adaptability to Changing Data: IL allows models to adapt and learn from new data as it arrives in real time. This adaptability ensures that the models remain up-to-date and accurate, even in dynamic and evolving environments.
2. Real-time Decision-Making: IL enables real-time decision-making based on recent data. It allows models to update and refine their predictions or classifications as new information becomes available, enabling timely and informed decisions.
3. Efficient Resource Utilization: IL optimizes resource utilization by focusing on relevant and informative data points. Instead of retraining the entire model on the whole dataset, IL selectively updates the model, saving computational resources and reducing training time.
4. Handling Concept Drift: Real-world data often exhibit concept drift, where the underlying relationships between variables change over time. IL is well-suited to handle concept drift as it can continuously update the model to capture these changes, ensuring that predictions remain accurate and relevant.
5. Scalability: IL techniques are scalable and can handle large volumes of data in real time. Models can be trained and updated incrementally, allowing for the efficient processing of massive datasets without retraining the entire model.
6. Cost Savings: IL reduces the need for frequent and extensive retraining of models. It leads to cost savings regarding computational resources, storage, and time required for model updates.
7. Improved Model Performance: IL allows models to leverage new information and adapt to changing conditions, improving performance over time. The latest trends and observations, which give rise to more exact predictions and more robust support for decisions, can be captured by models.

8. Continuous Learning: IL enables continuous learning from new data without discarding previous knowledge. Models can build on their prevailing knowledge and incorporate new information as it becomes available, leading to cumulative learning and improved performance over time.
9. Reduced Data Storage Requirements: IL methods typically store only the relevant information needed to update the model rather than storing the entire dataset. It reduces the storage requirements, making handling large volumes of data in real-time scenarios more efficient.
10. Rapid Model Deployment: IL allows for rapid model deployment and updates. As new data arrives, models can quickly adapt and incorporate the information, enabling organizations to deploy updated models faster and respond to changing circumstances promptly.
11. Personalization and Customization: IL facilitates personalized and customized experiences. Models can learn from user behavior, preferences, and feedback in real-time, enabling personalized recommendations, tailored services, and targeted interventions.
12. Early Detection of Anomalies: IL can detect anomalies or outliers in real-time data streams. By continuously monitoring and updating the model, it can quickly identify unexpected patterns or deviations from the norm, enabling early detection of anomalies and timely intervention.
13. Improved Data Efficiency: IL methods can extract maximum value from limited labeled data. By actively selecting and labeling the most informative instances for model updates, it optimizes the use of labeled data, reducing the need for extensive labeling efforts.
14. Privacy Preservation: IL can incorporate privacy-preserving techniques by selectively updating models without sharing sensitive or personal data. It ensures data privacy and compliance with privacy regulations while benefiting from the insights gained through IL.
15. Real-time Feedback Loops: IL enables the establishment of real-time feedback loops, where model predictions and decisions can be evaluated and, in turn, used to improve the learning process further. This iterative feedback loop allows for continuous model refinement and optimization.

These additional advantages highlight IL's flexibility, efficiency, adaptability, customization, and real-time responsiveness in various applications. Continuous learning harness power and adaptability, organizations can derive more value from their data and data-driven decisions in dynamic environments.

9.7.2 LIMITATIONS

While IL in real-time scenarios offers numerous advantages, there are particular limitations and challenges to consider during implementation [166, 168, 169]:

1. Computational Complexity: IL requires continuous updates and model adaptations, which can increase computational complexity. When new data arrives, the model will be modified and updated depending on its complexity and quantity of data received; it may need significant computing power.
2. Concept Drift Handling: Real-time data often exhibit concept drift, where the underlying relationships between variables change over time. IL algorithms must be designed to handle concept drift and adapt to the evolving data patterns. Please effectively address concept drift to avoid degraded model performance and inaccurate predictions.
3. Data Quality and Noise: Real-time data streams may contain noise, outliers, or erroneous observations. IL models must be robust to handle such noisy data and prevent them from adversely affecting learning. Ensuring data quality and incorporating techniques for noise handling is essential for accurate and reliable IL.
4. Labeling and Annotation: IL often requires labeled or annotated data for model updates. In real-time, labeling new data instances can be time-consuming and resource-intensive. Developing efficient strategies for AL, where the model actively selects the most informative instances for labeling, can help mitigate the labeling burden.
5. Scalability: The scalability of IL methods is crucial in real-time scenarios with large volumes of data. As the data grows, the model's capacity to handle the increasing data size, model updates, and computational requirements becomes essential. Scalable algorithms and architectures are necessary to ensure efficient and effective IL.

6. Overfitting and Catastrophic Forgetting: If adequately managed, IL models are prone to overfitting or catastrophic forgetting. Overfitting occurs when the model becomes too specialized to the previously seen data and fails to generalize well to new data. Catastrophic forgetting happens when the model forgets previously learned information when adapting to new data. Techniques such as regularization, model consolidation, and memory replay can help mitigate these challenges.
7. Data Privacy and Security: Real-time data streams may contain sensitive or personal information, and ensuring data privacy and security becomes crucial. IL should incorporate appropriate privacy-preserving techniques to protect sensitive data, such as differential privacy, secure multi-party computation, or federated learning.
8. Interpretability and Explainability: Some IL methods, namely, deep neural networks, can be complex and lack interpretability. In scenarios where interpretability is essential, such as in healthcare or finance, it becomes challenging to explain the model's decision-making process. Balancing the need for accuracy and interpretability is a consideration in implementing IL in real-time applications.
9. Model Monitoring and Maintenance: Continuous monitoring of the IL model's performance and regular model maintenance is necessary to ensure accuracy and reliability. Real-time model evaluation, performance tracking, and proactive model updates or retraining are crucial to address model degradation, concept drift, and evolving data patterns.

Addressing the limitations and challenges requires careful consideration of algorithm selection, model design, data quality management, privacy protection, scalability, and monitoring strategies. Organizations may effectively implement IL in real-time circumstances and exploit its benefits while reducing possible adverse outcomes by comprehending and overcoming the limitations.

9.8 CONCLUSION

An IL approach allows models to adapt and update on the basis of new data, without retraining the entire dataset. It overcomes the limitations of traditional batch learning and is particularly effective in handling dynamic data streams and concept drift. IL techniques allow models to improve performance, respond to new tasks or domains, and update them in real time when data is available, such as AL, TL, and OL.

There are many benefits to be gained from implementing IL on real-time scenarios, such as timely decisions, adaptation of data patterns, and optimal use of resources. It also provides personalized experiences and reduces the need for extensive retraining. IL methods are ascertained to be useful in diversified domains, for example, healthcare, finance, manufacturing, and telecommunications, which demonstrates its usefulness when addressing actual world problems.

There are challenges to using IL, such as managing concept drift, ensuring data quality, and resolving privacy concerns, which must be considered. The danger of overfitting, the need for interpretability, and ongoing model monitoring and maintenance should all be considered.

IL offers a strong foundation for adjusting to dynamic input and guiding real-time decision-making. Organizations may use IL's advantages to learn more, make better predictions, and innovate across various fields by utilizing AL, TL, and OL.

KEYWORDS

- **active learning**
- **incremental learning**
- **online learning**
- **real-time AI**
- **transfer learning**

REFERENCES

1. Domingos, P., & Hulten, G. (2000). Mining high-speed data streams. In *Proceedings of the Sixth ACM SIGKDD International Conference on Knowledge Discovery and Data Mining (KDD '00), New York, NY, USA* (pp. 71–80). Association for Computing Machinery. https://doi.org/10.1145/347090.347107
2. Hulten, G., Spencer, L., & Domingos, P. (2001). Mining time-changing data streams. In *Proceedings of the Seventh ACM SIGKDD International Conference on*

Knowledge Discovery and Data Mining (KDD '01), New York, NY, USA (pp. 97–106). Association for Computing Machinery. https://doi.org/10.1145/502512.502529

3. Bifet, A., & Gavaldà, R. (2007). Learning from time-changing data with adaptive windowing. In *Proceedings of the 7th SIAM International Conference on Data Mining, Minneapolis, Minnesota, USA*. DBLP. 10.1137/1.9781611972771.42.
4. Ganguly, K. K., & Hossain, B. M. M. (2018). Evaluating the effectiveness of conventional machine learning techniques for defect prediction: A comparative study. In *2018 Joint 7th International Conference on Informatics, Electronics & Vision (ICIEV) and 2018 2nd International Conference on Imaging, Vision & Pattern Recognition (icIVPR), Kitakyushu, Japan* (pp. 481–485). DOI: 10.1109/ICIEV.2018. 8641006.
5. Luu, S. T., Nguyen, H. P., Van Nguyen, K., & Luu-Thuy Nguyen, N. (2020). Comparison between traditional machine learning models and neural network models for vietnamese hate speech detection. In *2020 RIVF International Conference on Computing and Communication Technologies (RIVF), Ho Chi Minh City, Vietnam* (pp. 1–6). DOI: 10.1109/RIVF48685.2020.9140745.
6. Kaur, M., Shukla, A. K., & Kaur, S. (2021). An introduction to machine learning in a nutshell). In *10th International Conference on System Modeling & Advancement in Research Trends (SMART), Moradabad, India* (pp. 17–22). DOI: 10.1109/SMART 52563.2021.9676315.
7. Sarker, I. H. (2021). Machine learning: Algorithms, real-world applications and research directions. *SN Computer Science, 2*, 160. https://doi.org/10.1007/s42979-021-00592-x
8. Sarker, I. H. (2021). Deep learning: A comprehensive overview on techniques, taxonomy, applications and research directions. *SN Computer Science, 2*, 420. https://doi.org/10.1007/s42979-021-00815-1
9. Edwards, G. (2023). Machine learning - An introduction. https://towardsdatascience.com/machine-learning-an-introduction-23b84d51e6d0, [Online] (last accessed on June 07, 2023)
10. Usama, M., et al. (2019). Unsupervised machine learning for networking: Techniques, applications and research challenges. *IEEE Access*, *7*, 65579–65615. DOI: 10.1109/ACCESS.2019.2916648.
11. Casolla, G., Cuomo, S., Cola, V. S. d., & Piccialli, F. (2020). Exploring unsupervised learning techniques for the internet of things. *IEEE Transactions on Industrial Informatics,16*(4), 2621–2628. doi: 10.1109/TII.2019.2941142.
12. MacQueen, J. (1967). Some methods for classification and analysis of multivariate observations. In *Proceedings of the 5th Berkeley Symposium on Mathematical Statistics and Probability* (Vol. 1, pp. 281–297).
13. Ester, M., Kriegel, H. P., Sander, J., & Xu, X. (1996). A density-based algorithm for discovering clusters in large spatial databases with noise. In *Proceedings of the Second International Conference on Knowledge Discovery and Data Mining (KDD'96)* (pp. 226–231). AAAI Press.
14. Hartigan, J. A. (1972). Direct clustering of a data matrix. *Journal of the American Statistical Association, 67*(337), 123–129.
15. Breiger, R. L., Boorman, S. A., & Arabie, P. (1975). An algorithm for clustering relational data with applications to social network analysis and comparison with multi-dimensional scaling. *Journal of Mathematical Psychology, 12*(3), 328–383. https://doi.org/10.1016/0022-2496(75)90028-0.

16. Everitt, B. (1980). Cluster analysis. *Qualitative and Quantitative, 14*, 75–100. https://doi.org/10.1007/BF00154794
17. Hotelling, H. (1933). Analysis of a complex of statistical variables into principal components. *Journal of Educational Psychology, 24*(6), 417–441. https://doi.org/10.1037/h0071325
18. Van der Maaten, L., & Hinton, G. (2008). Visualizing data using t-SNE. *Journal of Machine Learning Research, 1*, 1–48.
19. Agrawal, R., Imieliński, T., & Swami, A. (1993). Mining association rules between sets of items in large databases. *SIGMOD Record, 22*(2), 207–216. https://doi.org/10.1145/170036.170072
20. Shannon, C. E. (1948). A mathematical theory of communication. *Bell System Technical Journal, 27*, 379–423. https://doi.org/10.1002/j.1538-7305.1948.tb01338.x
21. Chandola, V., Banerjee, A., & Kumar, V. (2009). Anomaly detection: A survey. *ACM Computer Surveys, 41*(3), 58. https://doi.org/10.1145/1541880.1541882
22. Goodfellow, I. J., Pouget-Abadie, J., Mirza, M., Xu, B., Warde-Farley, D., Ozair, S., Courville, A., & Bengio, Y. (2014). Generative adversarial nets. In *Proceedings of the 27th International Conference on Neural Information Processing Systems, Cambridge, MA, USA* (Vol. 2, pp. 2672–2680). MIT Press.
23. Fuller, W. A. (1975). Regression analysis for sample survey. *Sankhyā, Series C, 37*, 117–132.
24. Harrell, F. E. (2015). *Regression modeling strategies, with applications to linear models, logistic and ordinal regression, and survival analysis* (Second Ed.). Springer. https://doi.org/10.1007/978-3-319-19425-7
25. Hosmer Jr., D. W., Lemeshow, S., & Sturdivant, R. X. (2013). *Applied logistic regression. 3rd Ed.* John Wiley & Sons, Hoboken, NJ. https://doi.org/10.1002/9781118548387
26. Cox, D. R. (1958). The regression analysis of binary sequences. *Journal of the Royal Statistical Society: Series B (Methodological), 20*, 215–232. https://doi.org/10.1111/j.2517-6161.1958.tb00292.x
27. Breiman, L. (1984). *Classification and regression trees* (1st Ed.). Routledge. https://doi.org/10.1201/9781315139470
28. Quinlan, J. R. (1986). Induction of decision trees. *Machine Learning, 1*, 81–106. https://doi.org/10.1007/BF00116251
29. Cortes, C., & Vapnik, V. (1995). Support-vector networks. *Machine Learning, 20*, 273–297. https://doi.org/10.1007/BF00994018
30. Schölkopf, B., Smola, A., & Müller, K. R. (1998). Nonlinear component analysis as a Kernel Eigenvalue problem. *Neural Computation, 10*(5), 1299–1319. DOI: https://doi.org/10.1162/089976698300017467
31. Liu, Q., & Wu, Y. (2012). Supervised learning. In N. M. Seel (Ed.) *Encyclopedia of the sciences of learning*. Springer, Boston, MA. https://doi.org/10.1007/978-1-4419-1428-6_451
32. Bengio, Y., Delalleau, O., & Roux, N. L. (2006). Label propagation and quadratic criterion. In *Semi-supervised learning*. MIT Press.
33. Blum, A., & Mitchell, T. (1998). Combining labeled and unlabeled data with co-training. In *Proceedings of the Eleventh Annual Conference on Computational Learning Theory (COLT' 98), New York, NY, USA* (pp. 92–100). Association for Computing Machinery. https://doi.org/10.1145/279943.279962

34. Chapelle, O., Scholkopf, B., & Zien, A. (2006). *Semi-supervised learning.* MIT Press.
35. Zhu, X., & Lafferty, J. (2005). Harmonic mixtures: Combining mixture models and graph-based methods for inductive and scalable semi-supervised learning. In *Proceedings of the 22nd International Conference on Machine Learning (ICML'05), New York, NY, USA* (pp. 1052–1059). Association for Computing Machinery. https://doi.org/10.1145/1102351.1102484
36. Watkins, C. J., & Dayan, P. (1992). Q-learning. *Machine Learning, 8*(3-4), 279–292. DOI: 10.1007/BF00992698.
37. Mnih, V. et al. (2013). Playing Atari with deep reinforcement learning. In *NIPS'13: Proceedings of the 26th International Conference on Neural Information Processing Systems* (pp. 3381–3389). DOI: https://doi.org/10.48550/arXiv.1312.5602
38. Dietterich, T. G. (2000). Ensemble methods in machine learning. In *Multiple classifier systems* (pp. 1–15). Springer. DOI: 10.1007/3-540-45014-9_1.
39. Fernández-Delgado, M., et al. (2014). Do we need hundreds of classifiers to solve real world classification problems? *Journal of Machine Learning Research, 15*(1), 3133–3181.
40. Breiman, L. (2001). Random forests. *Machine Learning, 45*(1), 5–32. DOI:10.1023/A:1010933404324.
41. Friedman, J. H. (2001). Greedy function approximation: A gradient boosting machine. *Annals of Statistics, 29*(5), 1189–1232. DOI: 10.1214/aos/1013203451.
42. Lecun, Y., Bottou, L., Bengio, Y., & Haffner, P. (1998). Gradient-based learning applied to document recognition. *Proceedings of the IEEE, 86*(11), 2278–2324. DOI: 10.1109/5.726791.
43. Hochreiter, S., & Schmidhuber, J. (1997). Long short-term memory. *Neural Computation, 9*(8), 1735–1780. DOI: https://doi.org/10.1162/neco.1997.9.8.1735
44. Vaswani, A., et al. (2017). Attention is all you need. In *Advances in Neural Information Processing Systems* (pp. 5998–6008).
45. Byrd, R. H., Chin, G. M., Nocedal, J., & Wu, Y. (2012). Sample size selection in optimization methods for machine learning. *Mathematical Programming, 134*(1), 127–155. DOI: 10.1007/s10107-012-0572-5.
46. Hinton, G. E., & Salakhutdinov, R. R. (2006). Reducing the dimensionality of data with neural networks. *Science, 313*(5786), 504–507. DOI: 10.1126/science.1127647.
47. Konda, V., & Tsitsiklis, J. (1999). Actor-critic algorithms. In *Advances in Neural Information Processing Systems* (Vol. 12).
48. Graves, A., Wayne, G., & Danihelka, I. (2014). Neural Turing Machines. arXiv [cs.NE].
49. Deng, J., Dong, W., Socher, R., Li, L. J., Li, K., & Li, F.-F. (2009). ImageNet: A large-scale hierarchical image database. In *2009 IEEE Conference on Computer Vision and Pattern Recognition, Miami, FL, USA* (pp. 248–255). DOI: 10.1109/CVPR.2009.5206848.
50. Bengio, Y., Courville, A., & Vincent, P. (2013). Representation learning: A review and new perspectives. *IEEE Transactions on Pattern Analysis and Machine Intelligence, 35*(8), 1798–1828. DOI: 10.1109/TPAMI.2013.50.
51. Chen, T., & Guestrin, C. (2016). XGBoost: A scalable tree boosting system. In *Proceedings of the 22nd ACM SIGKDD International Conference on Knowledge Discovery and Data Mining (KDD'16), New York, NY, USA* (pp. 785–794). Association for Computing Machinery. https://doi.org/10.1145/2939672.2939785

52. L'Heureux, A., Grolinger, K., Elyamany, H. F., & Capretz, M. A. M. (2017). Machine learning with big data: Challenges and approaches. *IEEE Access, 5*, 7776–7797. DOI: 10.1109/ACCESS.2017.2696365.
53. Luu, S. T., Nguyen, H. P., Van Nguyen, K., & Nguyen, N. L. T. (2020). Comparison between traditional machine learning models and neural network models for vietnamese hate speech detection. In *2020 RIVF International Conference on Computing and Communication Technologies (RIVF), Ho Chi Minh City, Vietnam* (pp. 1–6). DOI: 10.1109/RIVF48685.2020.9140745.
54. Suyal, P., Bhatt, A. K., Pant, J., & Mohan, L. (2021). Comparative analysis between traditional learning and deep learning. In *3rd International Conference on Advances in Computing, Communication Control and Networking (ICAC3N), Greater Noida, India* (pp. 376–379). DOI: 10.1109/ICAC3N53548.2021.9725370.
55. Alfred, R., & Obit, J. H. (2021). The roles of machine learning methods in limiting the spread of deadly diseases: A systematic review. *Heliyon, 7*(6), e07371. DOI: 10.1016/j.heliyon.2021.e07371. Epub 2021 Jun 23. PMID: 34179541; PMCID: PMC 8219638.
56. Settles, B. (2012). Active learning. In *Synthesis lectures on artificial intelligence and machine learning*. Springer, Cham. DOI: https://doi.org/10.1007/978-3-031-01560-1
57. Lewis, D. D., & Gale, W. A. (1994). A sequential algorithm for training text classifiers. In W. Bruce Croft & C. J. van Rijsbergen, (Eds.), *SIGIR 94: Proceedings of the Seventeenth Annual International A CM-SIGIR Conference on Research and Development in Information Retrieval, London* (pp. 3–12). Springer-Verlag. DOI: https://doi.org/10.48550/arXiv.cmp-lg/9407020
58. Dasgupta, S., & Hsu, D. (2008). Hierarchical sampling for active learning. In *Proceedings of the 25th International Conference on Machine Learning (ICML'08), New York, NY, USA* (pp. 208–215). Association for Computing Machinery. https://doi.org/10.1145/1390156.1390183
59. Settles, B. (2009). Active learning literature survey. University of Wisconsin-Madison, Computer Sciences Technical Report 1648. https://minds.wisconsin.edu/bitstream/handle/1793/60660/TR1648.pdf?sequence=1&isAllowed=y
60. Settles, B., & Craven, M. (2008). An analysis of active learning strategies for sequence labeling tasks. In *Proceedings of the 2008 Conference on Empirical Methods in Natural Language Processing* (pp. 1070–1079). https://www.biostat.wisc.edu/~craven/papers/settles.emnlp08.pdf
61. Tong, S., & Koller, D. (2001). Support vector machine active learning with applications to text classification. *Journal of Machine Learning Research, 2*, 45–66. 10.1162/153244302760185243.
62. Tong, S., & Koller, D. (2000). Active learning for parameter estimation in Bayesian networks. In *Advances in Neural Information Processing Systems* (Vol. 13).
63. Ho, T. K., & Basu, M. (2002). Complexity measures of supervised classification problems. *IEEE Transactions on Pattern Analysis and Machine Intelligence, 24*, 289–300. 10.1109/34.990132.
64. Bordes, A., Ertekin, S., Weston, J., & Bottou, L. (2005). Fast kernel classifiers with online and active learning. *Journal of Machine Learning Research, 6*, 1579–1619. DOI: 10.5555/1046920.1194925.
65. Wu, J., Chen, J., & Huang, D. (2022). Entropy-based active learning for object detection with progressive diversity constraint. In *2022 IEEE/CVF Conference on Computer*

Vision and Pattern Recognition (CVPR), New Orleans, LA, USA (pp. 9387–9396). DOI: 10.1109/CVPR52688.2022.00918.

66. Holub, A., Perona, P., & Burl, M. C. (2008). Entropy-based active learning for object recognition. In *2008 IEEE Computer Society Conference on Computer Vision and Pattern Recognition Workshops, Anchorage, AK, USA* (pp. 1–8). DOI: 10.1109/CVPRW.2008.4563068.
67. Ferreira, P. M. (2016). Unsupervised entropy-based selection of data sets for improved model fitting. In *2016 International Joint Conference on Neural Networks (IJCNN), Vancouver, BC, Canada* (pp. 3330–3337). DOI: 10.1109/IJCNN.2016.7727625.
68. Cai, W., Zhang, Y., & Zhou, J. (2013). Maximizing expected model change for active learning in regression. In *2013 IEEE 13th International Conference on Data Mining, Dallas, TX, USA* (pp. 51–60). DOI: 10.1109/ICDM.2013.104.
69. Cai, W., Zhang, Y., & Zhou, J. (2013). Maximizing expected model change for active learning in regression. In *2013 IEEE 13th International Conference on Data Mining, Dallas, TX, USA* (pp. 51–60). DOI: 10.1109/ICDM.2013.104.
70. Qing, J., Knudde, N., Couckuyt, I., Spina, D., & Dhaene, T. (2020). Bayesian active learning for electromagnetic structure design. In *14th European Conference on Antennas and Propagation (EuCAP), Copenhagen, Denmark* (pp. 1–5). DOI: 10.23919/EuCAP48036.2020.9136051.
71. Arnavaz, K., Feragen, A., Krause, O., & Loog, M. (2021). Bayesian active learning for maximal information gain on model parameters. In *25th International Conference on Pattern Recognition (ICPR), Milan, Italy* (pp. 10524–10531). DOI: 10.1109/ICPR48806.2021.9411962.
72. Liu, P., Wang, L., Ranjan, R., He, G., & Zhao, L. (2022). A survey on active deep learning: From model driven to data driven. *ACM Computing Surveys, 54*(10s), 1–34. https://doi.org/10.1145/3510414
73. Settles, B., & Craven, M. W. (2008). An analysis of active learning strategies for sequence labeling tasks. In *Proceedings of the Conference on Empirical Methods in Natural Language Processing (EMNLP' 08), USA* (pp. 1070–1079). Association for Computational Linguistics.
74. Aggarwal, C. C., Kong, X., Gu, Q., Han, J., & Yu, P. S. (2014). Active learning: A survey. In *Data classification: Algorithms and applications* (pp. 571–605). CRC Press. https://doi.org/10.1201/b17320
75. El-Hasnony, I. M., Elzeki, O. M., Alshehri, A., & Marie, S. H. (2022). Multi-label active learning-based machine learning model for heart disease prediction. *Sensors, 22*(3), 1184. https://doi.org/10.3390/s22031184
76. Hu, L., Lu, S., & Wang, X.-Z. (2013). A new and informative active learning approach for support vector machine. *Information Sciences, 244*, 142–160. DOI: 10.1016/j.ins.2013.05.010.
77. Wang, X., Liu, B., Cao, S., et al. (2020). Important sampling based active learning for imbalance classification. *Science China Information Sciences, 63*, 182104. https://doi.org/10.1007/s11432-019-2771-0
78. Lu, X., Zhang, J., Li, T., & Zhang, Y. (2016). Incorporating diversity into self-learning for synergetic classification of hyperspectral and panchromatic images. *Remote Sensing, 8*(10), 804. https://doi.org/10.3390/rs8100804

79. Wang, L., Zhang, D., & Wang, L. (2014). Active learning-based ensemble for image classification. *IEEE Transactions on Pattern Analysis and Machine Intelligence, 36*(6), 1209–1222. DOI: 10.1109/TPAMI.2013.239.
80. Leng, Y., Xu, X., & Qi, G. (2013). Combining active learning and semi-supervised learning to construct SVM classifier. *Knowledge-Based Systems, 44*, 121–131. https://doi.org/10.1016/j.knosys.2013.01.032 https://www.sciencedirect.com/science/article/pii/S095070511300052X
81. Azab, A. M., Toth, J., Mihaylova, L., Arvaneh, M. (2018). A review on transfer learning approaches in brain–computer interface. In *Signal processing and machine learning for brain-machine interfaces* (pp. 81–101). DOI: 10.1049/PBCE114E_ch5IET Digital Library, https://digital-library.theiet.org/content/books/10.1049/pbce114e_ch5
82. Kanade, V. (2023). What is transfer learning? Definition, Methods, and Applications. https://www.spiceworks.com/tech/artificial-intelligence/articles/articles-what-is-transfer-learning/, (accessed June 11, 2023).
83. Zhuang, F., et al. (2021). A comprehensive survey on transfer learning. *Proceedings of the IEEE, 109*(1), 43–76. DOI: 10.1109/JPROC.2020.3004555.
84. Hosna, A., Merry, E., Gyalmo, J., et al. (2022). Transfer learning: A friendly introduction. *Journal of Big Data, 9*, 102. https://doi.org/10.1186/s40537-022-00652-w
85. Pan, S. J. &Yang, Q. (2010). A survey on transfer learning. *IEEE Transactions on Knowledge and Data Engineering, 22*(10), 1345–1359. DOI: 10.1109/TKDE.2009.191.
86. Fei-Fei, L., Fergus, R., & Perona, P. (2006). One-shot learning of object categories. *IEEE Transactions on Pattern Analysis and Machine Intelligence, 28*(4), 594–611. https://doi.org/10.1109/TPAMI.2006.79
87. Socher, R., Ganjoo, M., Sridhar, H., Bastani, O., Manning, C. D., & Ng, A. Y. (2013). Zero-shot learning through cross-modal transfer. arXiv [cs.CV].
88. Zhao, P., Hoi, S. C. H., Wang, J., & Li, B. (2014). Online Transfer Learning. *Artificial Intelligence, 216*, 76–102. DOI: https://doi.org/10.1016/j.artint.2014.06.003.
89. Alasbahi, R., & Zheng, X. (2022). An online transfer learning framework with extreme learning machine for automated credit scoring. *IEEE Access, 10*, 46697–46716. DOI: 10.1109/ACCESS.2022.3171569.
90. Geng, X., & Smith-Miles, K. (2009). IL. In S. Z. Li, & A. Jain (Eds), In *Encyclopedia of biometrics*. Springer, Boston, MA. https://doi.org/10.1007/978-0-387-73003-5_304
91. Luo, Y., Yin, L., Bai, W., & Mao, K. (2020). An appraisal of IL methods. *Entropy, 22*(11), 1190. https://doi.org/10.3390/e22111190
92. Belouadah, E., Popescu, A., Kanellos, I. (2021). A comprehensive study of class IL algorithms for visual tasks. *Neural Networks, 135*, 38–54. DOI: https://doi.org/10.1016/j.neunet.2020.12.003.
93. Zhang, Y., Wa, S., Zhang, L., & Lv, C. (2022). Automatic plant disease detection based on tranvolution detection network with GAN modules using leaf images. *Frontiers in Plant Science, 13*, 875693.
94. Saranya, P., & Nagarajan, N. (2020). Efficient agricultural yield prediction using metaheuristic optimized artificial neural network using Hadoop framework. *Soft Computing, 24*(16), 12659–12669. https://doi.org/10.1007/s00500-020-04707-z
95. Praveen, B., Talukdar, S., Shahfahad, et al. (2020). Analyzing trend and forecasting of rainfall changes in India using non-parametrical and machine learning approaches. *Scientific Reports, 10*, 10342. https://doi.org/10.1038/s41598-020-67228-7

96. Fayyaz, Z., Ebrahimian, M., Nawara, D., Ibrahim, A. F., & Kashef, R. (2020). Recommendation systems: Algorithms, challenges, metrics, and business opportunities. *Applied Sciences, 10*(21), 7748. https://doi.org/10.3390/app10217748
97. Lee, Y., Ahn, M. H., & Lee, S. J. (2022). IL with neural network algorithm for the monitoring pre-convective environments using geostationary imager. *Remote Sensing, 14*(2), 387. https://doi.org/10.3390/rs14020387
98. Bakthavatchalam, K., Karthik, B., Thiruvengadam, V., Muthal, S., Jose, D., Kotecha, K., & Varadarajan, V. (2022). IoT framework for measurement and precision agriculture: Predicting the crop using machine learning algorithms. *Technologies, 10*(13). https://doi.org/10.3390/technologies10010013
99. Shrivastava, V. K., Shrivastava, A., Sharma, N., Mohanty, S. N., & Pattanaik, C. R. (2023). Deep learning model for temperature prediction: A case study in New Delhi. *Journal of Forecasting,* 1– 16. https://doi.org/10.1002/for.2966
100. Kates, R. W., Travis, W. R., & Wilbanks, T. J. (2012). Transformational adaptation when incremental adaptations to climate change are insufficient. *Proceedings of the National Academy of Sciences, 109*(19), 7156–7161. DOI: https://doi.org/10.1073/pnas.111552110
101. Kim, W., & Hasegawa, O. (2018). Time series prediction of tropical storm trajectory using self-organizing incremental neural networks and error evaluation. *Journal of Advanced Computational Intelligence and Intelligent Informatics, 22*, 465–474. 10.20965/jaciii.2018.p0465.
102. Fernández, J. G., & Mehrkanoon, S. (2021). Broad-UNet: Multi-scale feature learning for nowcasting tasks. *Neural Networks, 144*, 419–427. https://doi.org/10.1016/j.neunet.2021.08.036.
103. Vincent, J. (2023). Google DeepMind will use machine learning to spot eye diseases early. https://www.theverge.com/2016/7/5/12095830/google-deepmind-nhs-eye-disease-detection, (last accessed June 16, 2023).
104. Jie, Z., Zhiying, Z., & Li, L. (2021). A meta-analysis of Watson for oncology in clinical application. *Scientific Reports, 11*, 5792. https://doi.org/10.1038/s41598-021-84973-5
105. Kent, J. (2023). Google uses artificial intelligence to detect breast cancer. https://healthitanalytics.com/news/google-uses-artificial-intelligence-to-detect-breast-cancer, (last accessed June 16, 2023).
106. Yim, J., Chopra, R., De Fauw, J., & Ledsam, J. (2023). Using AI to predict retinal disease progression. https://www.deepmind.com/blog/using-ai-to-predict-retinal-disease-progression, (last accessed June 16, 2023).
107. Smalley, E. (2017). AI-powered drug discovery captures pharma interest. *Nature Biotechnology, 35*(7), 604–605. link.gale.com/apps/doc/A498337436/HRCA?u=anon~3a6ded78&sid=googleScholar&xid=119e5753. Accessed 15 June 2023.
108. Michard, F., & Teboul, J. L. (2019). Predictive analytics: Beyond the buzz. *Annals of Intensive Care, 9*, 46. https://doi.org/10.1186/s13613-019-0524-9
109. Aresti, C. (2023). PayPal's use of machine learning to enhance fraud detection. https://d3.harvard.edu/platform-rctom/submission/paypals-use-of-machine-learning-to-enhance-fraud-detection-and-more/, (last accessed June 16, 2023).
110. Afriyie, J. K., Tawiah, K., Pels, W. A., Addai-Henne, S., Dwamena, H. A., Emmanuel, O. O., Ayeh, S. A., Eshun, J. (2023). A supervised machine learning algorithm for

detecting and predicting fraud in credit card transactions. *Decision Analytics Journal, 6*, 100163. https://doi.org/10.1016/j.dajour.2023.100163.

111. Hart, O., Edwards, J., Perez Leon, F., Johnson, M., Newman, K., Wu, A., Jason, J. (2020). Updating a machine learning fraud model based on third party transaction information. US10672005B1.
112. Chang, A. H., Yang, L. K., Tsaih, R. H., & Lin, S. K. (2022). Machine learning and artificial neural networks to construct P2P lending credit-scoring model: A case using Lending Club data. *Quantitative Finance and Economics, 6*(2), 303–325. DOI: 10.3934/QFE.2022013
113. Risk as a Service. https://www.jpmorgan.com/solutions/cib/securities-services/risk-as-a-service, (last Accessed June 16, 2023).
114. Risk Staff. (2023). Flow market-maker of the year: Citadel Securities. https://www.risk.net/derivatives/7934291/flow-market-maker-of-the-year-citadel-securities, (last Accessed June 17, 2023).
115. Budiharto, W. (2021). Data science approach to stock prices forecasting in Indonesia during Covid-19 using Long Short-Term Memory (LSTM). *Journal of Big Data, 8*, 47. https://doi.org/10.1186/s40537-021-00430-0
116. Ruby, M. (2023). 5 Machine learning techniques for sales forecasting - Comparing linear regression, random forest regression, XGBoost, LSTMs, and ARIMA time series forecasting in Python. https://towardsdatascience.com/5-machine-learning-techniques-for-sales-forecasting-598e4984b109, (last accessed June 17, 2023).
117. Woloszko, N. (2020). Adaptive trees: A new approach to economic forecasting. DOI: https://doi.org/10.1787/18151973
118. Reinforcement Learning, Control, and Optimization. (2023). https://www.bosch-ai.com/research/fields-of-expertise/reinforcement-learning-control-and-optimization/, (last accessed June 17, 2023).
119. Predictive Maintenance at Scale. (2023). https://www.siemens.com/global/en/products/services/digital-enterprise-services/analytics-artificial-intelligence-services/predictive-services/senseye-predictive-maintenance.html, (last accessed June 17, 2023).
120. How Big Data Analysis helped increase Walmarts Sales turnover? (2023). https://www.projectpro.io/article/how-big-data-analysis-helped-increase-walmarts-sales-turnover/109, (last accessed June 17, 2023).
121. Predictive Maintenance. (2023). https://www.ge.com/research/project/predictive-maintenance, (last accessed June 17, 2023).
122. Theissler, A., Pérez-Velázquez, J., Kettelgerdes, M., Elger, G. (2021). Predictive maintenance enabled by machine learning: Use cases and challenges in the automotive industry. *Reliability Engineering & System Safety, 215*, 107864. https://doi.org/10.1016/j.ress.2021.107864.
123. Demand Forecasting - AWS Whitepaper. (2023). https://docs.aws.amazon.com/pdfs/whitepapers/latest/demand-forecasting/demand-forecasting.pdf, (last accessed June 17, 2023).
124. Dodel, J. P., & McMahon, T. (2023). IL: Optimizing search relevance at scale using machine learning. https://aws.amazon.com/blogs/machine-learning/incremental-learning-optimizing-search-relevance-at-scale-using-machine-learning/, (last accessed June 17, 2023).

125. How personalisation is helping the retail brands? (2023). https://www.indianretailer.com/article/operations/marketing/How-Personalisation-is-helping-the-retail-brands.a5761?page=1, (last accessed June 17, 2023).
126. Chen, L., Thakuriah, P., & Ampountolas, K. (2021). Short-term prediction of demand for ride-hailing services: A deep learning approach. *Journal of Big Data Analytics in Transportation, 3*, 175–195. https://doi.org/10.1007/s42421-021-00041-4
127. Gallien, J., Mersereau, A., Garro, A., Mora, A., & Vidal, M. (2013). Initial shipment decisions for new products at Zara. *SSRN Electronic Journal, 63*. 10.2139/ssrn.2378859.
128. E-commerce Tech Trends: Reinforcement Learning for Dynamic Pricing. (2023). https://eleks.com/research/reinforcement-learning-for-dynamic-pricing/, (last accessed June 17, 2023).
129. Weiss, S. (2023). 3 secrets to increase retention for your e-commerce subscription business. https://mutesix.com/3-secrets-to-increase-retention-for-your-e-commerce-subscription-business/, (last accessed June 17, 2023).
130. Andrade, I. M. D., & Tumelero, C. (2022). Increasing customer service efficiency through artificial intelligence chatbot. *Revista de Gestão, 29*(3), 238–2251. https://doi.org/10.1108/REGE-07-2021-0120
131. Holmlund, M., Van Vaerenbergh, Y., Ciuchita, R., Ravald, A., Sarantopoulos, P., Ordenes, F. V., & Zaki, M. (2020). Customer experience management in the age of big data analytics: A strategic framework. *Journal of Business Research, 116*, 356–365. https://doi.org/10.1016/j.jbusres.2020.01.022.
132. Perlich, C., Dalessandro, B., Raeder, T., et al. (2014). Machine learning for targeted display advertising: Transfer learning in action. *Machine Learning, 95*, 103–127. https://doi.org/10.1007/s10994-013-5375-2
133. Inzaugaat, E. (2023). How machine learning can help with customer retention. https://towardsdatascience.com/how-machine-learning-can-help-with-customer-retention-6b5bf654e822, (last accessed June 17, 2023).
134. Tasar, O., Tarabalka, Y., & Alliez, P. (2019). IL for semantic segmentation of large-scale remote sensing data. *IEEE Journal of Selected Topics in Applied Earth Observations and Remote Sensing, 12*(9), 3524–3537. DOI: https://doi.org/10.48550/arXiv.1810.12448
135. Saias, J., Rato, L., & Gonçalves, T. (2022). An approach to churn prediction for cloud services recommendation and user retention. *Information, 13*(5), 227. https://doi.org/10.3390/info13050227
136. Amin, A., Adnan, A., & Anwar, S. (2023). An adaptive learning approach for customer churn prediction in the telecommunication industry using evolutionary computation and Naïve Bayes. *Applied Soft Computing, 137*, 110103. https://doi.org/10.1016/j.asoc.2023.110103.
137. Shao, Y., Zhao, Y., Yu, F., Zhu, H., & Fang, J. (2021). The traffic flow prediction method using the IL-based CNN-LTSM model: The solution of mobile application. *Mobile Information Systems,* 1–16. 10.1155/2021/5579451.
138. Bovenzi, G., Yang, L., Finamore, A., Aceto, G., Ciuonzo, D., Pescape, A., & Rossi, D. (2021). A first look at class IL in deep learning mobile traffic classification. In *5th Network Traffic Measurement and Analysis Conference, TMA*. Virtual.
139. Correani, A., De Massis, A., Frattini, F., Petruzzelli, A. M., & Natalicchio, A. (2020). Implementing a digital strategy: Learning from the experience of three digital

transformation projects. *California Management Review, 62*(4), 37–56. https://doi.org/10.1177/0008125620934864
140. Vrablecová, P., Ezzeddine, A. B., Rozinajová, V., Šárik, S., Sangaiah, A. K. (2018). Smart grid load forecasting using online support vector regression. *Computers & Electrical Engineering, 65*, 102–117. https://doi.org/10.1016/j.compeleceng.2017.07.006.
141. Mostafa, N., Ramadan, H. S. M., & Elfarouk, O. (2022). Renewable energy management in smart grids by using big data analytics and machine learning. *Machine Learning with Applications, 9*, 100363. https://doi.org/10.1016/j.mlwa.2022.100363.
142. Wang, Y., Chen, Q., Hong, T., & Kang, C. (2019). Review of smart meter data analytics: Applications, methodologies, and challenges. *IEEE Transactions on Smart Grid, 10*(3), 3125–3148. DOI: 10.1109/TSG.2018.2818167.
143. Mezni, H., Driss, M., Boulila, W., Atitallah, S. B., Sellami, M., & Alharbi, N. (2022). SmartWater: A service-oriented and sensor cloud-based framework for smart monitoring of water environments. *Remote Sensing, 14*(4), 922. https://doi.org/10.3390/rs14040922
144. Galić, E., & Höhfeld, M. (1996). Improving the generalization performance of multi-layer-perceptrons with population-based IL. In H. M. Voigt, W. Ebeling, I. Rechenberg, H. P. Schwefel (Eds.) *Parallel problem solving from nature — PPSN IV* (Vol. 1141). Springer, Berlin, Heidelberg. https://doi.org/10.1007/3-540-61723-X_1037
145. Cai, G., Gao, H., Yang, D., & Wang, L. (2023). Synchrophasor-based online load margin estimation using IL assisted LightGBM in smart grid. *IEEE Systems Journal, 17*(3), 4660–4670. DOI: 10.1109/JSYST.2023.3243128.
146. Puah, B. K., Chong, L. W., Wong, Y. W., Begam, K. M., Khan, N., Juman, M. A., & Rajkumar, R. K. (2021). A regression unsupervised IL algorithm for solar irradiance prediction. *Renewable Energy, 164*, 908–925. https://doi.org/10.1016/j.renene.2020.09.080.
147. Hu, T., Wu, W., Guo, Q., Sun, H., Shi, L., & Shen, X. (2020). Very short-term spatial and temporal wind power forecasting: A deep learning approach. *CSEE Journal of Power and Energy Systems, 6*(2), 434–443. DOI: 10.17775/CSEEJPES.2018.00010.
148. Belsare, K., & Goraya, M. (2022). Various frameworks for IoT-enabled intelligent waste management system using ML for smart cities. In *Mobile Computing and Sustainable Informatics* (pp. 797–817). 10.1007/978-981-19-2069-1_55.
149. Nemade, B., & Shah, D. (2022). An efficient IoT based prediction system for classification of water using novel adaptive IL framework. *Journal of King Saud University - Computer and Information Sciences, 34*(8), 5121–5131. https://doi.org/10.1016/j.jksuci.2022.01.009.
150. Eyre, T. J., Fisher, A., Hunt, L. P., & Kutt, A. S. (2011). Measure it to better manage it: a biodiversity monitoring framework for the Australian rangelands. *The Rangeland Journal, 33*, 239–253. DOI: https://doi.org/10.1071/RJ10071
151. Chen, Y., Lv, Y., Wang, X., & Wang, F. -Y. (2018). Traffic flow prediction with parallel data. In *21st International Conference on Intelligent Transportation Systems (ITSC), Maui, HI, USA* (pp. 614–619). DOI: 10.1109/ITSC.2018.8569276.
152. Kamble, S. J., & Kounte, M. R. (2022). Dynamic traveling route planning method for intelligent transportation using IL-based hybrid deep learning prediction model with fine-tuning. *Transport and Telecommunication Journal, 23*(4), 293–310. DOI: https://doi.org/10.2478/ttj-2022-0024

153. Wang, D., Yang, Y., & Ning, S. (2018). DeepSTCL: A deep spatio-temporal ConvLSTM for travel demand prediction. In *International Joint Conference on Neural Networks (IJCNN), Rio de Janeiro, Brazil* (pp. 1–8). DOI: 10.1109/IJCNN.2018.8489530.
154. Lourens, T. (2005). Using population-based IL to optimize feasible distribution logistic solutions (PhD diss., University of Stellenbosch, Stellenbosch).
155. Theissler, A., Pérez-Velázquez, J., Kettelgerdes, M., & Elger, G. (2021). Predictive maintenance enabled by machine learning: Use cases and challenges in the automotive industry. *Reliability Engineering & System Safety, 215*, 107864. https://doi.org/10.1016/j.ress.2021.107864.
156. Kalathas, I., & Papoutsidakis, M. (2021). Predictive maintenance using machine learning and data mining: A pioneer method implemented to Greek railways. *Designs, 5*(1), 5. https://doi.org/10.3390/designs5010005
157. Armacost, A. P., Barnhart, C., Ware, K. A., & Wilson, A. M. UPS optimizes its air network. *Interfaces, 34*(1), 15–25. http://www.jstor.org/stable/25062865.
158. Amatriain, X., & Basilico, J. (2015). Recommender systems in industry: A Netflix case study. In F. Ricci, L. Rokach, B. Shapira (Eds.) *Recommender systems handbook*. Springer, Boston, MA. https://doi.org/10.1007/978-1-4899-7637-6_11
159. Jaganathan, S., & Madhusudhanan, S. (2022). Polarity classification of social media feeds using IL—A deep learning approach. *IEICE Transactions on Fundamentals of Electronics, Communications and Computer Sciences, 105*(3), 584–593.
160. How Google Ads uses machine learning to improve Ad targeting and ROAS. https://www.bdex.com/blog/how-google-ads-uses-machine-learning-to-improve-ad-targeting-and-roas/, (last accessed June 17, 2023).
161. Wiggers, K. (2023). Facebook's improved AI isn't preventing harmful content from spreading. https://venturebeat.com/ai/facebooks-improved-ai-isnt-preventing-harmful-content-from-spreading/, (last accessed June 17, 2023).
162. Palmer, R. (2023). A Spotify Song and Playlist Recommendation Engine. https://www.mongodb.com/developer/code-examples/python/song-recommendations-example-app/, (last accessed June 17, 2023).
163. Dwivedi, P. (2023). YouTube analytics data: A comprehensive guide. https://hevodata.com/learn/leveraging-youtube-analytics-data/, (last accessed June 17, 2023).
164. Coenen, A. (2023). How the New York times is experimenting with recommendation algorithms. https://open.nytimes.com/how-the-new-york-times-is-experimenting-with-recommendation-algorithms-562f78624d26, (last accessed June 17, 2023).
165. Stern, D. H., Herbrich, R., & Graepel, T. (2009). Matchbox: Large scale online bayesian recommendations. In *Proceedings of the 18th International Conference on World Wide Web (WWW '09), New York, NY, USA* (pp. 111–120). Association for Computing Machinery. https://doi.org/10.1145/1526709.1526725
166. Luo, Y., Yin, L., Bai, W., & Mao, K. (2020). An appraisal of IL methods. *Entropy (Basel), 22*(11), 1190. DOI: 10.3390/e22111190. PMID: 33286958; PMCID: PMC7712976.
167. Pezoulas, V. C., De Vita, S., Tzioufas, A., Fotiadis, D., Kourou, K., Kalatzis, F., Exarchos, T., Zampeli, E., Gandolfo, S., Goules, A., Baldini, C., & Skopouli, F. N. (2020). Overcoming the barriers that obscure the interlinking and analysis of clinical data through harmonization and IL. *IEEE Open Journal of Engineering in Medicine and Biology, PP*, 1–1. 10.1109/OJEMB.2020.2981258.

168. Geng, X., & Smith-Miles, K. (2009). IL. In S. Z. Li,& A. Jain (Eds.), *Encyclopedia of biometrics*. Springer, Boston, MA. https://doi.org/10.1007/978-0-387-73003-5_304
169. Liu, J., Xie, Z., Nikolopoulos, D., & Li, D. (2020). {RIANN}: Real-time IL with approximate nearest neighbor on mobile devices. In *2020 {USENIX} Conference on Operational Machine Learning (OpML 20)*.

CHAPTER 10

Exploring of AI-Powered Application in Chatbots for Future Development

S. SASIKALA[1] and V. KAVITHA[2]

[1]*PSGR Krishnammal College for Women, Coimbatore, India*

[2]*Department of Computer Science with Cognitive Systems, Sri Ramakrishna College of Arts & Science, Coimbatore, Tamil Nadu, India*

ABSTRACT

In today's digital era, artificial intelligence (AI) has infused various surfaces of our lives, transforming the way we interact with technology. Among these innovations, chatbots have emerged as a prominent and versatile application of AI, transforming customer service, automating tasks, and enhancing user experiences across industries. "Exploring AI-Powered Applications in Chatbot Development for the Future," takes readers on a journey through the active landscape of chatbots, examining their current state, practical applications, and the exciting predictions they hold for the future. It offers a complete examination of chatbot technology, commencing with foundational concepts such as natural language processing (NLP) and dialogue management. It showcases real-world implementations of chatbots, highlighting their impact on diverse sectors, including e-commerce, healthcare, finance, and more. By analyzing case studies and industry-specific use cases, readers gain insights into how chatbots are reshaping customer interactions, improving operational efficiency, and driving innovation. It explores cutting-edge developments in chatbot design, such as voice-enabled user interfaces, multimodal interactions, and ethical implications. In order to assure privacy, fairness, and openness, it emphasizes responsible AI practices as a means of addressing the ethical issues

Real-Time Artificial Intelligence (AI): Key Motivations, Technologies, Platforms, and Use Cases. Pethuru Raj, B. Sundaravadivazhagan, V. Kavitha, B. Narendra Kumar Rao, & Hannah Vijaykumar (Eds.)

DOI: 10.1201/9781998511358-10

raised by chatbots that are driven by AI chatbots, emphasizing responsible AI practices to ensure privacy, fairness, and transparency. Chatbots are designed to be decisive in influencing the future of human–computer interactions as AI develops further. Exploring AI-powered applications in chatbot development for the future provides specialists, academics, programmers, and enthusiasts with the information and perceptions necessary to maximize the potential of AI-driven chatbots in a constantly evolving digital environment. This book lays the foundation for a future in which intelligent conversational agents are an essential part of our daily lives by illuminating the existing capabilities of chatbots and providing a look into their transformational capacity. AI is at the forefront of innovation in a time of accelerated technological growth. Among the many uses for AI, chatbots have become a fantastic link for efficient and natural communication between people and machines. Chatbots are poised to alter how we interact with technology as the world becomes more digital and networked by enhancing process efficiency, information accessibility, and responsiveness. It serves as the foundation for chatbot technology, highlighting how NLP makes it possible for machines and humans to communicate similarly. Additionally, it illustrates the revolutionary effects of chatbots across a variety of industries, including retail, healthcare, banking, and customer service, and provides actual instances to highlight their importance. It explores the evolving field of chatbot creation and looks at themes that will influence this technology's future. This book offers a thorough road map for comprehending and realizing the full potential of AI-powered chatbots, from voice-enabled interactions that close the gap between humans and computers to the ethical issues that lead responsible AI deployment. It provides a comprehensive picture of the transformational potential of chatbots and their vital role in determining the future of technology-driven dialogues, with a watchful eye on both the present and the horizon. AI is at the forefront of innovation in a time of accelerated technological growth. Among the many uses for AI, chatbots have become a fantastic link for efficient and natural communication between people and machines. Chatbots are poised to alter how we interact with technology as the world becomes more digital and networked by enhancing process efficiency, information accessibility, and responsiveness.

10.1 OVERVIEW OF AI AND ITS IMPACT ON THE DIGITAL ERA

Artificial intelligence, or AI, is a game-changing technology that has significantly changed the digital age and many facets of our life [1]. It refers to the creation of computer systems that are capable of carrying out operations like problem-solving, judgement, language comprehension, and visual perception that ordinarily demand for human intellect. Automation has been revolutionized by AI, enabling organizations and sectors to optimize operations, boost productivity, and minimize human involvement [2]. AI-powered systems can undertake labor-intensive, repetitive activities, which saves a lot of time and money.

Users of many digital services, like online purchasing, content recommendations, and targeted advertising, can have personalized experiences thanks to AI. In order to comprehend preferences and customize products for specific consumers, AI systems analyze user data. Computers already have the ability to understand and interpret human language thanks to natural language processing (NLP), a subfield of AI. Voice assistants, chatbots, and language translation services are examples of applications that are now widely used and have fully revolutionized how we interact with digital devices.

Neural networks are used in deep learning, a branch of machine learning, to analyze and learn from enormous datasets. By revolutionizing speech and picture recognition, this technology has enabled innovations like voice control and facial identification. Businesses can use AI-driven data analysis to extract valuable knowledge from massive datasets, allowing them to see patterns, forecast trends, and make data-driven decisions that would be challenging or impossible for humans to make. Industries are being upended by AI, creating new business models and opportunities. Businesses are using AI to gain a competitive edge, streamline processes, and offer cutting-edge goods and services.

10.2 INTRODUCTION TO CHATBOTS AND THEIR ROLE IN CUSTOMER INTERACTIONS

10.2.1 INTRODUCTION TO CHATBOTS

Chatbots are computer programs designed to simulate conversation with human users through messaging applications, websites, or mobile apps. NLP is used by them to comprehend and react to consumer inquiries or requests. Chatbots can be designed to perform a wide range of tasks, from providing customer support and handling routine inquiries to providing personalized recommendations and delivering news updates. They can also be used for marketing purposes, such as lead generation and user engagement.

There are two main types of chatbots: Rule-based chatbots and AI-powered chatbots. While AI-powered chatbots employ machine learning algorithms to learn from user interactions and improve their responses over time, rule-based chatbots utilize predefined rules to respond to user inputs.

Chatbots have become increasingly popular in recent years, as businesses look for ways to automate routine tasks and improve customer engagement. They offer several advantages over traditional customer support channels, including 24/7 availability, faster response times, and the ability to handle multiple conversations simultaneously.

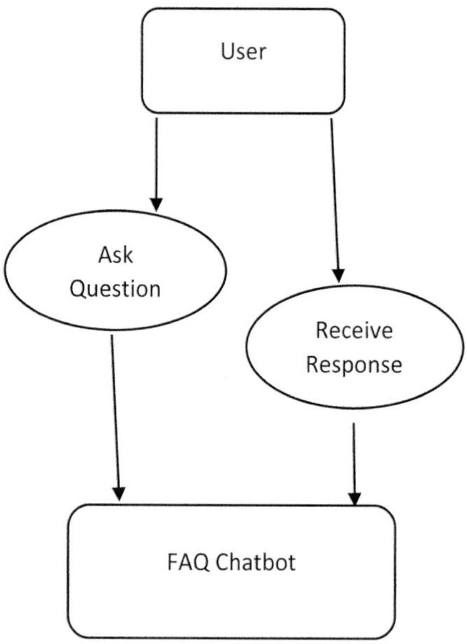

FIGURE 10.1 Chatbot.

1. ***Customer interactions:*** Chatbots can provide automated customer support by answering frequently asked questions, handling routine inquiries, and resolving issues. AI, machine learning, and natural language understanding (NLU) are all used by customer care chatbots to simulate human communication. Businesses have been incorporating chatbots into their service models over time to automate

repetitive tasks and respond to customer enquiries. Chatbots can be effective when used for the correct tasks, despite some of their limitations. Businesses may be hesitant about how customer care chatbots might affect the customer experience, and with good reason. Here is an example of how the Azure chatbot works in customer service.

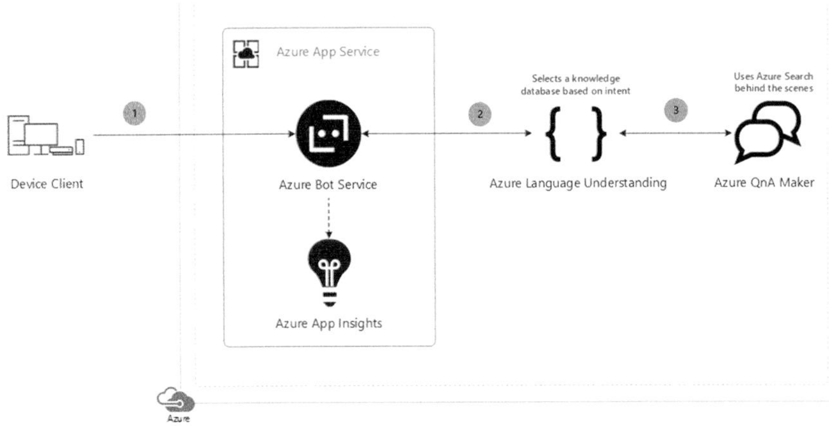

FIGURE 10.2 Commerce chatbot for customer service in Azure.

Note: Commerce chatbot for customer service in Azure involves using Azure's services and tools to build, deploy, and manage a chatbot that can assist customers with various commerce-related inquiries and tasks.

Source: https://learn.microsoft.com/en-us/gaming/azure/reference-architectures/cognitive-css-bot

2. ***E-commerce:*** Chatbots can assist customers with shopping by providing product recommendations, answering questions about products, and processing orders. The interaction with a pleasant and competent sales assistant is replicated in the digital realm by an e-commerce bot. The e-commerce chatbot may converse with customers, determine their needs from the interaction, and recommend products that are appropriate. The chatbot responds to the customer's questions regarding an item they are thinking about buying, much like a persistent human sales assistant. It responds to follow-up queries and frequently asked questions, retrieves user evaluations, and performs any other action necessary to persuade the buyer to buy. Additionally, ecommerce chatbots can cross-sell and upsell pertinent products.

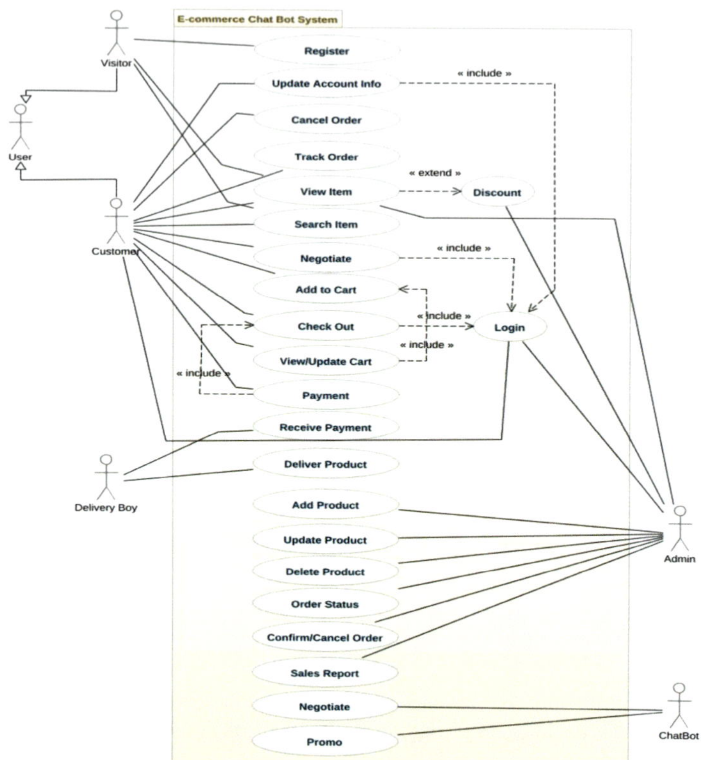

FIGURE 10.3 Use case of an e-commerce chatbot.

Note: E-commerce chatbot can serve various use cases to enhance the customer experience, streamline operations, and boost sales.

Source: https://app.genmymodel.com/api/repository/asifalidosani/Ecommerce%20chatbot%20Use%20Case

10.3 EVOLUTION OF CHATBOT TECHNOLOGY AND ITS CURRENT STATE

NLP, machine learning, and advancements in AI have all contributed to the intriguing evolution of chatbot technology. Here is a concise summary of the key turning moments in the evolution of chatbot technology:

1. Early Rule-Based Systems (1960s–1990s):

Early rule-based systems were among the first attempts at creating chatbot-like programs. These systems were simple and relied on predefined rules and patterns to generate responses to user inputs. Here is a closer look at these early rule-based chatbot systems.

- *ELIZA (1966):*

ELIZA, created by Joseph Weizenbaum, was one of the earliest and most famous chatbots. It encouraged users to have text-based dialogues while simulating a Rogerian psychotherapist. ELIZA identified keywords in user input and responded with preprogrammed templates that reflected the input back as questions or statements. Though it appeared to understand to some extent, it merely followed predetermined scripts.

- *PARRY (1972):*

PARRY, developed by Kenneth Colby, simulated a paranoid patient and interacted with users through text-based conversations. It employed a set of rules and heuristics to mimic the behavior of a person with paranoid schizophrenia. The guarded and suspicious tone of PARRY's comments was intentional, making the exchange more difficult and interesting.

- *SHRDLU (1970s):*

SHRDLU, developed by Terry Winograd, operated in a simulated block world environment. Users could communicate with SHRDLU using natural language to manipulate blocks within the simulated world. Although SHRDLU was limited to this specific task, it demonstrated the potential for using rule-based systems in interactive environments.

- *RACTER (1984):*

Racter, created by William Chamberlain and Thomas Etter, generated creative and surreal text using a combination of predefined grammatical rules and random structures. The output was often nonsensical, but it occasionally produced intriguing and poetic language.

- *ALICE and AIML (Late 1990s):*

ALICE (Artificial Linguistic Internet Computer Entity) was a chatbot developed by Richard Wallace, powered by AIML (Artificial Intelligence Markup Language). It used a large rule set to enable more contextually aware interactions with users. ALICE's responses became more sophisticated than its predecessors, but it still relied heavily on predefined rules.

2. RISE of AIML (1995)

The "RISE of AIML" refers to the emergence and popularity of AIML in the late 1990s and early 2000s. Early chatbot technology benefited greatly from the use of AIML, especially when developing rule-based systems with enhanced functionality.:

- *AIML Development:*

AIML was created by Richard Wallace in 1995. It is an XML-based markup language specifically designed for programming chatbots and NLP. The main purpose of AIML was to provide a standard format for defining patterns and responses in rule-based chatbot systems.

- *ALICE—A Notable Implementation:*

AIML powered ALICE, a chatbot created by Richard Wallace in the mid-1990s. Comparing user input to predefined patterns in AIML lessons, it was created to hold contextually relevant conversations with users and provide appropriate responses.

- *Pattern-Response Paradigm:*

AIML is based on a simple pattern-response paradigm. The patterns are the input text or questions provided by the user, and the responses are the appropriate replies generated by the chatbot. AIML files consist of rules that map input patterns to desired output responses.

- *Expansion and Adoption:*

AIML's straightforward syntax and ease of use made it accessible to developers, even those without advanced programming skills. As a result, many enthusiasts and developers started creating their own chatbots using AIML. AIML-based chatbots quickly gained popularity, leading to a rise in the number of chatbot implementations during this period.

- *Transition to Machine Learning:*

As technology advanced, machine learning and NLP techniques started to take precedence over rule-based systems. While AIML-based chatbots remained popular for some time, more sophisticated chatbots that incorporated machine learning algorithms and neural networks eventually surpassed them in terms of NLU and response generation.

3. Chatterbots and Smarter Systems (Late 1990s–Early 2000s)

Chatterbots and Smarter Systems represent the next stage in the evolution of chatbot technology, following the early rule-based systems. During the late 1990s and the early 2000s, developers began exploring more sophisticated approaches to create chatbots that could generate responses based on learning from past interactions rather than just relying on predefined rules. Here is a brief explanation of

- *Chatterbots:*

Chatterbots are chatbots that use machine learning and statistical techniques to generate responses to user inputs. Unlike early rule-based chatbots, which used fixed patterns to match user input, chatterbots attempt to identify patterns and associations in large datasets of conversation history to predict appropriate responses.

- *Cleverbot (1997) and Jabberwacky (1988):*

Cleverbot and Jabberwacky were early examples of chatterbots that gained popularity during this period. Cleverbot, developed by Rollo Carpenter in 1997, learned from conversations with users and other chatterbots, while Jabberwacky, created by British programmer Rollo Carpenter, used a similar approach to learn from its interactions with users.

- *Learning from Data:*

Chatterbots used statistical algorithms to learn from vast amounts of conversation data, which allowed them to generate responses that seemed more contextually relevant and natural. However, their understanding was still limited compared to modern AI language models.

- *Limitations:*

Chatterbots were not truly "intelligent" in the way we understand AI today. They relied on statistical patterns and lacked deep contextual understanding, often leading to responses that seemed sensible but could be nonsensical or incorrect.

- *Smarter Systems:*

Over time, researchers and developers improved chatterbot technology by incorporating more sophisticated machine learning algorithms, such as support vector machines (SVMs) and neural networks. Smarter chatbot systems were created as a result of these developments, able to handle a wider range of conversational subjects and give more precise responses. Neural network algorithms have found applications in various domains, including building smarter systems. The choice of a neural network architecture depends on the specific problem and requirements [3]. Here are some neural network algorithms that can be effective in the development of smarter systems for various applications such as convolutional neural network (CNNs), recurrent neural networks (RNNs), long short-term memory (LSTM) networks, gated recurrent unit (GRU) networks.

- *Transition to AI Language Models:*

While chatterbots were a step forward, the real breakthrough in chatbot technology came with the development of AI language models like GPT (generative pretrained transformer) and BERT (bidirectional encoder representations from transformers). These models, based on deep learning, demonstrated remarkable language understanding and generation abilities, enabling more human-like and contextually aware conversations.

4. Modern NLP and AI Advancements (Mid-2000s–Early 2010s):

Modern NLP and AI advancements have propelled the field of chatbot technology to new heights. These advancements, which took place from the mid-2000s to the early 2010s, revolutionized the way chatbots understand and respond to human language. Here is a brief explanation of modern NLP and AI advancements:

- *Statistical NLP:*

Modern NLP techniques moved away from rule-based systems and embraced statistical approaches. Instead of relying on handcrafted rules, statistical NLP utilized large datasets to identify patterns and probabilities in language usage. This led to significant improvements in language understanding and the ability to handle more complex language structures.

- *Machine Learning Algorithms:*

Machine learning algorithms, such as SVMs and conditional random fields (CRFs), played a crucial role in modern NLP. These algorithms were applied to tasks including named entity recognition (NER), sentiment analysis, and part-of-speech tagging, which helped chatbots better comprehend the context and intent of human input.

- *Sentiment Analysis:*

Sentiment analysis became a powerful tool in modern NLP, enabling chatbots to discern the emotional tone and sentiment of user messages. This allowed chatbots to respond more empathetically and appropriately to users' emotions, leading to improved user experiences. The emotion of user can be analyzed in NLP through text input, preprocessing, machine learning models, emotion detection, sentiment score, response generation, and feedback loop.

- *Semantic Understanding:*

Modern NLP advancements focused on semantic understanding, moving beyond just matching keywords to capturing the meaning and intent behind user queries. Techniques like word embeddings and deep learning-based approaches allowed chatbots to understand the relationships between words and concepts, improving their ability to comprehend natural language.

- *Introduction of Deep Learning:*

The introduction of deep learning, particularly neural networks, marked a significant breakthrough in NLP. Deep learning models like RNNs and transformers brought about a paradigm shift, enabling chatbots to learn from vast amounts of data and generate more coherent and contextually relevant responses.

- *IBM's Watson and Jeopardy:*

IBM's Watson, introduced in 2011, demonstrated the power of AI in language understanding by competing on the quiz show "Jeopardy" against human champions and winning. Watson's success showcased the potential of AI for complex language tasks and brought NLP and chatbot technology into the public eye.

- *Rise of Virtual Assistants:*

Modern NLP advancements paved the way for the rise of virtual assistants like Apple's Siri, Google Assistant, and Amazon's Alexa. These virtual assistants incorporated NLP and AI to provide users with personalized and interactive experiences, handling tasks ranging from voice commands to setting reminders and answering questions.

5. Siri and Personal Assistants (2011):

Siri and Personal Assistants represent a pivotal moment in the evolution of chatbot technology, bringing conversational AI to the mainstream. Siri, introduced by Apple in 2011, was one of the first widely adopted personal assistant chatbots, showcasing the potential of AI-powered virtual assistants. Here is a brief explanation of Siri and Personal Assistants:

- *Siri's Introduction:*

Siri is a virtual assistant that uses NLP and speech recognition to interpret voice commands and provide responses to users. Apple acquired Siri Inc. in 2010 and integrated the technology into its iOS devices, officially launching Siri as a key feature with the release of the iPhone 4S in 2011.

- *Voice-Activated Interaction:*

Siri's unique selling point was its ability to interact with users through voice commands, making it more user-friendly and accessible. Instead of typing queries, users could simply speak to Siri, and the AI-powered assistant would understand and respond appropriately.

- *Natural Language Processing:*

Siri's success was attributed to its advanced NLP capabilities. It could understand context, handle conversational interactions, and answer a wide range of queries, from simple informational questions to setting reminders, sending messages, and even making restaurant reservations.

- *Rise of Personal Assistants:*

Following Siri's introduction, other tech giants, such as Google and Amazon, quickly developed their own virtual assistants, including Google Assistant and Amazon Alexa. These personal assistants expanded beyond smartphones and found their way into smart speakers, smart TVs, and other smart devices.

- *Integration with Ecosystems:*

Personal assistants like Siri were tightly integrated into their respective ecosystems. Siri leveraged Apple's ecosystem to access apps, calendar events, contacts, and other system services. Similarly, Google Assistant and Alexa were integrated into their respective ecosystems, utilizing Google's search capabilities and Amazon's vast e-commerce services.

- *Continual Improvement:*

Siri and other personal assistants have undergone continuous improvement over the years. These virtual assistants are now more precise, contextually aware, and responsive because of developments in NLP, machine learning, and cloud computing.

- *Expanding Functionality:*

With each iteration, personal assistants have expanded their functionality to perform a wide range of tasks, from controlling smart home devices to providing weather updates, news, and navigation assistance. They have become an integral part of the daily lives of millions of users worldwide.

6. Growth of Messaging Apps and Chatbots (Mid-2010s):

- *ChatGPT:*

The growth of messaging apps and chatbots has been a significant trend in recent years, transforming the way businesses and individuals interact and communicate. This growth is fueled by advancements in technology, changes in user behavior, and the increasing demand for more personalized and convenient experiences. Here is a brief explanation of the growth of messaging apps and chatbots:

- *Popularity of Messaging Apps:*

Messaging apps have become immensely popular due to their ease of use, real-time communication capabilities, and the ability to connect with friends, family, and colleagues across the globe. Apps like WhatsApp, Facebook Messenger, WeChat, and Telegram have amassed billions of users worldwide.

- *Shift to Conversational Interfaces:*

As messaging apps gained traction, users started expecting businesses and services to be accessible through these platforms. This shift led to the rise of conversational interfaces, where businesses began offering customer support, information, and services through chat-based interactions.

- *Rising Demand for Convenience:*

Consumers increasingly prefer faster and more convenient ways to interact with businesses. Messaging apps and chatbots provide a seamless and efficient means of communication, allowing users to get answers to their queries, make purchases, and access services without leaving the messaging app.

- *Chatbots for Customer Support:*

Businesses embraced chatbots to handle customer support inquiries effectively. Chatbots, powered by AI and NLP, can handle a large volume of queries simultaneously, provide instant responses, and offer 24/7 support, enhancing customer satisfaction.

- *E-commerce and Transactions:*

Chatbots have been integrated into e-commerce platforms, enabling users to browse products, place orders, and make payments within the messaging app. This streamlined process has led to increased sales and improved user experiences.

- *Personalization and Contextual Understanding:*

AI-powered chatbots leverage user data and contextual information to provide personalized recommendations and responses. They can remember past interactions, understand user preferences, and tailor their responses accordingly.

- *Expansion Across Industries:*

Chatbots have expanded beyond customer support and e-commerce. They are now used in various industries, including healthcare, banking, travel, and education, providing valuable services such as appointment scheduling, financial advice, travel recommendations, and educational assistance [4].

- *Multilingual Support:*

Advanced chatbots can support multiple languages, enabling businesses to reach and engage with diverse global audiences.

- *Integration with Voice Assistants:*

Chatbots have also integrated with voice assistants like Siri, Google Assistant, and Alexa, allowing users to interact with businesses through voice commands.

7. Chatbot Platforms and APIs (Late 2010s):

In the late 2010s, there was a significant development in the chatbot ecosystem with the emergence of chatbot platforms and APIs. These platforms and APIs provided developers and businesses with powerful tools and services to build, deploy, and manage chatbots more efficiently and effectively. Here is a brief explanation of Chatbot Platforms and APIs in the late 2010s.

- *Chatbot Platforms:*

Chatbot platforms are comprehensive software frameworks that offer a suite of tools, resources, and services to create, train, and deploy chatbots. These platforms abstract much of the underlying complexity, making it easier for developers, regardless of their AI expertise, to build and integrate chatbots into their applications or services.

- *Developer-Friendly Tools:*

Chatbot platforms typically come with user-friendly graphical interfaces, drag-and-drop builders, and prebuilt components, allowing developers to design and configure chatbot interactions without extensive coding knowledge.

- *Natural Language Processing (NLP) Integration:*

Many chatbot platforms include built-in NLP capabilities, enabling chatbots to understand and interpret user input more accurately. These NLP engines help extract intents, entities, and context from user messages, facilitating more natural and context-aware conversations.

- *Dialog Management:*

Advanced chatbot platforms offer dialog management features that allow developers to create dynamic and context-driven conversations. To create more engaging encounters, they can design conversation flows, deal with user interruptions, and manage numerous turns of discourse.

- *Multichannel Support:*

Chatbot platforms often provide support for various messaging channels, such as websites, mobile apps, social media platforms, and messaging apps. This versatility allows businesses to deploy their chatbots on multiple channels and reach a wider audience.

- *Analytics and Insights:*

To improve performance and optimize user interactions, chatbot platforms usually offer analytics and reporting features. Developers can track user engagement, identify bottlenecks, and gather insights into user behavior to fine-tune their chatbots [5].

- *Cloud-Based Deployment:*

Chatbot platforms often support cloud-based deployment, enabling easy scalability and maintenance. The cloud infrastructure ensures that chatbots can handle varying loads of user interactions and remain available 24/7.

- *Integration with APIs:*

Many chatbot platforms allow integration with third-party APIs, enabling chatbots to access external services and data, expanding their capabilities and potential use cases.

- *AI Language Model Integration:*

Some advanced chatbot platforms can integrate with large-scale AI language models like OpenAI's GPT, giving developers access to state-of-the-art language processing capabilities.

8. AI Language Models (Recent Years):

AI language models have been at the forefront of recent advancements in NLP and AI. These models are made to comprehend and produce text that resembles human speech, opening up a variety of applications and use cases. Here is a brief explanation of AI language models in recent years:

- *Introduction of AI Language Models:*

In recent years, researchers and engineers have made significant strides in developing large-scale AI language models. These models are based on deep learning techniques, particularly transformer architectures, which allow them to process and understand vast amounts of text data.

- *Pretrained and Fine-Tuned Models:*

AI language models are typically pretrained on massive datasets to learn the statistical patterns and relationships in language. These pretrained models are then fine-tuned on specific tasks to make them more contextually relevant and useful for various applications.

- *Contextual Word Embeddings:*

AI language models create contextual word embeddings, which represent words in a way that captures their meaning and context within a given sentence or text. This contextual understanding enables the model to generate more coherent and contextually relevant responses.

- *GPT Series by OpenAI*

One of the most prominent AI language model series is the GPT by OpenAI. GPT-2, GPT-3, and subsequent versions have shown remarkable capabilities in various language tasks, such as text completion, translation, question-answering, and even creative writing.

- *Multimodal Capabilities:*

Some advanced AI language models can process not only text but also other media, such as images and videos. These multimodal models can perform tasks that involve both language understanding and visual perception, opening up new possibilities for AI applications.

- *Improved Natural Language Understanding:*

AI language models have greatly improved NLU. They can grasp the nuances of language, handle complex sentence structures, and comprehend context, leading to more accurate and context-aware responses.

- *Transfer Learning and Few-Shot Learning:*

AI language models excel at transfer learning, meaning they can apply knowledge learned from one domain to another [6]. Additionally, some models support few-shot learning, where they can perform tasks with minimal training examples, making them more adaptable and efficient.

- *Real-World Applications:*

AI language models have been integrated into various applications, including virtual assistants, chatbots, language translation services, content generation, sentiment analysis, and more. Their versatility and adaptability make them valuable tools across industries. Some examples for contextual word embeddings that are used in real-world applications are search engines, chatbot and virtual assistants, sentimental analysis, machine translation system, NER, and information retrieval [7].

- *Ethical Considerations:*

With the growing capabilities of AI language models, ethical considerations around potential misuse and misinformation have come to the forefront. Researchers and developers strive to build responsible AI systems and mitigate potential risks associated with biased or harmful outputs.

9. Multi-Modal Chatbots (Ongoing):

Multi-modal chatbots are the next frontier in chatbot technology, combining the understanding and generation of not only text but also other forms of media, such as images, videos, and audio. These chatbots leverage advanced AI models that can process and interpret multiple modalities, providing more immersive and interactive user experiences. Here is a brief explanation of multi-modal chatbots and their ongoing development:

- *Integration of Multiple Modalities:*

Traditional chatbots primarily work with text-based inputs and responses. Multi-modal chatbots, on the other hand, go beyond text and can understand and generate content from various sources, such as images, videos, audio clips, and other visual or auditory data.

- *Improved Contextual Understanding:*

By incorporating multiple modalities, these chatbots gain a deeper understanding of user inputs. For example, when presented with an image or video, the chatbot can use visual data to provide more contextually relevant and accurate responses [8].

- *Enhanced Interactivity:*

Multi-modal chatbots enable richer and more interactive conversations. Users can provide input in different formats, such as describing an image or asking questions related to a video, and the chatbot can respond accordingly.

- *Real-World Applications:*

Multi-modal chatbots have promising applications in various domains. For instance, in e-commerce, a user can share an image of a product they are interested in, and the chatbot can retrieve information, provide reviews, and even facilitate purchases.

- *Visual Question Answering (VQA):*

One of the key research areas in multi-modal chatbots is visual question answering (VQA). This involves combining visual content (images or videos) with natural language questions to produce accurate and informative responses.

- *Cross-Modal Understanding:*

Developing multi-modal chatbots involves building models that can understand and align information from different modalities. For example, a chatbot must be able to relate a description of an image to text-based responses.

- *Data Challenges:*

Training multi-modal chatbots requires large and diverse datasets that contain both textual and multimedia information. Acquiring and annotating such datasets is a complex task and remains an ongoing challenge for researchers.

- *Ongoing Research and Advancements:*

Multi-modal chatbots are an active area of research, and ongoing advancements in AI and machine learning are driving their development. Researchers continue to improve the performance and capabilities of these chatbots, making them more versatile and robust.

- *Ethical Considerations:*

With the integration of multiple modalities, new ethical considerations arise, such as privacy concerns related to image and video processing. Ensuring the responsible use of multi-modal chatbots is crucial as they become more prevalent in various applications.

10.4 UNDERSTANDING CHATBOT ARCHITECTURE AND DESIGN

Different types of chatbot architectures: Rule-based, retrieval-based, and generative models:

1. **Rule-based:** One kind of chatbot design uses predetermined rules and patterns to operate: rule-based chatbots. These chatbots have particular instructions placed into them to react to human input using if-then statements or conditional logic. A rule-based chatbot analyzes user input when they communicate with it and compares it to a set of predefined rules to identify the best course of action.
    - ***Rule Definition:*** The chatbot developer defines a set of rules that cover different user inputs and potential responses. These rules are created based on the expected user queries and the chatbot's intended functionality.
    - ***User Interaction:*** When a user sends a message or query to the chatbot, the chatbot processes the input and tries to find a matching rule for that particular input.
    - ***Rule Matching:*** The chatbot checks the user input against its predefined rules. If a matching rule is found, the corresponding response associated with that rule is retrieved.
    - ***Response Generation:*** The chatbot generates a response based on the rule that matched the user input. The response can be a prewritten text or a dynamic message created using place-holders for specific information from the user input [9].
    - ***Sending the Response:*** The chatbot sends the generated response back to the user.

- ***Iterative Process:*** If the user provides additional information or asks follow-up questions, the chatbot goes through the same process again, attempting to match the new input with its rules and providing relevant responses accordingly.

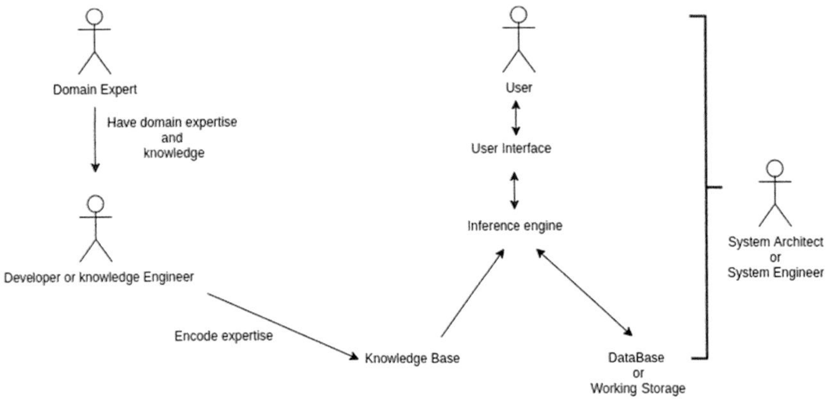

FIGURE 10.4 Rule-based architecture.

Note: A rule-based architecture, often referred to as a rule-based system or rule-based expert system, is a computational framework that relies on a set of explicit, predefined rules to make decisions or perform tasks.

Source: https://www.oreilly.com/library/view/python-natural-language/9781787121423/846638a9-a253-4924-9f8c-0bf99104d6b7.xhtml

2. **Retrieval-based:** One of the typical methods for creating chatbots is a retrieval-based architecture. Instead of creating entirely new responses using language generation models like GPT-3.5, the chatbot generates responses by obtaining pre-existing responses from a database or knowledge base. The retrieval-based chatbot typically consists of the following components:

    - ***Input Processing:*** The chatbot processes the user's input, which could be in the form of text or speech. It may involve tasks like tokenization, spell checking, and other preprocessing steps.
    - ***Matching and Ranking:*** After preprocessing the user's input, the chatbot compares it to the entries in the knowledge base or database. It uses techniques like TF-IDF (term frequency-inverse document frequency), BM25, or word embeddings to

measure the similarity between the user's input and the candidate responses [10].

- ***Knowledge Base:*** The knowledge base contains a collection of predefined responses to different user inputs. These responses can be curated by human experts or generated through other methods. The knowledge base can be as simple as a set of predefined rules or as complex as a database with a vast array of information.
- ***Response Selection:*** Once the candidate responses are ranked based on their similarity to the user's input, the chatbot selects the most appropriate response from the ranked list. The chosen response is then returned to the user as the chatbot's reply.

Advantages of Retrieval-Based Chatbots:

1. Better Control over Responses.
2. Reduced Risk of Generating Inappropriate Content.
3. Faster Response Times.

Disadvantages of Retrieval-Based Chatbots:

1. Limited Creativity.
2. Dependency on Knowledge Base.
3. Inflexibility.

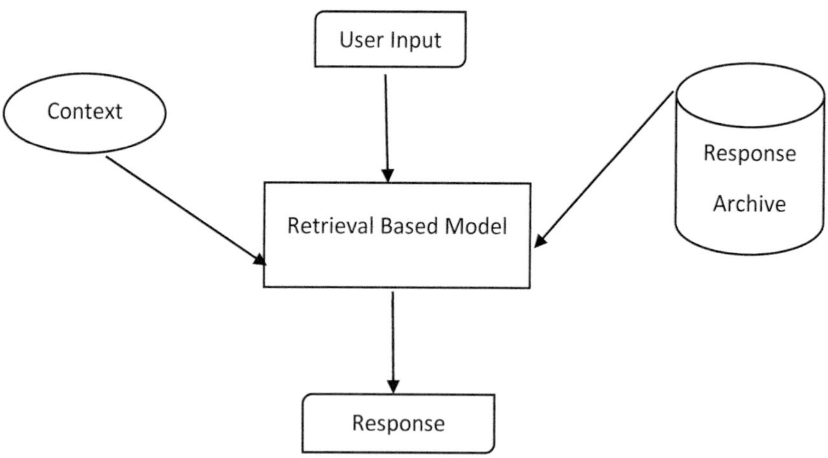

FIGURE 10.5 Retrieval-based architecture.

Generative models: Another widely used method for creating chatbots is generative modeling. Generative models can create responses entirely from scratch based on the input they receive, in contrast to retrieval-based chatbots that rely on pre-existing responses. GPT-3.5, the fundamental model that drives me as an AI language model, is one of the most well-known generative language models. Below is an overview of the generative model's architecture in chatbots:

- ***Transformer-based Architecture:*** Generative chatbots are often built using transformer-based architectures. The transformer architecture was introduced in the "Attention is All You Need" paper by Vaswani et al., and it has revolutionized NLP tasks.
- ***Encoder–Decoder Framework:*** The chatbot's generative model typically follows an encoder–decoder framework. In this architecture, the input text is encoded into a hidden representation by the encoder, and then the decoder generates the response based on that hidden representation.
- ***Self-Attention Mechanism:*** The transformer's key component is the self-attention mechanism, which allows the model to weigh the importance of different words in the input sequence when making predictions. This mechanism enables the model to capture long-range dependencies and contextual information effectively.
- ***Pretraining and Fine-tuning:*** Generative chatbots like GPT-3.5 are pretrained on a large corpus of text from the internet. During pretraining, the model learns to predict the next word in a sentence given the context of the previous words. This process helps the model acquire a vast amount of world knowledge and linguistic patterns [11].
- ***Transfer Learning:*** After pretraining, the model is fine-tuned on specific tasks or domains to adapt it to the chatbot's target use case. Fine-tuning allows the model to learn task-specific responses and behavior based on labeled examples provided during the training process.
- ***Sampling Techniques:*** When generating responses, chatbots can use various sampling techniques, such as greedy sampling, beam search, or nucleus sampling (also known as top-p sampling). These techniques control the diversity and randomness of the generated responses [12].

14. **Testing and Iteration:** Regularly test the chatbot with real users, collect feedback, and iterate on the design to address any shortcomings.
15. **Scalability and Performance:** Design the chatbot with scalability in mind, considering potential increases in user load and performance requirements.

10.6 CONVERSATIONAL FLOW AND USER EXPERIENCE CONSIDERATIONS

Building upon the previous response, let us delve further into conversational flow and user experience considerations:

1. **Fallback Plan for Unknown Queries:** Prepare the chatbot to gracefully handle queries that it cannot understand or provide a relevant response to. Offer alternative options, redirect users to relevant resources, or suggest rephrasing their question.
2. **Avoid Repetition:** Minimize repetition in responses to prevent users from feeling frustrated or like they are stuck in a loop. Acknowledge previous responses but try to offer new information or insights when needed [15].
3. **Confirmation and Verification:** When the chatbot is about to execute a critical action or provide sensitive information, seek confirmation from the user before proceeding. This enhances user trust and reduces the chances of errors.
4. **Progress Indicators:** If the chatbot is performing a complex task or retrieving information, use progress indicators to keep users informed about the ongoing process. It helps manage user expectations and avoids perceived delays.
5. **Dynamic Responses:** Create responses that adjust based on the user's input. For example, if a user mentions they are a beginner, tailor the chatbot's responses to be more informative and beginner-friendly.
6. **User Initiative Recognition:** Allow users to take the lead in conversations. Give them the option to explore further or ask questions on their terms. Avoid an overly restrictive or rigid conversational structure [16].

7. **Offering Help and Commands:** Make it clear to users what commands they can use to get assistance or access specific features. For example, providing a "help" command that lists available options can be beneficial.
8. **Minimizing Wait Time:** Optimize the chatbot's response time to be as fast as possible. If there is an unavoidable delay, use interactive messages to entertain or engage users during the wait.
9. **Exit and Restart Points:** Allow users to exit the chatbot gracefully or restart the conversation at any point. Provide a clear path for users to end the conversation when they have completed their tasks.
10. **Visual Cues and Formatting:** Use proper formatting, such as bold or italic text, to highlight important information or specific options. Visual cues can help users navigate the conversation more efficiently.
11. **A/B Testing:** Conduct A/B testing with different conversational flows to identify which version yields better user engagement and satisfaction. Use data-driven insights to refine the chatbot's design.
12. **Handling Incomplete Inputs:** Account for situations when the user's input is incomplete or ambiguous. Ask follow-up questions or provide suggestions to help users provide the required information.
13. **Learning from User Interactions:** Continuously learn from user interactions to improve the chatbot's responses, understand common user intents, and refine the conversational flow.
14. **Personas and Scenarios:** Create personas and scenarios to map out potential user journeys. Understanding different user personas and their objectives will help tailor the conversation more effectively.

10.7 BUILDING INTELLIGENT CHATBOT CONVERSATIONS

Designing chatbot conversations using dialogue management systems when designing chatbot discussions with dialogue management systems, users and the chatbot must communicate in an organized and dynamic way. In order to guarantee a seamless and natural conversation experience, dialogue management technologies are essential. The following procedures can be used to create efficient chatbot dialogues when using dialogue management systems:

1. **Define the chatbot's purpose and scope:** Outline the chatbot's objectives and the precise tasks it is capable of carrying out. To avoid confusion and establish user expectations, decide on the chatbot's domain and capabilities.
2. **Identify user intents:** Recognize the many intentions or goals people may have when communicating with the chatbot. Asking for information, booking a reservation, troubleshooting, or looking for advice are examples of common intentions.
3. **Design an intent-to-action mapping:** Define the activities that the chatbot should take in response to each user intent. These could entail giving a clear response, seeking clarification, presenting alternatives, or carrying out a particular duty.
4. **Create a conversation flow:** Establish the actions the chatbot should do in response to each user intent. These could require providing a succinct response, requesting further explanation, offering solutions, or doing a specific responsibility.
5. **Use Natural Language Understanding (NLU):** Implement NLU capabilities to accurately interpret user messages and extract important entities or parameters required to fulfill user requests.
6. **Implement context and memory:** Give the chatbot the ability to keep the dialogue in context. To avoid asking the same questions repeatedly and to deliver meaningful answers, keep in mind earlier exchanges.
7. **Personality and tone:** Choose the right persona and timbre for the chatbot. This will make for a more interesting and relatable dialogue.
8. **Test and iterate:** To find any problems or potential improvements, extensively test the chatbot. Obtain user feedback and make any necessary adjustments to the dialogue management and conversation flow [17].
9. **Monitor and update:** Keep an eye on how the chatbot is performing and how users are interacting. Gather information to pinpoint typical user problems or bottlenecks and release updates often to enhance the chat experience.
10. **Feedback Mechanism:** The agent asks the user for input to determine whether the discussion is going well and whether the user is happy with the bot's response. This encourages the bot to learn from its errors and improve in subsequent discussions.

11. **Policy Learning:** Policy learning is a higher-level framework that teaches the bot positive language to improve overall end-user satisfaction. Broadly, it creates the network of "happy paths" and routes the conversation to end-user satisfaction. The bot then tries to learn from the interaction and follow the conversation flow it had with similar users in the past. This makes use of reinforcement learning technique.

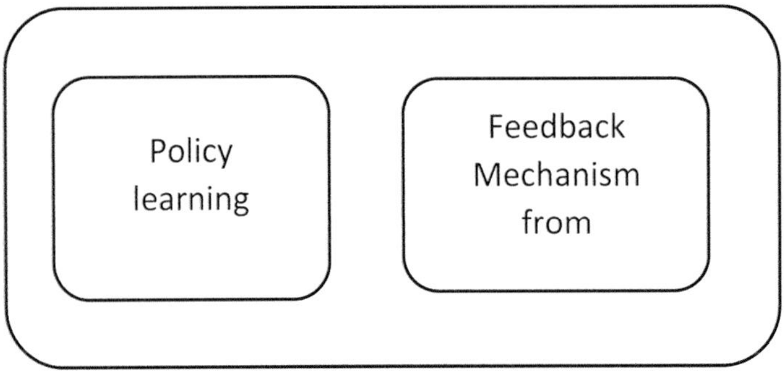

FIGURE 10.7 Choosing the best action to perform based on conversation.

10.8 DIALOG MANAGER CONTROL FLOW

The dialogue must control the following general procedure in order to provide a proper response:

1. Gather user responses from all the response models.
2. Check if any priorities are defined in the list of user responses and pick the response that holds the highest priority.
3. Some dialogue managers choose the responses based on the model selection policy when there is no priority specified. The model selection policy in this case may score each user response and select the one with the highest score.

10.9 IMPLEMENTING CONTEXT-AWARE CHATBOTS FOR PERSONALIZED INTERACTIONS

Context-aware chatbots have the capacity to recall and comprehend prior encounters with users, resulting in more distinctive and interesting dialogues.

Here is a step-by-step tutorial for creating personalized interactions with context-aware chatbots:

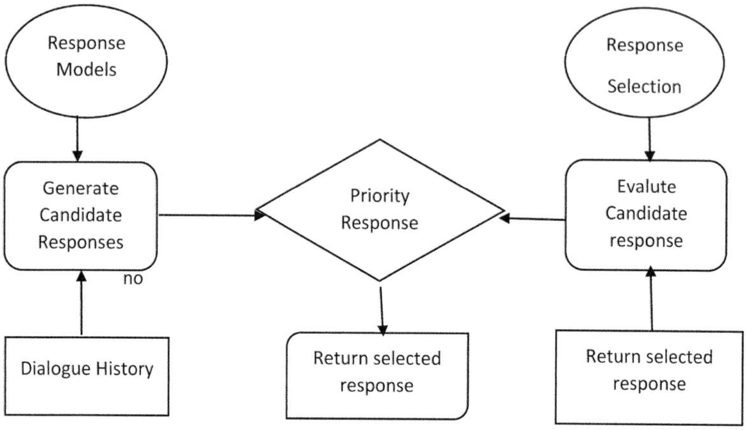

FIGURE 10.8 Dialogue manager control flow of a chatbot.

1. **Data Collection and User Profiling:** Gather relevant data about users, such as their preferences, past interactions, and any other contextual information that can help personalize the conversation. Create user profiles or maintain a user context database to store and update this information.
2. **Context Management:** Implement a context management system to keep track of ongoing conversations with each user. Maintain conversation history and session context, including user queries, bot responses, and any relevant variables or parameters. Use this context to understand the current state of the conversation and respond accordingly [18].
3. **Natural Language Understanding (NLU):** Utilize NLU techniques to extract entities and intents from user inputs accurately. Identify key elements in the conversation that are relevant to maintain context, such as dates, locations, preferences, or user-specific data.
4. **Contextual Responses:** Train the chatbot to respond contextually based on user history and preferences. Use the context and user profiles to tailor responses and recommendations to each individual user.

5. **Personalized Recommendations:** Leverage user profiles and historical data to provide personalized recommendations or suggestions. For example, if it is an e-commerce chatbot, recommend products based on the user's past purchases or browsing behavior.
6. **Adaptability and Learning:** Implement a learning mechanism that allows the chatbot to adapt its responses based on user feedback and new information. Continuously update and improve the user profiles and context based on user interactions.
7. **Seamless Handover:** If the chatbot cannot handle a specific query, ensure a seamless handover to a human agent while providing them with the relevant conversation history. Maintain context during the handover process to avoid repetition when the human agent takes over.
8. **Multi-turn Interactions:** Design the chatbot to handle multi-turn interactions where users may have complex requests that span several conversational turns. Use context to maintain continuity and avoid losing track of the conversation's flow.
9. **Privacy and Security:** Ensure that user data is handled securely and comply with data privacy regulations. Allow users to control their data and provide options for data deletion if desired.
10. **Testing and User Feedback:** Thoroughly test the context-aware chatbot to ensure that it performs accurately and delivers personalized responses. Gather feedback from users to identify areas for improvement and refine the chatbot's context management.

10.10 CONCLUSION

The significance of AI in today's world. It might highlight how AI has revolutionized various industries, enhanced efficiency, and enabled the development of innovative technologies. For companies and organizations looking to improve consumer interaction and expedite numerous processes, exploring the usage of AI-powered apps in chatbots presents an exciting and dynamic field. Chatbots that are driven by AI have the potential to greatly enhance consumer interactions. They can offer round-the-clock assistance, individualized interactions, and quick responses, all of which increase client happiness and loyalty. It can be incorporated into a variety of commercial operations, including customer service, marketing, and sales. They are valuable resources for the entire organization since they can help with lead generation, order processing, and problem solving. Future chatbots are probably going to be able to communicate with users via text, speech, and visual interfaces. The evolution of chatbots is from their early stages to their current sophisticated form. This would involve mentioning the advancements in NLP and machine learning that have contributed to the improved capabilities of chatbots. It could summarize some of the diverse applications of AI and chatbots across different domains. These applications may include customer support, virtual assistants, healthcare, finance, e-commerce, and more. There are many benefits of using AI and chatbots, such as enhanced customer experience, cost savings, and increased productivity. It might also acknowledge the challenges in the field, like ethical concerns, data privacy, and potential biases in AI algorithms. This will make user interactions more adaptable and accessible. Strong privacy and security measures will be essential as chatbots are increasingly used to handle sensitive data. Regulations governing data protection must be followed. The goal of chatbots in the future is to improve human talents, not to completely replace them. Chatbots and human agents working together can improve customer service and problem-solving efficiency.

KEYWORDS

- **artificial intelligence**
- **chatbots**
- **dialog management**
- **neural networks**
- **pretrained**

REFERENCES

1. Abu-Alsaad, H. A. (2019). Agent applications in e-learning systems and current development and challenges of adaptive e-learning systems. In *2019 11th International Conference on Electronics, Computers and Artificial Intelligence (ECAI)* (pp. 1–6). IEEE.
2. de Barcelos Silva, A., Gomes, M. M., da Costa, C. A., da Rosa Righi, R., Barbosa, J., Pessin, G., De Doncker, G., Federizzi, G. (2020). Intelligent personal assistants: A systematic literature review. *Expert Systems with Applications, 147*, 113193.

3. Benotti, L., Martnez, M. C., Schapachnik, F. (2017). A tool for introducing computer science with automatic formative assessment. *IEEE Transactions on Learning Technologies, 11*(2), 179–192.
4. Brandtzaeg, P. B., & Følstad, A. (2017). Why people use chatbots. In *Inter-national Conference on Internet Science*. Springer.
5. Ciechanowski, L., Przegalinska, A., Magnuski, M., & Gloor, P. (2019). In the shades of the uncanny valley: An experimental study of human–chatbot interaction. *Future Generation Computer Systems, 9*, 539–548.
6. Clarizia, F., Colace, F., Lombardi, M., Pascale, F., Santaniello, D. (2018). Chatbot: An education support system for student. In *International Symposium on Cyberspace Safety and Security*. Springer.
7. Dsouza, R., Sahu, S., Patil, R., & Kalbande, D.R. (2019). Chat with bots intelligently: A critical review & analysis. In *International Conference on Advances in Computing, Communication and Control (ICAC3)* (pp 1–6). IEEE.
8. Hien, H. T., Cuong, P.-N., Nam, L. N. H., Nhung, H. L. T. K., & Thang, L.D. (2018). Intelligent assistants in higher-education environments: The fit-ebot, a chatbot for administrative and learning support. In *Proceedings of the Ninth International Symposium on Information and Communication Tech- nology.*
9. Ho, C. C., Lee, H. L., Lo, W. K., Lui, K. F. A. (2018). Developing a chatbot for college student programme advisement. In *International Symposium on Educational Technology (ISET)* (pp. 52–56). IEEE.
10. Ismail, M., Ade-Ibijola, A. (2019). Lecturer's apprentice: A chatbot for assisting novice programmers. In *International Multidisciplinary Information Technology and Engineering Conference (IMITEC)* (pp. 1–8). IEEE.
11. Kumar, M. N., Chandar, P. L., Prasad, A. V., & Sumangali, K. (2016). An- droid based educational chatbot for visually impaired people. In *IEEE International Conference on Computational Intelligence and Computing Re- search (ICCIC)* (pp. 1–4). IEEE.
12. Lee, K., Jo, J., Kim, J., & Kang, Y. (2019). Can chatbots help reduce the workload of administrative officers? -implementing and deploying faq chatbot service in a university. In *International Conference on Human-Computer Interaction*. Springer.
13. Lin, M. P.-C., & Chang, D. (2020). Enhancing post-secondary writers' writing skills with a chatbot. *Journal of Educational Technology & Society, 23*(1), 78–92.
14. Lin, Y.-H., & Tsai, T. (2019). A conversational assistant on mobile devices for primitive learners of computer programming. In *IEEE International Conference on Engineering, Technology and Education (TALE)* (pp. 1–4). IEEE.
15. Mabunda, K., Ade-Ibijola, A. (2019). Pathbot: An intelligent chatbot for guiding visitors and locating venues. In *6th International Conference on Soft Computing & Machine Intelligence (ISCMI)* (pp. 160–168). IEEE.
16. Mckie, I. A. S., & Narayan, B. (2019). Enhancing the academic library experience with chatbots: An exploration of research and implications for practice. *Journal of the Australian Library and Information Association, 68*(3), 268–277.
17. Molnar, G., & Szuts, Z. (2018). The role of chatbots in formal education. In *IEEE 16th International Symposium on Intelligent Systems and Infor Matics (SISY)* (pp. 000197–000202). IEEE.

18. Murtarelli, G., Gregory, A., & Romenti, S. (2021). A conversation-based perspective for shaping ethical human–machine interactions: The challenge of chatbots. *Journal of Business Research, 129*, 927–935.
19. Nguyen, H. D., Pham, V. T., Tran, D. A., & Le, T. T. (2019). Intelligent tutoring chatbot for solving mathematical problems in high school. In *11th International Conference on Knowledge and Systems Engineering (KSE)* (pp. 1–6). IEEE.
20. Ondas, S., Pleva, M., & Hladek, D. (2019). How chatbots can be involved in the education process. In *17th International Conference on Emerging eLearning Technologies and Applications (ICETA)* (pp. 575–580). IEEE.

CHAPTER 11

AX-YOLOV8-Based Indian Traffic Sign Detection and Classification in Autonomous Vehicles

ANTONY VIJAY J.[1*], MYTHILY M.[2], and BEAULAH DAVID[3]

[1]*Department of Computer Science and Engineering, R. M. K. College of Engineering and Technology, Puduvoyal, Thiruvallur District, Chennai, Tamil Nadu, India*

[2]*Department of Computer Science and Engineering, Karunya Institute of Technology and Sciences, Coimbatore, Tamil Nadu, India*

[3]*Department of Information Technology, Hindustan College of Engineering and Technology, Coimbatore, Tamil Nadu, India*

ABSTRACT

The field of autonomous vehicles has been rapidly developing in recent years, and deep learning technology has played an increasingly important role in this development. With the growing number of vehicles on roads, traffic sign recognition technology has become increasingly important in mitigating the risk of accidents. It is essential for autonomous vehicles to understand the environment and obey traffic rules and regulations in order to prevent accidents, and traffic sign recognition plays a key role in this process. In this chapter, we discuss traffic sign detection works by analyzing real-time video data from a camera mounted on the vehicle to detect road signs such as stop sign, speed limit signs, no entry sign, cautionary traffic signs, and informative traffic signs. To detect the traffic signs effectively, every traffic sign is scanned and pre-processed to

Real-Time Artificial Intelligence (AI): Key Motivations, Technologies, Platforms, and Use Cases.
Pethuru Raj, B. Sundaravadivazhagan, V. Kavitha, B. Narendra Kumar Rao, & Hannah Vijaykumar (Eds.)

DOI: 10.1201/9781998511358-11

extract key points using deep learning-based AX-YOLOv8 Algorithm. It generates dynamic anchor boxes around the traffic signs and non-maximum suppression technique produces confidence scores for each box, based on that traffic sign prediction is performed. AX-YOLOv8 is used to recognize the various sizes of traffic sign images more effectively and this feature works perfectly for removing noises as well as helps in dimensionality reduction by removing the less important features from the image. To classify the traffic sign, the CNN model acts as the backbone network of AX-YOLOv8 to predict the 43 classes of traffic sign images. This model is analyzed and compared with various deep learning algorithms, in which AX-YOLOv8 achieves the maximum precision of 98.9 and 99.2% in detection and classification for TRID, Indian traffic sign datasets. This system alerts about the road traffic signs to the driver before the driver reaches new roads. This greatly reduces drivers' stress levels and helps them respond proactively to changes in the driving environment even when traveling on new roads.

11.1 INTRODUCTION

Automation technology helps to incorporate computerized platforms with sensors, artificial intelligence (AI), and data analysis capabilities. This collects data from various sources, and accordingly, it makes decisions on its own [1]. Nowadays, businesses in fields, such as automobile, production, minerals extraction, and energy are particularly adept at utilizing autonomous technologies in order to increase productivity, security, and sustainable development [2]. These technologies have become more prevalent as a result of various circumstances. The primary purpose of the technology has changed and it is now more than just a tool for accomplishing tasks more quickly, effectively, and efficiently [3]. Rather, it has interfered with and changed the whole thing we do. These technologies have also benefited from the growth of related technologies, such as enhanced sensing, networking, artificial intelligence, machine learning (ML), and the Internet of Things [4, 5].

Road safety in metropolitan areas faces issues, such as being drunken, cell phone usage which distracts while driving and the person visuals are susceptible to drowsiness whenever looking at traffic sign images of various sizes for an extended period in complicated scenes, and our cognitive ability is restricted. This frequently results in wrong decisions and significantly lowers effectiveness [6, 7]. To overcome these issues, traffic sign identification technologies in self-driving cars can help drivers identify traffic signs. In order to evaluate driving hazards and ensure safety, autonomous vehicles must be able to recognize the tiny, unclear, and faded traffic signs accurately [8, 9]. To improve the efficiency and dependability of vehicle systems, various researches were conducted to increase the detection accuracy of various traffic sign images [10]. Nowadays, traffic sign detection and classification in real time is becoming a challenging task. To achieve this, deep learning model requires a lot of training and testing for better classification. There are various neural networks are there for processing the dataset, such as EfficientNet, ResNet-50, Inception-v4 [11, 12].

The object recognition algorithms are divided into two categories: (a) one-step recognition and (b) two-step recognition process. The one-step object recognition algorithms will recognize and classify the object effectively in a single step which are You Only Look One (YOLO) and Single Shot Multi-Box Detector (SSD) [13, 14]. The two-step object detection algorithms perform object detection in the first stage and classification in the second stage. This category of algorithms requires two steps to recognize the traffic signs such as feature pyramid networks (FPN), Fast R-CNN, spatial pyramid pooling (SPP-Net), and so forth [15, 16]. There are a lot of algorithms that were released among those YOLO sounds good in all aspects such as Precision, Accuracy, mAPm, and F1-Score [17, 18]. Recently, YOLO version 8 has been released which is one of the best algorithms for recognizing and classifying traffic signs effectively [19, 20]. The architecture of YOLOv8 is represented in Figure 11.1.

This architecture is mainly focusing only on bigger size objects, it is difficult to recognize the smaller images, and the performance metrics of YOLOv8 really could recognize and classify all the sizes of traffic sign images [21].

The research covered in this chapter is arranged consequently: The detailed review of the literature survey is mentioned in Section 11.2. The traffic sign detection and recognition process is represented in Section 11.3. The discussion about the TRID and ITSD datasets is in Section 11.4. The various performance metrics calculation is represented in Section 11.5. The comparative results of various performance metrics based on neural networks are presented in Section 11.8. The conclusion of this chapter is discussed in Section 11.9, and finally, references are presented.

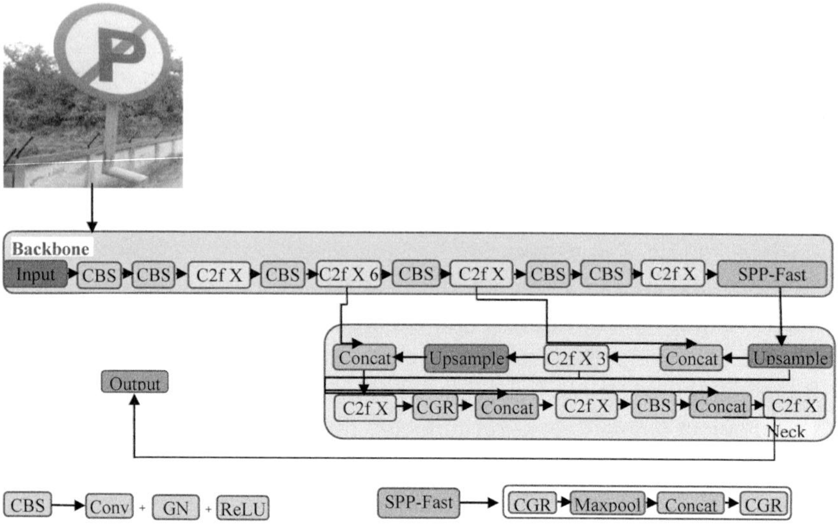

FIGURE 11.1 YOLOv8 network architecture.

11.2 RELATED WORK

11.2.1 TWO-STAGE OBJECT DETECTORS

Gudigar et al. [22] stated that there are various algorithms that are useful for detecting and classifying the traffic sign objects accurately. In this, R-CCN creates region proposal regions in the form of bounding boxes, which locate the features from the chosen bounding boxes. Finally, classification takes place with the help of support vector machine (SVM). The performance of this approach is a little bit lag in processing. So the response time of R-CNN is very slow.

Prabhakar et al. [23] proposed a spatial pyramid pooling algorithm which is more efficient in detecting and classifying the traffic sign images because it includes the spatial pyramid pooling layer prior to the fully connected layer, which handled all the images without resizing it. It creates and processes more than 2000 region proposals per image, so it reduces the performance of SPP-Net. Even though, it reduces the performance of SPP-Net which is three times faster than the RCCN algorithm.

Liu et al. [24] proposed Fast R-CNN modifies SPP to RoI pooling. This follows many patches from one image, to train efficiently, it requires only one forward and backward pass through the convolution layers. Zeng et al. [25] proposed a region proposal network (RPN) to select the exact bounding boxes in the image. Which required a lot of training to predict certain regions. The performance of Faster R-CNN is three times faster than Fast R-CNN.

Gudigar and Chokkadi [26] proposed the FPN model to supply the multiscale features to extract any size of images. This model uses a convolution neural network as hidden layers to handle the various sizes of images and recognize them effectively. Winkle [27] proposed the Mask R-CNN algorithm to solve the issue of instance segmentation. Which resolved the RoI pooling layer alignment issue and its performance is much better than Faster R-CNN.

Mammeri [28] presented Cascade R-CNN solved the issue that happened in fixing the correct range of Intersection over Union (IoU) in predicting the precision and recall of the model. This ensures the confidence score for all the bounding boxes, so the best RoI can be identified effectively.

11.2.2 SINGLE-STAGE OBJECT DETECTORS

In the single-stage object detector, the region proposal process is skipped and recognition of the object and classification has happened in the single stage. These techniques perform better than two-stage detectors. Tijerina et al. [29] represented an SSD approach that creates bounding boxes on different aspect ratios based on the size of feature maps. The neural network generates mAP% for each bounding box. So it predicts the various scales of images effectively. Färber [30] proposed YOLOv1 algorithm, which is a single-stage object detection algorithm. This algorithm applied the sliding window technique to analyze the bounding boxes in the image and predict the traffic sign effectively.

Zhang et al. [31] proposed YOLOv2 incorporates Darknet-19 as a backbone network, so it performs better than YOLOv1. Novak et al. [32] proposed YOLOv3 which includes the new neural network Darknet-53, which extracts the traffic signs more perfectly than YOLOv2.

Dewi et al. [33] proposed YOLOv4 which contains various backbone networks such as CSPResNext50, CSPDarknet53, and EfficientNet-B3. The copy of the features extracted from each layer is added to the Dense layer, so the feature maps will not get lost and show better performance than YOLOv4.

Wang et al. [34] proposed YOLOv5 which has four categories of processing the images, such as Small, Medium, Large, and Extra Large. Processing small-sized images becomes easy compared with Extra Large images. That is, based on the sizes of the images the performance can vary. YOLOv6 model is also known as MT-YOLOv6. Dang et al. [35] proposed YOLOv6 realizes higher mAP with YOLOv5 on various frames per second and its inference speed is higher than YOLOv5.

Zhang et al. [36] proposed YOLOv7 was trained to detect real-time object detection, with the MS COCO dataset for 80 generic classes. There are six variants of models in YOLOv7, according to the processing of various sizes of images the performance rate has been changed. When compared with YOLOv6, it is 1.2% better in performance.

Lou et al. [37] proposed YOLOv8 supported with anchor-free object recognition system, that is, it can identify the center position of the objects and draw bounding boxes around them in various aspect ratios, and based on threshold with IoU, it can go for classification. To train up this model, mosaic augmentation technique is used and this model achieves better results in the Mean Average Precision for five versions of YOLOv8 and this approach is far better than all the previous versions of YOLO. The Position of the Bounding box (POB) is calculated by the Equation 11.1.

$$POB_{(\sigma_i,\sigma_{i+1})} = -((x_{i+1} - x)\, log\, log\, (\sigma_i) + (x - x_i)\, log\, log\, (\sigma_{i+1})) \quad (11.1)$$

In the above equation, σ is the Outcome of the sigmoid function, where x_{i+1}, x_i represent the Neural Network range of parameters and x is the label.

An alignment for the bounding boxes is to be automatically calculated by the Equation 11.2.

$$T = C^{w1}\, x\, IoU^{w2} \quad (11.2)$$

where C is the Model Classification value, IoU is the Intersection over Union value, and $w1$ and $w2$ are the various weight metrics.

11.3 PROPOSED SYSTEM

The family of YOLO models provides better results in various performance metrics but this could not perform well for small or arbitrary-sized images. In order to solve this issue, we proposed AX-YOLOv8. This approach performs well in detecting the classifying the traffic sign in any size of image. To achieve this, few changes have been made to YOLOv8 model that are listed as follows:

1. The multiple description convolution neural network (MD-CNN) module is incorporated into the proposed module to collect the lost description of the features. This feature collects information during the downsampling of the max pooling process. It is represented in Figure 11.2.
2. Some of the C2f modules is replaced by Depth-wise Convolution (DWC) Module mainly used to combine various modules to collect the associated information about the overall architecture. The C2f Module is represented in Figure 11.3A and the DWC Module is presented in Figures 11.3B and 11.3C.
3. The adaptive feature fusion (AFF) process takes the various parameters to collect the initial and global features to compare the performance of the AX-YOLOv8 model. Finally, it chooses the best model out of various methods which is represented in Figure 11.4.

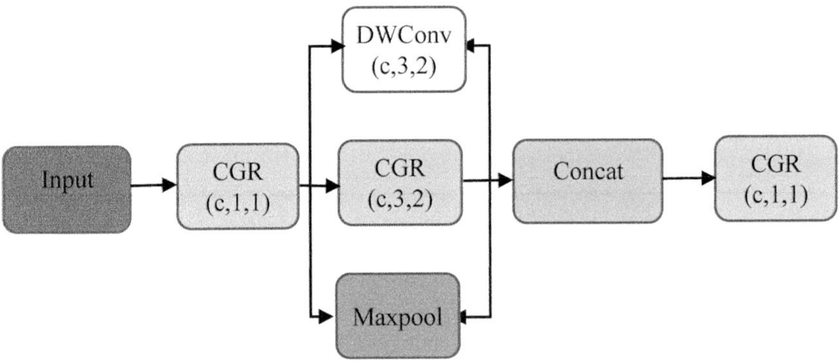

FIGURE 11.2 Downsampling process.

The downsampling process combines the following three approaches, such as group normalization (GN), ReLU, and convolution.

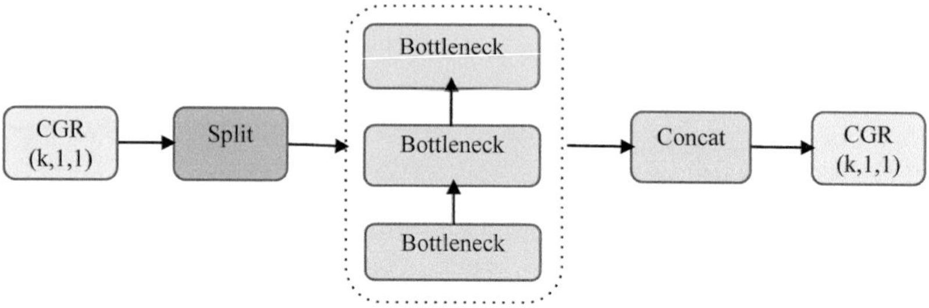

FIGURE 11.3a C2f.

Actually, the C2f module, that is, credit to RoboFlow is replaced by the DWC module. In C2f, all the outcomes from the bottlenecks are concatenated and applied to the dimensionality reduction process.

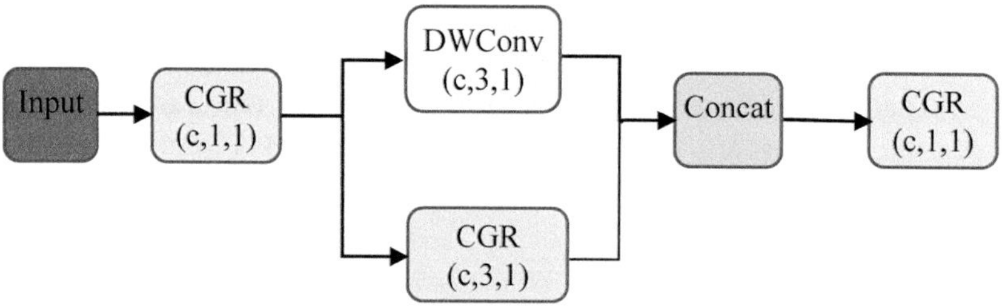

FIGURE 11.3b DWS.

Depth-wise convolution (DWC) module is mainly used to concatenate all the necessary associated information, which will help in the data analysis process.

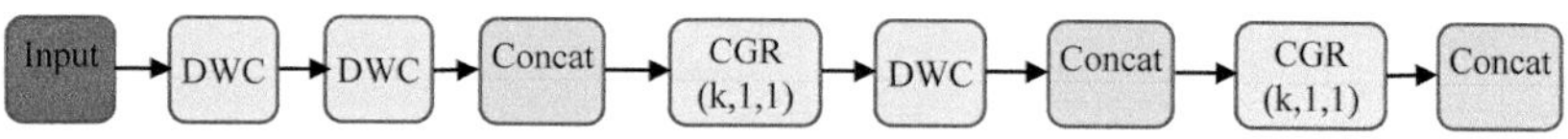

FIGURE 11.3c DWS structure.

The DWS structure makes use of DWS to concatenate all the contextual information about the layers which will be better managing spaces than dense layers.

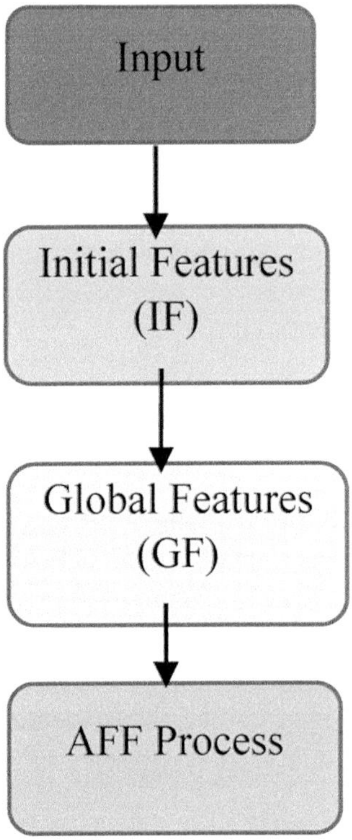

FIGURE 11.4 Adaptive feature fusion process.

After combining all those newly added features and modules to AX-YOLOv8, the system architecture of AX-YOLOv8 is represented in Figure 11.5.

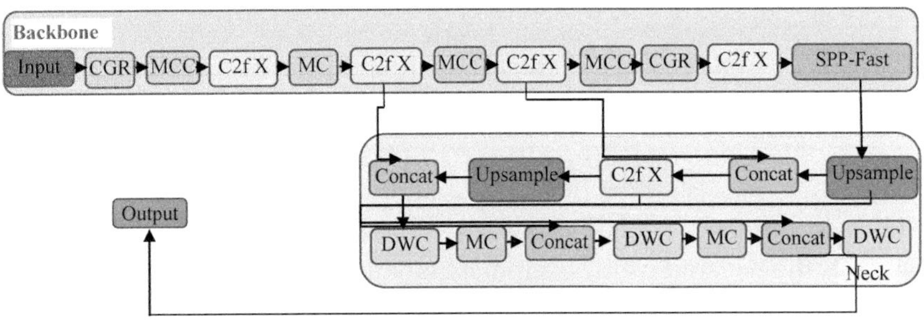

FIGURE 11.5 AX-YOLOv8 system architecture.

11.4 DATASETS

This chapter mainly concentrates on two major datasets such as TRID, Indian Traffic Sign Datasets (ITSD). In total, 1200 images were taken for the training and testing process from both datasets. This dataset is categorized into three major types, such as Mandatory Signs, cautionary signs, and informatory signs, which have the resolution ranges from 400 * 400 to 1920 * 1080. So the model can detect and analyze any size of traffic sign images. Samples of two datasets are represented in Figure 11.6.

11.5 PERFORMANCE METRICS

It is categorized into two metrics which are regression and classification metrics. Regression metrics usually deal with continuous values in the relationship between predicted and actual. Likewise, classification metrics only consider the discrete values. These two categorized metrics are applicable only for supervised deep learning models. From these metrics, traffic sign recognition errors can be easily identified.

11.6 REGRESSION METRICS

11.6.1 MEAN ABSOLUTE ERROR (MAE)

This metric finds the absolute error from the predicted and actual values. In this metric, a comparison of values between predicted and actual image frames is calculated for predicting the bounding boxes of traffic sign images. This can be measured by Equation 11.3.

$$MAE = \frac{1}{M}\sum_{j=1}^{M}\left|x_{j-}\hat{x}_{j}\right| \tag{11.3}$$

11.6.2 MEAN SQUARED DEVIATION (MSD)

This metric finds the relationship between actual and predicted values. It removes the negative values automatically using the squaring function. If the model has errored in prediction, then the outcome abruptly shows an error. This is another form of error spotting technique in predicting the bounding boxes of traffic sign images. This can be measured by Equation 11.4.

$$MSD = \frac{\sum (x_j - \hat{x}_j)^2}{M} \tag{11.4}$$

FIGURE 11.6 TRID and ITSD dataset samples.

11.6.3 ROOT MEAN SQUARED ERROR (RMSE)

This metric predicts the error using Euclidean distance, which is the square root of Mean Square Deviation. Boundary detection of traffic sign images can be easily monitored by the Euclidean distance formula. Which can be measured by Equation 11.5.

$$RMSE = \sqrt{(MSD)} \tag{11.5}$$

11.7 CLASSIFICATION METRICS

Classification metrics yields only discrete output in the form of True or False. Various metrics are available under classification to evaluate the performance of a model.

11.7.1 ACCURACY

Accuracy (A) is a metric categorized under classification, which accurately predicts the correct location of traffic signs, which can be implemented by predicting the true predictions from the total observations. Which can be measured by Equation 11.6.

$$A = \frac{True\,Predictions}{Total\,Observations} \tag{11.6}$$

where True Predictions = TN + TP, Total Observations = TP + TN + FP + FN.

Metrics of all the input images and ground truth are manually created in order to assess the program's results. Each image's results from the approach were contrasted with the actual data. The following parameters create the confusion matrixes and corresponding performance measures:

1. **TP:** Count of correctly predicting the traffic sign images in bounding boxes.
2. **TN:** Count of correctly predicting the negative cases of traffic sign images in bounding boxes.
3. **FP:** Count of wrongly predicting the traffic sign images in bounding boxes.

4. **FN:** Count of wrongly predicting the negative cases of traffic sign images in bounding boxes.

From the aforementioned factors, many performance targets are derived, including TP, FP, accuracy, precision, and recall.

11.7.2 INTERSECTION OVER UNION (IOU)

IoU is an evaluation metric used to measure the accuracy in predicting traffic sign object detection, which is measured by dividing the Area of Intersection of boxes and the Area of Union Boxes. This method is widely used in object detection. Which can be measured by Equation 11.7.

$$IoU = \frac{Area\,of\ Intersection\,of\ boxes}{Area\,of\ Union\ Boxes} \tag{11.7}$$

The Threshold (T) is set to be 0.5 and IoU = 1 is the maximum value.

If the value of IoU $\geq T$, then the prediction is correct else the prediction is incorrect, that is, traffic sign images are not predicted correctly.

11.7.3 POSITIVE PREDICTIVE VALUE (PPV)

Usually, Positive Predicted Value can be determined by the Error Matrix. PPV also known as precision, which is used to classify only the positive observations from the dataset. This metric helps to locate the road sign images effectively. Which can be measured by the Equation 11.8.

$$PPV = \frac{True\ Positive}{Predicted\ Positive} \tag{11.8}$$

Where True Positive = TP, Predicted Positive = TP + FP.

11.7.4 AVERAGE PRECISION (AP)

Average Precision (AP) is calculated from the average of deep learning model accuracy. Which can be evaluated by Equation 11.9.

$$AP = \sum_{T=0}^{T=n-1} [True\,Positive(T) - True\,Positive(T+1) * Precision(T)] \tag{11.9}$$

Here n is the Number of Categories of Traffic Signs and T is the threshold.

11.7.5 MEAN AVERAGE PRECISION (MAP)

Mean Average Precision is used to analyze the throughput and segmentation of the traffic sign model. This can be calculated by Equation 11.10.

$$mAP = \frac{1}{n}\sum_{i=1}^{n} AP_i \tag{11.10}$$

Performance evaluated based on Mean Average Precision (mAP) with various values of IoU ranges between 0.55 and 0.97. In 95 h of the training process, AX-YOLOv8 was trained with 15,000 traffic sign images with various ranges of IoU values.

Positive Predicted Value of the model with various sizes of mAP values in the form of bounding boxes, such as mAP (s), mAP (m), mAP (l), and mAP (xl). Here, s indicates Small, m indicates Medium, l denotes Large, and xl denotes Extra Large. These are classified based on IoU values ranging between 0.5 and 1.0, which is mentioned in Table 11.1.

TABLE 11.1 PPV Values of VGG NET-19 Based AX-YOLOv8 with Various IoU.

S. No.	Parameters Evaluation Metrics	mAP Score
1	Bounding Box PPV-mAP (s)	0.674412755
2	Bounding Box PPV-mAP (m)	0.674412755
3	Bounding Box PPV-mAP (l)	1.0
4	Bounding Box PPV-mAP (xl)	0.87124484

The comparison of mAP metrics with input size is represented in Table 11.2.

The mAP metrics with a number of iterations start with 520 to 15k. This table clearly reveals that larger-size traffic sign images produced the maximum IoU values ranging from 0.8915 to 1.0. The conclusion from the table is mAP (l) which predicts the region effectively at the size of 15k.

TABLE 11.2 mAP Metrics vs Input Size.

Metrics	520	2.5k	5k	7.5k	10k	12.5k	15k
mAP (s)	0.5015	0.5976	0.7256	0.7566	0.7587	0.7887	0.7987
mAP (m)	0.5015	0.5977	0.6792	0.7092	0.7556	0.7890	0.7987
mAP (l)	0.8915	0.8999	0.9204	0.9489	0.9845	0.9967	1.0
mAP (xl)	0.4957	0.5967	0.8233	0.8345	0.8768	0.9180	0.9456

11.7.6 PREVALENCE

Prevalence is one of the metrics of Error Matrix, which covers only the portion of positive outcomes by the total outcomes. This metric only takes the traffic sign portion from the image. This can be measured by Equation 11.11.

$$P = \frac{Actual\ Positive}{Total\ Observations} \tag{11.11}$$

where Actual Positive = FN + TP, Total Observations = TP + TN + FP + FN.

11.7.7 SENSITIVITY

Sensitivity is also called True Positive Rate or Recall. This metric is suitable for analyzing the positive test cases correctly. This is easy to locate the position of the road sign perfectly. Which can be measured by Equation 11.12.

$$S = \frac{True\ Positives}{Actual\ True} \tag{11.12}$$

Where True Positives = TP, Actual True = TP + FN.

In this, it focuses only on sensitive areas with different sizes of pixels such as Small, Medium, Large, and Extra Large. The iteration process starts from 520 to 15k input size. True Positive Rate of the model with various sizes of mAP values in the form of bounding boxes, such as mAP (s), mAP (m), mAP (l), and mAP (xl). Here Small, Medium, Large, and Extra Large are classified based on IoU values ranging between 0.5 and 1.0, which is mentioned in Table 11.3.

TABLE 11.3 TPR Values of VGG NET-19 Based AX-YOLOv8 with Various IoU.

S. No.	Parameters Evaluation Metrics	mAP Score
1	Bounding Box TPR-mAP (s)	0.7991085
2	Bounding Box TPR-mAP (m)	0.7991085
3	Bounding Box TPR-mAP (l)	0.7991085
4	Bounding Box TPR-mAP (xl)	0.7991085

The comparison of TPR metrics with input size is represented in Table 11.4.

TABLE 11.4 TPR Metrics vs Input Size.

Metrics	520	2.5k	5k	7.5k	10k	12.5k	15k
TPR (s)	0.7838	0.7957	0.8230	0.8828	0.8891	0.8962	0.9228
TPR (m)	0.7838	0.7957	0.8230	0.8828	0.8891	0.8962	0.9228
TPR (l)	0.7838	0.7957	0.8230	0.8828	0.8891	0.8962	0.9228
TPR (xl)	0.7838	0.7957	0.8230	0.8828	0.8768	0.8962	0.9228

11.7.8 FALSE-NEGATIVE RATE (FNR)

FNR metric only calculates the False-Negative, that is, wrongly predicted cases. Usually, in the traffic sign detection process, some percentage of flaws or errors happen, that can be identified by FNR. Which can be measured by Equation 11.13.

$$FNR = \frac{False\ Negatives}{Actual\ True} \tag{11.13}$$

Where False-Negatives = FN, Actual True = TP + FN.

11.7.9 FALSE-POSITIVE RATE (FPR)

The FPR metric is also called miss rate. It is used to predict the outcome wrongly, that is, No Overtaking sign instead of No Left Turn sign. This will calculate the error rate of AX-YOLOv8 model. This can be calculated by Equation 11.14.

$$FPR = \frac{False\ Positive}{Actual\ False} \tag{11.14}$$

Where False-Positive = FP, Actual False = TN + FP.

11.7.10 SPECIFICITY

Specificity is a metric for predicting the negative cases correctly. This metric is very useful in classifying the not relevant to traffic sign image regions effectively. Which can be measured by Equation 11.15.

$$SP = \frac{True\ Negatives}{Actual\ False} \tag{11.15}$$

Where True-Negatives = TN, Actual False = TN + FP.

11.7.11 *F1-SCORE*

The F1 Score metric is the combination of Positive Predictive Value (PPV) and Sensitivity (S). This metric can be used instead of accuracy when the data are uneven. It is represented by the harmonic mean of the above two metrics. This can be effectively predicting the traffic signs even if the values are uneven. This can be measured by Equation 11.16.

$$F1 - Score = \frac{PPV * S}{PPV + S} \tag{11.16}$$

The AX-YOLOv8 model-based VGG NET-19 shows better performance in detecting the traffic sign portions than other neural networks.

11.8 COMPARATIVE RESULTS

Surveying the results of various deep learning models from the various research papers, the following result is consolidated. This result is represented in Table 11.5.

TABLE 11.5 Comparison of Various NN with PPV, TPR, and F1-Score Metrics.

Model	Neural Network	PPV	TPR	F1-Score
AX-YOLOv8	VGG NET-19	84.1%	89.2%	89.7%
	Darknet53	79.5%	77.4%	79.82%
	ResNet-50	82.5%	81.4%	80.06%

A Comparison of training time and input size with various Neural Networks is discussed in Table 11.6. The comparison of the Mean Average Precision metric with the present work is discussed in Table 11.7 and the Comparison of Performance Analysis of various algorithms is represented in Table 11.8.

TABLE 11.6 Training of Various Neural Networks.

Model	Neural Network	Training Time (hrs)	Input Size
AX-YOLOv8	VGG NET-19	95	15K
	Darknet53	40	9K
	ResNet-50	62	13K

TABLE 11.7 Mean Average Precision of Various Neural Networks.

Model	Neural Network	mAP
AX-YOLOv8	VGG NET-19	82.3%
	Darknet53	74.2%
	ResNet-50	76.1%

TABLE 11.8 Comparison of Performance Analysis of Various Algorithms.

Algorithms	Performance Analysis		
	mAP	Precision	Recall
YOLOv8	39.5	52.1	40
AX-YOLOv8	42.5	54.3	42.2

The performance analysis of YOLOv8 and AX-YOLOv8 with 10 levels of thresholds in Figure 11.7. Here threshold values range between 0.5 and 0.95.

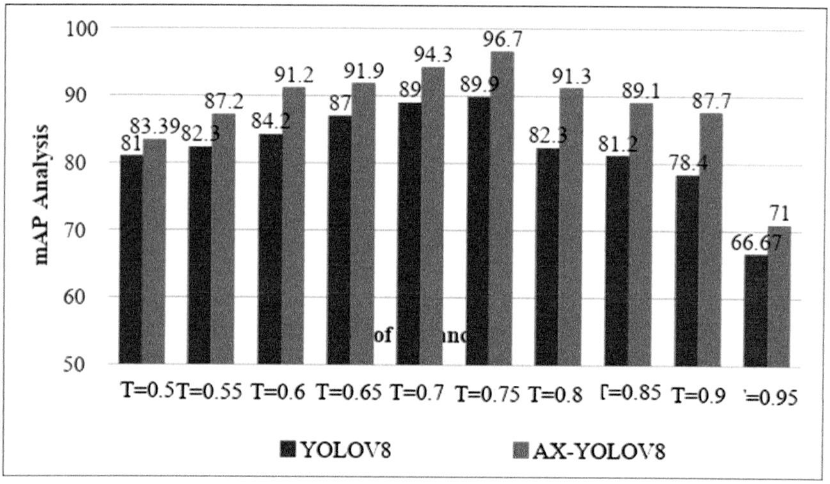

FIGURE 11.7 Performance analysis of YOLOv8 and AX-YOLOv8.

11.9 CONCLUSION

Arbitrary Extra Large You Only Look Once version 8 (AX-YOLOv8) is one of the best object detection variants of the YOLO family.

This version works better for detecting the adaptive sizes of objects effectively in an image frame. The proposed system continuously monitors the road traffic signs effectively from the IP camera. This methodology is applied to various performance metrics in terms of regression and classification. In this system, three neural networks such as VGG NET-19, Darknet53, ResNet-50 accompanied separately by AX-YOLOv8 model to identify the performance of the Neural Network based on four metrics (PPV, TPR, F1-Score, mAP). Based on experimental results AX-YOLOv8 with VGG NET-19 produces the highest PPV (84.1%), TPR (89.2%), F1-Score (89.7%), and mAP (82.3%). When VGG NET-19 is compared with the other two neural networks, it performs well in detecting the traffic sign images effectively.

KEYWORDS

- **AX-YOLOv8**
- **convolution neural network**
- **deep learning dynamic anchor boxes**
- **multi-scale feature residual aggregation module**
- **non-maximum suppression**

REFERENCES

1. Khaleel, S., Bayzeed, K. I., Khaleel, S. O., & Hashmi, M. J. (2023). traffic sign detection and recognition for driverless vehicles using convolutional neural network in deep learning. *Mathematical Statistician and Engineering Applications*, *72*(1), 1609–1623.
2. Singal, G., Singhal, H., Kushwaha, R., Veeramsetty, V., Badal, T., & Lamba, S. (2023). RoadWay: Lane detection for autonomous driving vehicles via deep learning. *Multimedia Tools and Applications*, *82*(4), 4965–4978.
3. Megalingam, R. K., Thanigundala, K., Musani, S. R., Nidamanuru, H., & Gadde, L. (2022). Indian traffic sign detection and recognition using deep learning. *International Journal of Transportation Science and Technology*, *12*(3), 683–699.
4. Sanyal, B., Kumar Mohapatra, R., & Dash, R. (2022). Traffic sign recognition on Indian database using wavelet descriptors and convolutional neural network ensemble. *Concurrency and Computation: Practice and Experience*, *34*(10), e6827.

5. Mishra, J., & Goyal, S. (2022). An effective automatic traffic sign classification and recognition deep convolutional networks. *Multimedia Tools and Applications*, *81*(13), 18915–18934.
6. Kanagaraj, N., Hicks, D., Goyal, A., Tiwari, S., & Singh, G. (2021). Deep learning using computer vision in self driving cars for lane and traffic sign detection. *International Journal of System Assurance Engineering and Management*, *12*(6), 1011–1025.
7. Satti, S. K., Maddula, P., & Ravipati, N. V. (2021). Unified approach for detecting traffic signs and potholes on Indian roads. *Journal of King Saud University-Computer and Information Sciences*, *34*(10), 9745–9756.
8. Manasa, R., Karibasappa, K., & Singh, M. K. (2021). Differential evolution evolved RBFNN based automated recognition of traffic sign images. *Results in Control and Optimization*, *5*, 100056.
9. Hasan, N., Anzum, T., & Jahan, N. (2021). Traffic sign recognition system (TSRS): SVM and convolutional neural network. In *Inventive Communication and Computational Technologies: Proceedings of ICICCT 2020* (pp. 69–79). Springer: Singapore.
10. Alam, A., & Jaffery, Z. A. (2020). Indian traffic sign detection and recognition. *International Journal of Intelligent Transportation Systems Research*, *18*, 98–112.
11. Vincent, M. A., Vidya, K. R., & Mathew, S. P. (2020, December). Traffic sign classification using deep neural network. In *2020 IEEE Recent Advances in Intelligent Computational Systems (RAICS)* (pp. 13–17). IEEE.
12. Kumar, K. N., Vishnu, C., Mitra, R., & Mohan, C. K. (2020, October). Black-box adversarial attacks in autonomous vehicle technology. In *2020 IEEE Applied Imagery Pattern Recognition Workshop (AIPR)* (pp. 1–7). IEEE.
13. Cococcioni, M., Rossi, F., Ruffaldi, E., Saponara, S., & de Dinechin, B. D. (2020). Novel arithmetics in deep neural networks signal processing for autonomous driving: Challenges and opportunities. *IEEE Signal Processing Magazine*, *38*(1), 97–110.
14. Tiron, G. Z., & Poboroniuc, M. S. (2019, October). Neural network based traffic sign recognition for autonomous driving. In *2019 International Conference on Electromechanical and Energy Systems (SIELMEN)* (pp. 1–5). IEEE.
15. Song, Y., Fan, R., Huang, S., Zhu, Z., & Tong, R. (2019). A three-stage real-time detector for traffic signs in large panoramas. *Computational Visual Media*, *5*, 403–416.
16. Jain, A., Mishra, A., Shukla, A., & Tiwari, R. (2019). A novel genetically optimized convolutional neural network for traffic sign recognition: A new benchmark on Belgium and Chinese traffic sign datasets. *Neural Processing Letters*, *50*, 3019–3043.
17. Vennelakanti, A., Shreya, S., Rajendran, R., Sarkar, D., Muddegowda, D., & Hanagal, P. (2019, January). Traffic sign detection and recognition using a CNN ensemble. In *2019 IEEE international conference on consumer electronics (ICCE)* (pp. 1–4). IEEE.
18. Natarajan, S., Annamraju, A. K., & Baradkar, C. S. (2018). Traffic sign recognition using weighted multi-convolutional neural network. *IET Intelligent Transport Systems*, *12*(10), 1396–1405.
19. Yang, T., Long, X., Sangaiah, A. K., Zheng, Z., & Tong, C. (2018). Deep detection network for real-life traffic sign in vehicular networks. *Computer Networks*, *136*, 95–104.

20. Varma, V. S. K. P., Adarsh, S., Ramachandran, K. I., & Nair, B. B. (2018). Real time detection of speed hump/bump and distance estimation with deep learning using GPU and ZED stereo camera. *Procedia Computer Science*, *143*, 988–997.
21. Martínez-Díaz, M., & Soriguera, F. (2018). Autonomous vehicles: Theoretical and practical challenges. *Transportation Research Procedia*, *33*, 275–282.
22. Gudigar, A., Chokkadi, S., Raghavendra, U., & Acharya, U. R. (2017). Local texture patterns for traffic sign recognition using higher order spectra. *Pattern Recognition Letters*, *94*, 202–210.
23. Prabhakar, G., Kailath, B., Natarajan, S., & Kumar, R. (2017, July). Obstacle detection and classification using deep learning for tracking in high-speed autonomous driving. In *2017 IEEE Region 10 Symposium (TENSYMP)* (pp. 1–6). IEEE.
24. Liu, S., Tang, J., Wang, C., Wang, Q., & Gaudiot, J. L. (2017). A unified cloud platform for autonomous driving. *Computer*, *50*(12), 42–49.
25. Zeng, Y., Xu, X., Shen, D., Fang, Y., & Xiao, Z. (2016). Traffic sign recognition using kernel extreme learning machines with deep perceptual features. *IEEE Transactions on Intelligent Transportation Systems*, *18*(6), 1647–1653.
26. Gudigar, A., Chokkadi, S., & U, R. (2016). A review on automatic detection and recognition of traffic sign. *Multimedia Tools and Applications*, *75*, 333–364.
27. Winkle, T. (2016). Safety benefits of automated vehicles: Extended findings from accident research for development, validation and testing. In M. Maurer, J. Gerdes, B. Lenz, & H. Winner (Eds.), *Autonomous driving: Technical, legal and social aspects* (pp. 335–364). Springer: Berlin, Heidelberg.
28. Mammeri, A., Zhou, D., & Boukerche, A. (2016). Animal-vehicle collision mitigation system for automated vehicles. *IEEE Transactions on Systems, Man, and Cybernetics: Systems*, *46*(9), 1287–1299.
29. Tijerina, L., Blommer, M., Curry, R., Swaminathan, R., Kochhar, D. S., & Talamonti, W. (2016). An exploratory study of driver response to reduced system confidence notifications in automated driving. *IEEE Transactions on Intelligent Vehicles*, *1*(4), 325–334.
30. Färber, B. (2016). Communication and communication problems between autonomous vehicles and human drivers. In M. Maurer, J. Gerdes, B. Lenz, & H. Winner (Eds.), *Autonomous Driving: Technical, Legal and Social Aspects* (pp. 125–144). Springer: Berlin, Heidelberg.
31. Zhang, J., Huang, M., Jin, X., & Li, X. (2017). A real-time Chinese traffic sign detection algorithm based on modified YOLOv2. *Algorithms*, *10*(4), 127.
32. Novak, B., Ilić, V., & Pavković, B. (2020, May). YOLOv3 Algorithm with additional convolutional neural network trained for traffic sign recognition. In *2020 Zooming Innovation in Consumer Technologies Conference (ZINC)* (pp. 165–168). IEEE.
33. Dewi, C., Chen, R. C., Jiang, X., & Yu, H. (2022). Deep convolutional neural network for enhancing traffic sign recognition developed on YOLOV4. *Multimedia Tools and Applications*, *81*(26), 37821–37845.
34. Wang, J., Chen, Y., Dong, Z., & Gao, M. (2023). Improved YOLOv5 network for real-time multi-scale traffic sign detection. *Neural Computing and Applications*, *35*(10), 7853–7865.

35. Dang, T. P., Tran, N. T., To, V. H., & Tran Thi, M. K. (2023). Improved YOLOv5 for real-time traffic signs recognition in bad weather conditions. *The Journal of Supercomputing*, *79*(10), 1–19.
36. Zhang, H., Ruan, Y., Huo, A., & Jiang, X. (2023, May). Traffic sign detection based on improved YOLOv7. In *2023 5th International Conference on Intelligent Control, Measurement and Signal Processing (ICMSP)* (pp. 71–75). IEEE.
37. Lou, H., Duan, X., Guo, J., Liu, H., Gu, J., Bi, L., & Chen, H. (2023). DC-YOLOv8: Small-size object detection algorithm based on camera sensor. *Electronics*, *12*(10), 2323.

CHAPTER 12

Empowering Women's Health: Artificial Intelligence Applications in Pre-Eclampsia Screening and Monitoring

B. KANNAKI DEVI and D. RAMYACHITRA

Department of Computer Science, Bharathiyar University, Coimbatore, Tamil Nadu, India

ABSTRACT

Pre-eclampsia is a very risky pregnancy syndrome that endangers mother and fetus. Early detection and treatment of pre-eclampsia are essential to reducing the likelihood of negative outcomes. Artificial intelligence (AI) may improve the accuracy and efficacy of pre-eclampsia diagnosis and therapy. Predictive detection is based on medical history and maternal characteristics biomarkers and may be created using AI techniques like ML and image analysis to predict women who are at the highest risk of having pre-eclampsia. AI algorithms may also help medical personnel evaluate lab results, gauge the progression of an illness, and select the most appropriate treatment. Remotely monitoring pre-eclampsia patients with wearable technology and other technologies enables prompt response in the event of symptoms or changes in vital signs. Overall, the application of AI in pre-eclampsia has the ability to improve outcomes for women and newborns by enabling early detection and personalized therapy of this potentially lethal illness. AI can assist in detecting pre-eclampsia, a dangerous pregnancy complication, by analyzing patient data such as blood pressure, urine protein levels, and medical history.

Real-Time Artificial Intelligence (AI): Key Motivations, Technologies, Platforms, and Use Cases. Pethuru Raj, B. Sundaravadivazhagan, V. Kavitha, B. Narendra Kumar Rao, & Hannah Vijaykumar (Eds.)

DOI: 10.1201/9781998511358-12

Machine learning algorithms can identify patterns and risk factors, enabling early diagnosis and timely intervention and reducing maternal and fetal health risks. These algorithms can provide healthcare providers with early warning signs, enabling timely intervention and preventive measures to reduce maternal and fetal health risks. AI-driven prediction models offer improved accuracy in pre-eclampsia diagnosis, facilitating better patient care, reducing healthcare costs, and ultimately saving lives by identifying high-risk pregnancies more effectively than traditional methods.

12.1 INTRODUCTION

This chapter explores the profound impact of artificial intelligence (AI) on the healthcare industry. AI technology integration into various healthcare domains, such as medical imaging, diagnostics, personalized medicine, and patient care has revolutionized how healthcare is delivered [1]. The chapter delves into the potential benefits, challenges, and ethical considerations surrounding the implementation of AI in healthcare. It also highlights specific use cases and success stories to demonstrate the effectiveness of AI-driven solutions in improving patient outcomes and optimizing medical workflows [2].

AI has emerged as a transformative technology in the field of healthcare [1]. It includes a variety of methods and algorithms that let computers interpret and analyze enormous volumes of medical data, act intelligently, or anticipate the future [2, 3].

Each year, over half a million pregnant women lose their lives due to pregnancy-related complications, predominantly in low- and middle-income nations where medical resources are often scarce. Pre-eclampsia, a condition characterized by high blood pressure typically becomes the world's top cause of maternal and child morbidity and death after 20 weeks of pregnancy [1–3]. The elevated blood pressure associated with pre-eclampsia directly impacts the functioning of the cardiovascular, renal, and hepatic systems [6, 7].

The link between hypertension and pre-eclampsia is intricate and bidirectional. Pre-eclampsia can occur in people who have chronic hypertension. On the other hand, pre-eclampsia raises the risk of prolonged cardiovascular issues, such as BP, in the mother, which can result in increased morbidity and mortality [4, 5].

Moreover, infants born from pregnancies complicated by pre-eclampsia are at a heightened risk of being born prematurely, experiencing restricted growth within the uterus, and facing an increased likelihood of mortality. As a result, up to 900,000 infant deaths occur annually due to these factors [8, 9].

12.1.1 OVERVIEW OF PRE-ECLAMPSIA AND ITS IMPACT ON MATERNAL AND FETAL HEALTH

Pre-eclampsia is a serious pregnancy complication that typically occurs after the 20th week of pregnancy and is characterized by high BP (hypertension) and signs of damage to organs, often the liver and kidneys. It is a leading cause of maternal and fetal morbidity and mortality worldwide. Pre-eclampsia affects about 5–8% of pregnancies and can have significant implications for both the mother and the unborn baby [4, 5].

12.1.1.1 OVERVIEW OF PRE-ECLAMPSIA

1. **Hypertension:** The hallmark feature of pre-eclampsia is high BP. BP readings regularly more than 140/90 mmHg are considered hypertensive in pregnancy.
2. **Proteinuria:** Another key characteristic of pre-eclampsia is the presence of protein in the urine (proteinuria), which indicates kidney involvement and damage.
3. **Organ Dysfunction:** Pre-eclampsia can affect various organs, including the liver, kidneys, brain, and blood clotting system. In severe cases, it can lead to organ failure.
4. **Symptoms:** Common symptoms of pre-eclampsia include persistent headaches, vision changes, and abdominal pain, swelling of the hands and face (edema), and rapid weight gain.
5. **Risk Factors:** Several risk factors increase the likelihood of developing pre-eclampsia, including first-time pregnancies, multiple pregnancies (twins or more), a history of pre-eclampsia in previous pregnancies, chronic hypertension, obesity, diabetes, and certain autoimmune disorders [6].

12.1.1.2 IMPACT ON MATERNAL HEALTH

1. **Maternal Mortality:** Pre-eclampsia is a significant contributor to maternal mortality. In severe cases, it can lead to complications, such as eclampsia (seizures), stroke, and organ failure, putting the mother's life at risk.
2. **Premature Birth:** Pre-eclampsia frequently requires that the baby be delivered earlier than expected in order to save the health of the mother. This can lead to premature birth, which carries its own set of risks for the baby.
3. **Cardiovascular Risk:** Pre-eclampsia during pregnancy doubles the risk of cardiovascular disease in later life.
4. **Postpartum Complications:** Some women may experience complications related to pre-eclampsia after giving birth, such as postpartum pre-eclampsia, which occurs within days or weeks after delivery [6].

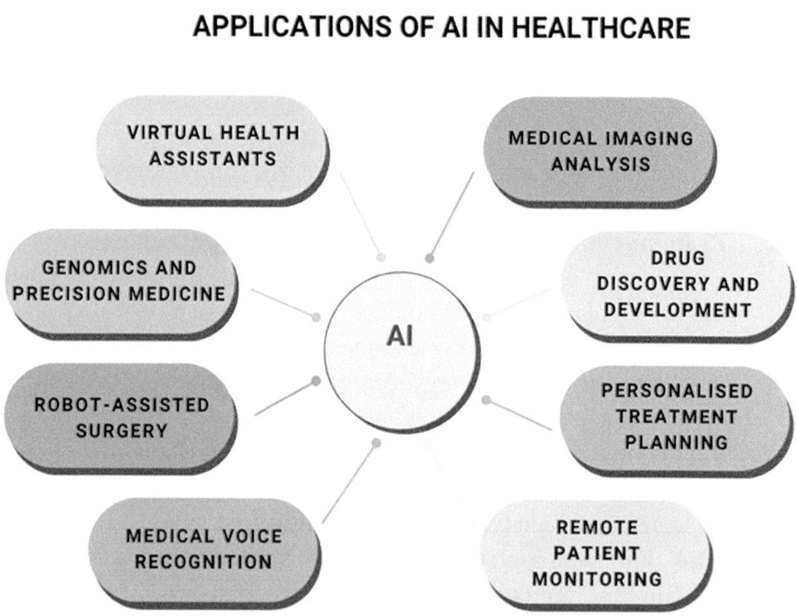

FIGURE 12.1 Risk factors of pre-eclampsia. It illustrates all the applications of artificial intelligence in healthcare industry, AI plays a very major role in all the medical fields such as gene editing, drug development, early disease detection, and treatment, etc.

Source: Adapted and modified from [12, 15].

12.1.1.3 IMPACT ON FETAL HEALTH

1. **Intrauterine Growth Restriction:** Pre-eclampsia can lead to inadequate blood flow to the placenta, resulting in reduced oxygen and nutrient supply to the fetus. This can lead to restricted fetal growth, known as intrauterine growth restriction (IUGR).
2. **Preterm Birth:** The risk of preterm birth increases with pre-eclampsia, which can result in various health challenges for the baby due to underdeveloped organs and systems.
3. **Fetal Distress:** Reduced oxygen supply and blood flow to the fetus can lead to fetal distress, affecting the baby's well-being during pregnancy and labor.
4. **Stillbirth:** In severe cases of pre-eclampsia, there is an increased risk of stillbirth, wherein the baby dies in the womb before delivery.

Early detection, close monitoring, and timely intervention are crucial in managing pre-eclampsia and reducing its impact on both maternal and fetal health. Regular prenatal check-ups, blood pressure monitoring, and urine testing are essential components of prenatal care to identify and manage pre-eclampsia effectively. Healthcare providers play a critical role in ensuring the best possible outcomes for the mother and the fetus through careful management and appropriate medical interventions [7].

12.1.2 IMPORTANCE OF EARLY DETECTION AND CURRENT SCREENING METHODS FOR PRE-ECLAMPSIA

The significance of early detection of pre-eclampsia cannot be overstated, as it significantly influences the management and outcome of the condition for both the mother and the baby. Early identification allows healthcare providers to implement appropriate interventions, closely monitor the pregnancy, and take necessary actions to prevent complications. Here are some key reasons highlighting the significance of early detection:

1. **Prevention of Severe Complications:** Early detection allows healthcare providers to intervene before pre-eclampsia progresses to a severe form, such as eclampsia (seizures) or HELLP syndrome (a serious liver disorder). Timely management can reduce the risk of life-threatening complications for the mother and the fetus.

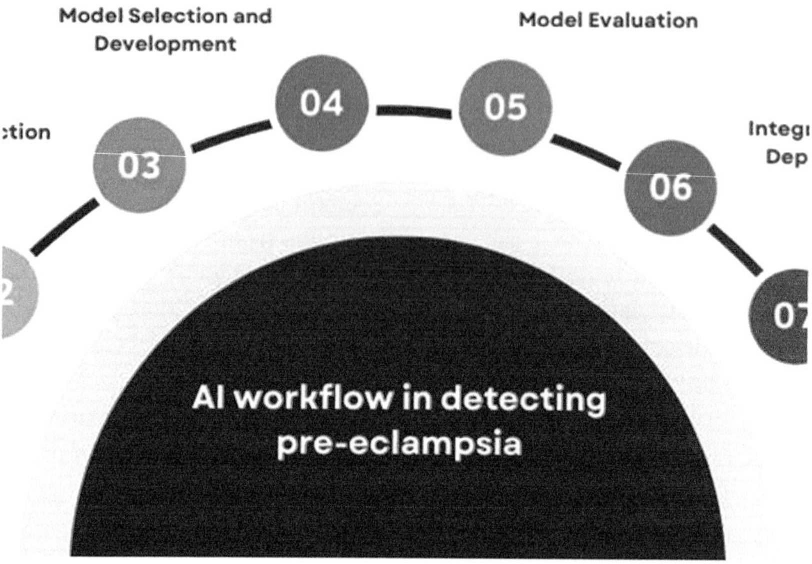

FIGURE 12.2 AI workflow in detecting pre-eclampsia [31]. It illustrates the roadmap of medical data generation during natural language processing and machine learning data investigation.

Source: Adapted and modified from [31].

2. **Improved Maternal and Fetal Outcomes:** Early detection and timely intervention contribute to better maternal and fetal outcomes. Close monitoring of blood pressure, kidney function, and fetal growth helps healthcare providers make informed decisions about the timing and mode of delivery to optimize both maternal and neonatal health.
3. **Prenatal Planning and Counselling:** Early identification of pre-eclampsia allows for better prenatal planning and counseling. Healthcare providers can discuss the condition, its potential risks, and the importance of compliance with medical recommendations with the pregnant woman and her family.
4. **Optimizing Maternal Health:** Early detection enables healthcare providers to initiate appropriate treatments and lifestyle modifications to manage hypertension and reduce the risk of complications related to high blood pressure, such as strokes and cardiovascular diseases.

5. **Preterm Birth Prevention:** Identifying pre-eclampsia early helps in the timely initiation of interventions to prevent preterm birth or manage preterm delivery, which can have significant health implications for the newborn [8].

12.1.2.1 CURRENT SCREENING METHODS FOR PRE-ECLAMPSIA

1. **Blood Pressure Monitoring:** Regular blood pressure measurements during prenatal check-ups are essential for identifying hypertension, which is a key feature of pre-eclampsia.
2. **Urine Protein Testing:** A routine urine test can detect the presence of proteinuria, which is a crucial indicator of kidney involvement in pre-eclampsia.
3. **Symptom Assessment:** Women are asked about symptoms, such as headaches, visual disturbances, and sudden weight gain, which can be early signs of pre-eclampsia.
4. **Fetal Monitoring:** Regular fetal growth assessments and monitoring of fetal well-being through methods like ultrasound and fetal heart rate monitoring are important to detect potential complications in the baby due to pre-eclampsia.
5. **Biomarker Testing:** Some Studies have looked at the use of biomarkers for pre-eclampsia risk prediction and early diagnosis, including placental growth factor (PlGF) and soluble fms-like tyrosine kinase 1 (sFLT-1).
6. **Doppler Ultrasound:** This specialized ultrasound can assess blood flow in the uterine and fetal blood vessels, providing insights into placental health and potential risks.

It is critical to recognize that while these screening methods are valuable, none of them are definitive in diagnosing pre-eclampsia. Diagnosis is based on the combination of clinical findings and the presence of specific criteria. Healthcare providers need to interpret these results in the context of a woman's overall health and medical history to make accurate assessments and provide appropriate care. Early detection and regular monitoring are key components of prenatal care that contribute to ensuring a safe and healthy pregnancy for both the mother and the baby [9].

12.2 PRE-ECLAMPSIA RISK FACTORS AND CLINICAL PRESENTATION

Pre-eclampsia is a complex and potentially life-threatening pregnancy complication that affects both the mother and the baby. Numerous risk factors that raise the possibility of having pre-eclampsia have been found, despite the fact that the precise reason is not entirely understood. Additionally, the clinical presentation of pre-eclampsia can vary, and early detection is crucial for timely management. The key risk factors and clinical signs of pre-eclampsia are the following:

12.2.1 RISK FACTORS

1. **First Pregnancy (Primigravida):** Compared with women who have had prior pregnancies, those who are pregnant for the first time are more likely to have pre-eclampsia.
2. **Multiple Pregnancies (Twins or More):** Women carrying multiple fetuses, such as twins or triplets are at higher risk of pre-eclampsia.
3. **Chronic Hypertension:** Women with chronic hypertension have an increased risk of developing pre-eclampsia during pregnancy.
4. **Age:** Women under 20 and over 40 years old are at higher risk of pre-eclampsia.
5. **Family History:** A family history of pre-eclampsia in first-degree relatives, such as a mother or sister increases the risk.
6. **Obesity:** Women who are obese or have a high body mass index (BMI) have an elevated risk of pre-eclampsia.
7. **Diabetes:** women with a history of diabetes or gestational diabetes (Diabetes in pregnancy) are at higher risk.
8. **Autoimmune Disorders:** Certain autoimmune conditions such as lupus or antiphospholipid syndrome can increase the risk of pre-eclampsia [11].
9. **Medical History:** A woman's medical history plays a crucial role in assessing her risk. Previous pregnancies with pre-eclampsia, chronic hypertension, diabetes, or autoimmune disorders are important risk indicators.
10. **Diabetes:** Pre-existing diabetes or gestational diabetes (diabetes during pregnancy) raises the risk of pre-eclampsia.

11. **Renal Disease:** Women who already have kidney illness or anomalies in the kidneys are more vulnerable.
12. **In Vitro Fertilization (IVF):** Pre-eclampsia risk among women who conceive through IVF may be increased.
13. **Interval between Pregnancies:** Short intervals between pregnancies may increase the risk of pre-eclampsia.
14. **Race and Ethnicity:** According to certain research, some racial and ethnic groups, such as African Americans may be more susceptible to pre-eclampsia [13].

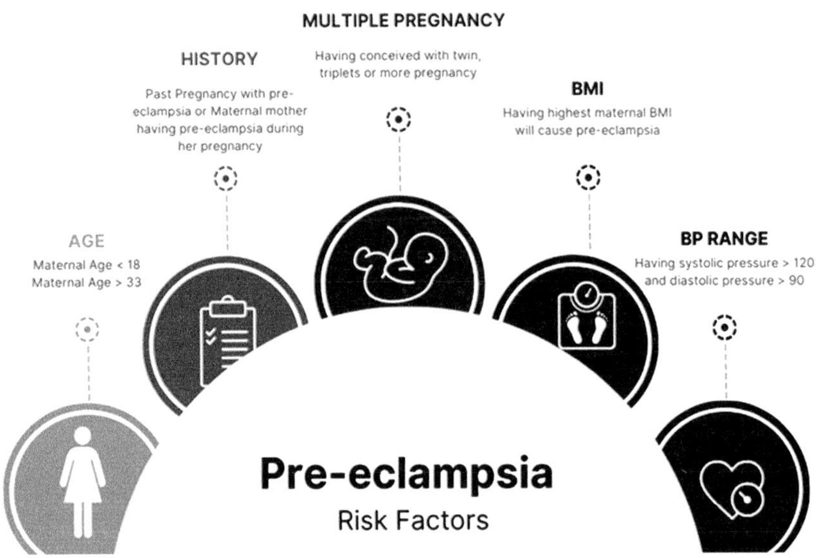

FIGURE 12.3 Risk factors of pre-eclampsia. It illustrates the risk factors of developing pre-eclampsia such that the above-mentioned factors play an important role in detecting the persons who would have pre-eclampsia in the future in advance.

Source: Adapted and modified from [5, 6].

12.2.2 CLINICAL PRESENTATION

The clinical presentation of pre-eclampsia can vary, and some women may not experience any symptoms. However, common signs and symptoms may include:

1. **High Blood Pressure (Hypertension):** Persistent high BP (140/90 mmHg or higher) is a key characteristic of pre-eclampsia.
2. **Proteinuria:** Discovering protein in the urine, detected through a urine test, is another hallmark of pre-eclampsia.
3. **Edema:** Swelling, especially in the hands, face, and legs, is a common symptom, although it can be seen in normal pregnancies as well.
4. **Headaches:** headaches that are persistent or severe and that are frequently not cured by over-the-counter drugs.
5. **Visual Disturbances:** Visual alterations such as hazy vision, flashing lights, or light sensitivity.
6. **Abdominal Pain:** Upper abdominal pain, especially on the right side, under the ribs.
7. **Sudden Weight Gain:** Rapid weight gain, especially in the later stages of pregnancy.

It's important to note that some symptoms of pre-eclampsia, such as BP and proteinuria, can only be detected through regular prenatal check-ups and urine tests. Additionally, not all women with pre-eclampsia will exhibit the same symptoms, making early detection through routine prenatal care critical for timely complications for both the mother and the child are managed and reduced. If any concerning signs or symptoms arise during pregnancy, it is essential to promptly inform healthcare providers for evaluation and appropriate care [12].

12.2.3 IDENTIFYING POPULATIONS AT HIGH RISK FOR PRE-ECLAMPSIA

Identifying populations at high risk for pre-eclampsia is essential to implement targeted interventions and closely monitor pregnancies to prevent or manage complications effectively. Healthcare providers use various factors to assess the possibility of pre-eclampsia patients. Some of the key methods and risk factors used to identify high-risk populations include:

1. Healthcare providers typically use a combination of these risk factors to assess a woman's overall risk of pre-eclampsia. Women identified as high-risk may receive more frequent prenatal visits, specialized monitoring, and targeted interventions to manage

and prevent complications associated with pre-eclampsia. Additionally, close communication and shared decision-making between healthcare providers and pregnant women are essential in providing personalized care based on individual risk profiles. Early identification and appropriate management of high-risk populations are critical to improving maternal and fetal outcomes in pregnancies affected by pre-eclampsia [13].

12.3 ROLE OF AI IN DETECTING PRE-ECLAMPSIA

The role of AI in pre-eclampsia detection is becoming increasingly important in improving maternal and fetal outcomes. AI-driven approaches leverage ML algorithms to analyze massive amounts of data and identify patterns and risk factors associated with pre-eclampsia. AI is increasingly being utilized to aid in the detection and prediction of pre-eclampsia, a serious hypertensive disorder during pregnancy [1, 2].

Here are some of the ways where AI Algorithms being applied for particular disease

The choice of algorithm for disease prediction varies on a number of variables, including the illness kind, the data available, and the precise predictions' objectives. Here are some common types of disease prediction and the algorithms commonly used for them:

1. **Cancer Prediction:**
   a) Logistic Regression
   b) Support Vector Machines (SVM)
   c) Random Forest
   d) Deep Learning (e.g., Convolutional Neural Networks for image-based cancer detection)

2. **Diabetes Prediction:**
   a) Logistic Regression
   b) Decision Trees
   c) k-Nearest Neighbors (k-NN)
   d) Support Vector Machines
   e) Naive Bayes
   f) Neural Networks

3. **Heart Disease Prediction:**
   a) Logistic Regression
   b) Decision Trees
   c) Random Forest
   d) Support Vector Machines
   e) Neural Networks

4. **Infectious Disease Outbreak Prediction:**
   a) Time Series Forecasting (e.g., ARIMA, LSTM)
   b) Epidemic Models (e.g., SIR, SEIR)
   c) Machine Learning Models for Spread Prediction

5. **Rare Disease Diagnosis:**
   a) Ensemble Methods (Random Forest, Gradient Boosting)
   b) Deep Learning (Convolutional Neural Networks, Recurrent Neural Networks)
   c) Transfer Learning (using pre-trained models)

6. **Mental Health Prediction:**
   a) Logistic Regression
   b) Support Vector Machines
   c) Neural Networks (for sentiment analysis)
   d) Natural Language Processing (NLP) models for text-based analysis

7. **Genetic Disease Prediction:**
   a) Genetic Algorithms
   b) Bayesian Networks
   c) Random Forest (for feature selection in genetic data)

8. **Chronic Disease Progression Prediction:**
   a) Survival Analysis (e.g., Cox Proportional-Hazards Model)
   b) Longitudinal Data Analysis (e.g., Linear Mixed-Effects Models)

9. **Drug Interaction and Adverse Event Prediction:**
   a) Association Rule Mining
   b) Bayesian Networks
   c) Deep Learning for analyzing structured and unstructured data

10. **Environmental Disease Risk Prediction:**
   a) Geospatial Analysis and GIS (Geographic Information Systems)
   b) Environmental Modeling

It is significant to note that the choice of algorithm may also depend on the quantity and quality of available data, as well as the specific features and characteristics of the disease being studied. Often, a combination of multiple algorithms or ensemble methods is used to improve prediction accuracy and robustness. Additionally, the field of disease prediction is constantly evolving, with new algorithms and techniques emerging as research progresses.

The integration of AI in pre-eclampsia detection and prediction can lead to earlier diagnosis, improved risk stratification, and more personalized care for pregnant women. However, itis important to remember that AI should complement the expertise of healthcare professionals and not replace their judgment. It should be used as a tool to support decision-making and improve patient outcomes, with appropriate validation and integration into clinical practice guidelines.

12.3.1 UTILIZING ML ALGORITHMS FOR RISK PREDICTION

Utilizing ML algorithms for risk prediction in pre-eclampsia is a powerful approach that leverages data-driven models to assess a woman's likelihood of developing the condition. Machine learning algorithms analyze large and complex datasets, including maternal demographics, medical history, vital signs, biomarkers, and other relevant factors to create predictive models that can identify high-risk pregnancies. Here's how machine learning algorithms are utilized for risk prediction in pre-eclampsia:

1. **Data Collection and Pre-processing:** Various sources, including wearable technology, medical imaging, and electronic health records (EHRs), are used to gather pertinent data. The data are then pre-processed to handle missing values, normalize the data, and remove any noise or outliers. There are several data collections repositories are available some of them are as follows GenBank, UniProt, PubMed, Ensembl, The Cancer Genome Atlas (TCGA), 1000 Genomes Project, European Nucleotide Archive (ENA), Proteomics Identifications Database (PRIDE), Gene Expression Omnibus (GEO), Exome Aggregation Consortium (ExAC), Sequence Read Archive (SRA), ArrayExpress, Model Organism Databases (MODs), Protein Data Bank (PDB), DrugBank.
2. **Feature Selection and Engineering:** To create precise predictions, machine learning algorithms need pertinent characteristics

(variables). The most informative variables are found using feature selection approaches, and feature engineering may entail developing new features using domain expertise [8].

3. **Model Training:** The pre-processed data are divided into training sets and validation sets. Many algorithms for machine learning, such as logistic regression, support vector machines (SVM), decision trees, random forests, and neural networks, are trained on the training data using a labeled dataset (previous cases of pre-eclampsia and non-pre-eclampsia pregnancies).
4. **Model Validation:** On the validation set, the trained model is assessed for performance and generalizability to fresh data. Cross-validation techniques are often used to estimate model performance effectively.
5. **Risk Prediction:** Once the model is validated and deemed accurate, it can be applied to new patient data to identify the risk of pre-eclampsia for individual pregnancies. The model generates a risk score or probability, indicating the likelihood of developing pre-eclampsia.
6. **Alerting and Clinical Decision Support:** Based on risk prediction, healthcare providers can be alerted to high-risk pregnancies, allowing for early intervention, closer monitoring, and personalized care plans.
7. **Iterative Improvement:** Machine learning models are iteratively improved by incorporating new data and retraining the model periodically. Continuous learning helps keep the model up-to-date and ensures it captures emerging risk factors [9].

Benefits of Utilizing Machine Learning for Risk Prediction in Pre-Eclampsia:

1. **Precision and Accuracy:** Machine learning algorithms can analyze a wide range of data and identify subtle patterns that may not be apparent using traditional approaches.
2. **Early Detection:** Machine learning models can detect pre-eclampsia risk earlier in pregnancy, allowing for timely interventions and improved outcomes.
3. **Personalized Care:** Risk prediction models facilitate personalized care plans, tailoring interventions to individual patient needs.

4. **Scalability:** Machine learning models can handle large and diverse datasets, making them scalable for population-wide risk prediction.
5. **Research and Insights:** Machine learning can reveal novel risk factors and associations, advancing our understanding of pre-eclampsia and potentially leading to new prevention and treatment strategies [10].

It is important to note that while machine learning holds significant promise, rigorous validation and clinical studies are crucial to ensure its accuracy, safety, and ethical considerations in real-world healthcare settings. Collaboration between AI experts, healthcare providers, and researchers is essential to successfully implement ML for risk prediction in pre-eclampsia detection.

12.3.2 AI-DRIVEN ANALYSIS OF PATIENT DATA AND MEDICAL RECORDS

Analysis of patient information and medical records using AI is a transformative approach that harnesses the power of AI to extract important insights and patterns from enormous amounts of healthcare data. This data-driven analysis has the potential to improve clinical decision-making, optimize patient care, and enhance medical research. Here is how AI is utilized for the analysis of patient data and medical records:

1. **Data Integration:** AI algorithms can integrate diverse types of patient information, including wearable device data, genetic information, medical imaging, laboratory results, and EHRs. By combining these sources, AI can create a comprehensive view of a patient's health status and medical history.
2. **Natural Language Processing (NLP):** NLP is a crucial component of AI-driven analysis of medical records. NLP techniques allow AI to understand and process unstructured clinical notes and medical literature. This enables AI to extract relevant information, such as patient symptoms, diagnoses, treatment plans, and outcomes, from free-text documents [11].
3. **Diagnosis and Prognosis Support:** AI can assist healthcare providers in diagnosing and predicting patient outcomes by analyzing patient data. AI algorithms can learn from historical

patient cases and identify patterns that correlate with specific conditions or predict disease progression.

4. **Risk Stratification:** AI can stratify patients based on their risk profiles for certain diseases or complications. This enables health-care providers to prioritize interventions for high-risk patients, improving patient outcomes and resource allocation.
5. **Treatment Recommendations:** AI can analyze patient data and medical literature to suggest evidence-based treatment options, including personalized therapies tailored to individual patient characteristics [12].
6. **Clinical Decision Support:** AI may help medical personnel make choices in real-time by analyzing patient data and suggesting the right tests, treatments, or interventions in accordance with the most recent recommendations and best practices.
7. **Identifying Clinical Trials Candidates:** AI-driven analysis can identify potential candidates for clinical trials based on specific criteria, facilitating recruitment and accelerating the development of new treatments [13].
8. **Healthcare Quality and Efficiency:** AI can analyze healthcare data to identify inefficiencies, optimize workflows, and improve the overall quality of healthcare delivery.

Benefits of AI-driven Analysis of Patient Data and Medical Records:

1. **Data-driven Insights:** AI can analyze large datasets and reveal hidden patterns that may not be apparent through traditional methods.
2. **Personalized Medicine:** AI enables tailored treatment plans based on individual patient characteristics and response to therapies.
3. **Early Detection and Prevention:** AI can aid in early disease detection and risk prediction, enabling timely interventions and preventive measures.
4. **Improved Patient Outcomes:** AI-driven decision support can enhance clinical decision-making, resulting in improved patient outcomes and lower healthcare expenses.
5. **Efficiency and Time Savings:** AI can automate data analysis tasks, freeing up healthcare providers' time for more direct patient care [14].

Challenges:

1. **Data Quality and Standardization:** For successful AI analysis, ensuring data quality and standardization is essential.
2. **Interoperability:** Integrating data from different sources can be challenging due to interoperability issues.
3. **Ethical and Privacy Concerns:** AI analysis of patient data raises concerns about data privacy, security, and ethical considerations.

AI-driven analysis of patient data and medical records holds immense the ability to completely transform healthcare by data-driven insights, improving patient care, and advancing medical knowledge. As AI technology continues to evolve, its impact on healthcare is expected to grow exponentially. However, it is important to address challenges and establish robust governance frameworks to guarantee the ethical and appropriate application of AI in healthcare [12].

12.3.3 AI-POWERED ANALYSIS OF PHYSIOLOGICAL PARAMETERS AND BIOMARKERS

AI-powered analysis of physiological parameters and biomarkers is a cutting-edge approach in healthcare that leverages AI to interpret complex data related to a patient's health status. Physiological parameters include various vital signs and measurements, while biomarkers are specific indicators that can be measured to assess a person's health condition. AI algorithms can analyze these data points to gain valuable insights, aid in diagnosis, predict disease progression, and personalize treatment plans. Here is how AI is utilized for the analysis of physiological parameters and biomarkers:

1. **Physiological Data Collection:** AI can process and analyze data from various physiological measures, including respiration rate, temperature, heart rate, and blood pressure, oxygen saturation, and electrocardiograms (ECGs). Additionally, data from wearable devices and continuous monitoring systems provide a wealth of real-time physiological information.
2. **Biomarker Analysis:** AI algorithms can analyze data from specific biomarkers, such as proteins, hormones, genetic markers, or other molecular indicators present in blood, urine, or tissue samples.

AI can identify patterns and correlations between biomarker levels and health conditions [15].

3. **Early Disease Detection:** AI has the ability to identify minor alterations in physiological parameters and biomarker levels that might be signs of a disease or other health issue in its early stages, allowing for early identification and treatment.
4. **Predictive Analytics:** By analyzing historical patient data and biomarker trends, AI can predict disease progression or identify patients at higher risk of developing certain conditions.
5. **Personalized Medicine:** AI-driven analysis of physiological parameters and biomarkers can support the customization of treatment regimens based on unique patient features, ensuring more targeted and effective therapies [16].
6. **Drug Response Prediction:** AI can analyze biomarker data to predict a patient's response to specific medications, helping healthcare providers select the most suitable treatment options.
7. **Clinical Decision Support:** Clinical decision assistance in real-time by AI algorithms analyzing physiological parameters and biomarkers to aid in diagnosis and treatment planning.
8. **Monitoring Disease Progression:** AI can continuously monitor changes in physiological parameters and biomarkers to assess disease progression and treatment efficacy.

Benefits of AI-powered Analysis of Physiological Parameters and Biomarkers:

1. **Precision Medicine:** AI enables personalized treatment plans based on an individual's unique physiological and biomarker profile.
2. **Early Intervention:** AI can identify early warning signs of disease, allowing for timely interventions and improved patient outcomes.
3. **Data Integration:** AI can efficiently analyze large and diverse datasets from multiple sources, providing comprehensive insights for healthcare providers.
4. **Data-driven Insights:** AI algorithms can uncover hidden patterns and associations within physiological and biomarker data, enhancing medical knowledge and research.
5. **Real-time Monitoring:** AI-driven continuous monitoring facilitates real-time patient assessment and timely clinical interventions [17].

Challenges:

1. **Data Quality and Availability:** The quality and availability of physiological and biomarker data may vary, affecting the accuracy of AI analysis.
2. **Interoperability and Integration:** Integrating data from different sources and healthcare systems can be challenging due to interoperability issues.
3. **Ethical and Privacy Concerns:** AI analysis of sensitive patient data raises ethical considerations about data privacy, security, and consent.
4. **Validation and Regulation:** Validating AI algorithms for medical use and obtaining regulatory approval are crucial to ensure their safety and efficacy.

As AI continues to advance, its role in analyzing physiological parameters and biomarkers is expected to grow, transforming healthcare by enabling more precise and personalized medical interventions. AI professionals, healthcare providers, academics, and regulatory agencies must work together to integrate AI in healthcare responsibly and effectively [17].

12.4 AI-BASED MEDICAL IMAGING FOR PRE-ECLAMPSIA

AI-based medical imaging is a promising approach for the early detection and diagnosis of pre-eclampsia. Medical imaging modalities, such as ultrasound, Doppler imaging, and magnetic resonance imaging (MRI), play a crucial role in evaluating placental and maternal vascular function, which are key factors in pre-eclampsia. AI-driven analysis of medical images can aid in identifying subtle changes and patterns that may indicate the development of pre-eclampsia. Here is how AI is utilized in medical imaging for pre-eclampsia [18].

12.4.1 DOPPLER ULTRASOUND ANALYSIS

AI algorithms can analyze Doppler ultrasound scans to assess blood flow in the uterine and fetal blood vessels. Changes in blood flow patterns, such as abnormal resistance in the uterine arteries can be indicative of impaired placental function and increased pre-eclampsia risk. Doppler ultrasound analysis employs sound waves as a noninvasive imaging tool to assess blood flow via blood vessels and other bodily structures. It is commonly used in various medical specialties, including obstetrics to assess blood flow in the maternal and fetal circulatory systems. In the context of pre-eclampsia, Doppler ultrasound is particularly valuable in evaluating the placental and maternal vascular function, which can provide crucial information for early detection and monitoring of the condition.

Here is how Doppler ultrasound analysis is used in obstetrics and its significance in the context of pre-eclampsia:

1. **Principle of Doppler Ultrasound:** Doppler ultrasound works based on the shifting in frequency of sound waves reflected by moving objects, or the Doppler effect. When sound waves encounter moving blood cells, the frequency of the reflected sound waves changes, and this shift is used to calculate the velocity and direction of blood flow.
2. **Fetal Middle Cerebral Artery (MCA) Doppler:** Doppler ultrasound can also assess blood flow in the fetal middle cerebral artery. Abnormalities in MCA Doppler, such as an increased peak systolic velocity (PSV), can indicate fetal anemia or other issues related to placental function.
3. **Umbilical Artery Doppler:** Doppler ultrasound of the umbilical artery provides information about blood flow between the placenta and the fetus. Abnormalities in umbilical artery Doppler, such as an increased pulsatility index, can be indicative of fetal growth restriction, which is often associated with pre-eclampsia.
4. **Fetal Ductus Venosus Doppler:** Doppler ultrasound of the fetal ductus venosus assesses blood flow in the fetal liver. Abnormalities in this parameter can provide insights into fetal well-being and potential complications related to pre-eclampsia [19].

12.4.2 PLACENTAL EVALUATION

AI can be applied to analyze MRI or ultrasound images of the placenta to identify abnormalities or signs of placental insufficiency, which is associated with pre-eclampsia. Pre-eclampsia is a major pregnancy problem marked by high BP and damage to other organs that commonly develops after the 20th week of pregnancy. Placental examination can be very helpful in identifying pre-eclampsia. Pre-eclampsia can be dangerous for both the mother and the baby if not diagnosed and managed promptly.

Here are some ways in which placental evaluation can help in detecting pre-eclampsia:

1. **Uterine Artery Doppler:** Uterine artery Doppler is a specialized ultrasound evaluation that assesses blood flow in the uterine arteries. Changes in the uterine artery blood flow pattern can be indicative of poor placental function and placental insufficiency. In pre-eclampsia, there is often impaired blood flow to the placenta, resulting in abnormal Doppler waveforms. Abnormal uterine artery Doppler findings can help identify women at risk of developing pre-eclampsia or other complications related to impaired placental function [20].
2. **Placental Size and Appearance:** During routine prenatal ultrasounds, the size and appearance of the placenta are assessed. In cases of pre-eclampsia, the placenta may appear larger or smaller than expected, and it may have abnormal features, such as calcifications or infarcts. These changes in the placental appearance can raise suspicion for pre-eclampsia and may prompt further evaluation and monitoring.
3. **Histopathological Examination:** After delivery, the placenta may undergo histopathological examination by a pathologist. The examination can reveal signs of placental dysfunction, such as abnormalities in the placental structure, inflammation, or evidence of inadequate blood flow. These findings can provide valuable insights into the underlying cause of pre-eclampsia and help guide future management and care.
4. **Clinical Assessment:** Placental evaluation is just one component of the overall assessment for pre-eclampsia. Other factors, such as the mother's blood pressure, proteinuria (presence of protein in the urine), and symptoms such as headaches or vision changes are also crucial in diagnosing pre-eclampsia. Combining clinical evaluation with placental assessment can improve the accuracy of pre-eclampsia detection [20].

12.4.3 FETAL GROWTH ASSESSMENT

AI can assist in evaluating fetal growth and detecting signs of IUGR, which is often linked to pre-eclampsia. Fetal growth assessment is an important component in the detection and management of pre-eclampsia during pregnancy. Monitoring fetal growth helps healthcare providers identify potential issues and make informed decisions about the management of the pregnancy [20].

Here is how fetal growth assessment can aid in the detection of pre-eclampsia:

1. **Ultrasound Monitoring:** Regular ultrasound examinations during pregnancy allow healthcare providers to assess the fetus's growth and development. By measuring various fetal biometric parameters, such as head circumference, abdominal circumference, and femur length, the estimated fetal weight can be calculated. Comparing these measurements to standard growth charts helps identify if the fetus is growing appropriately for its gestational age.
2. **Customized Growth Charts:** In some cases, customized growth charts may be used based on maternal characteristics, such as height, weight, and ethnicity. Customized growth charts can provide a more personalized assessment of fetal growth and help detect deviations from the expected growth patterns.
3. **Serial Measurements:** Monitoring fetal growth through serial ultrasound examinations allows healthcare providers to observe changes over time. Slower than expected growth or a sudden deceleration in growth may indicate potential issues with placental function, which is a key factor in pre-eclampsia.
4. **Monitoring High-Risk Pregnancies:** In pregnancies with a higher risk of developing pre-eclampsia, such as those with a history of pre-eclampsia in previous pregnancies, chronic hypertension, or other risk factors, healthcare providers may opt for more frequent fetal growth assessments to detect any growth issues early.

Early detection of abnormal fetal growth can prompt further evaluation and monitoring, allowing healthcare providers to take appropriate actions to manage pre-eclampsia and optimize the care of both the mother and the baby. In some cases, if the fetus is not growing adequately or if pre-eclampsia poses significant risks to the mother or the baby, early delivery may be recommended to prevent further complications [21].

12.4.4 PREDICTION MODELS

AI can develop predictive models based on medical imaging and other clinical data to estimate a woman's risk of developing pre-eclampsia. A prediction model, also known as a predictive model, is a mathematical or statistical tool used to make predictions or forecasts about future outcomes based on historical data and known variables. The primary purpose of a prediction model is to estimate the likelihood of a particular event or to predict a specific value based on the information available from past observations.

1. **PIERS (Pre-eclampsia Integrated Estimate of RiSk) Model:** This model combines clinical and biochemical factors to predict the risk of developing pre-eclampsia. The variables included in the model are maternal age, gestational age at recruitment, maternal race/ethnicity, mean arterial pressure, presence of chronic hypertension, BMI, previous preterm delivery, and the ratio of soluble fms-like tyrosine kinase 1 (sFlt-1) to PlGF in maternal serum.
2. **fullPIERS Model:** An expanded version of the PIERS model, the fullPIERS model includes additional variables such as the number of fetuses, any antihypertensive drug use before admission, and liver enzymes (AST and ALT) in maternal bloodwork.
3. **GLOWM Model:** Developed by the Global Library of Women's Medicine, this model uses variables, such as maternal age, parity, BMI, mean arterial pressure, and uterine artery pulsatility index to predict the risk of pre-eclampsia [21].
4. **Fetal Medicine Foundation (FMF) Model:** This model combines maternal factors, such as mean arterial pressure, BMI, and history of chronic hypertension with uterine artery Doppler measurements to assess the risk of pre-eclampsia.
5. **RAPID (Risk Prediction for Early Pre-eclampsia Detection) Model:** This model uses a combination of maternal characteristics, mean arterial pressure, and biomarkers (PlGF, pregnancy-associated plasma protein A, and inhibin A) to predict the risk of early-onset pre-eclampsia.
6. **ASPRE (Combined Multimarker Screening and Randomized Patient Treatment with Aspirin for Evidence-Based Pre-eclampsia Prevention) Model:** Designed to identify women at high risk for preterm pre-eclampsia, this model includes maternal

factors, mean arterial pressure, and biochemical markers (PAPP-A and PlGF).

It is important to note that each prediction model may have different validation and performance characteristics. Additionally, the use of these models in clinical practice may vary based on regional guidelines and healthcare settings [20].

As research and medical advancements continue, new prediction models or updates to existing ones may have emerged. Therefore, consult the latest literature and guidelines to ensure you have access to the most up-to-date information on pre-eclampsia prediction models. Healthcare professionals, particularly those specializing in maternal-fetal medicine, can also provide valuable insights and recommendations for risk assessment in individual cases.

12.4.5 DECISION SUPPORT FOR HEALTHCARE PROVIDERS

AI-powered medical imaging tools can provide healthcare providers with decision support, assisting in the interpretation of complex imaging data and aiding in early diagnosis. AI in decision support for healthcare providers in pre-eclampsia offers several benefits, aiding in early detection, risk assessment, and personalized care. Here are some specific ways AI can be leveraged:

1. **Risk Prediction Models:** AI can develop and enhance predictive models that consider various patient factors, biomarkers, and clinical data to assess the risk of pre-eclampsia for individual pregnant women. These models can help identify high-risk patients, enabling timely interventions and closer monitoring [20].
2. **Continuous Monitoring and Alerts:** AI-powered monitoring systems can continuously analyze vital signs, lab results, and other relevant data in real-time. If any concerning patterns or deviations are detected, the system can generate alerts, prompting healthcare providers to take immediate action.
3. **Image Analysis:** AI algorithms can analyze ultrasound and Doppler scans to detect early signs of pre-eclampsia, such as abnormal blood flow patterns in the placenta or uterine arteries. This assists healthcare providers in making more accurate and timely diagnoses.

4. **Personalized Care Plans:** AI can assist in developing personalized care plans for pregnant women based on their individual risk profiles. By considering various factors, such as medical history, age, and biomarker levels, AI systems can recommend tailored monitoring schedules and interventions [21].
5. **Clinical Decision Support Systems:** AI-driven decision support systems can be integrated into EHRs to provide evidence-based guidelines and best practices for pre-eclampsia management. This ensures that healthcare providers have access to the latest information and treatment recommendations.
6. **Telemedicine and Remote Monitoring:** AI-enabled telehealth solutions and wearable devices can facilitate remote monitoring of pregnant women, especially those at higher risk of pre-eclampsia. AI algorithms can analyze data collected from these devices to provide insights and alerts to both patients and healthcare providers [22].
7. **Research and Insights:** AI can analyze vast amounts of medical literature and research data to identify new trends, risk factors, and treatment approaches related to pre-eclampsia. This can support evidence-based decision-making and keep healthcare providers informed about the latest advancements.
8. **Data Integration:** AI can integrate data from multiple sources, such as EHRs, lab reports, and imaging studies, to create a comprehensive view of a patient's health status. This holistic approach can help healthcare providers make well-informed decisions.
9. **Clinical Trials and Research Recruitment:** AI can aid in identifying eligible candidates for clinical trials and research studies focused on pre-eclampsia. By matching patients with specific criteria, AI can streamline the recruitment process and accelerate research efforts [23].

By leveraging AI in decision support, healthcare providers can improve patient outcomes, enhance efficiency, and make more informed and data-driven decisions in pre-eclampsia detection and management. It is important to note that AI is a tool that complements clinical expertise and judgment, and healthcare providers remain central to the decision-making process.

12.5 REMOTE MONITORING OF VITAL SIGNS AND BLOOD PRESSURE TRENDS

Remote monitoring of vital signs and blood pressure trends in pre-eclampsia is an essential aspect of antenatal care for pregnant women with this condition. Pre-eclampsia is a serious pregnancy complication characterized by high blood pressure and signs of damage to other organ systems, often affecting the liver and kidneys. It is crucial to closely monitor the mother and the baby's well-being during this condition to ensure timely interventions and reduce the risk of complications.

Remote monitoring allows healthcare providers to keep a constant check on the patient's health without requiring her to be physically present at the medical facility all the time. This is particularly beneficial for women with pre-eclampsia, as they may need to be monitored more frequently and closely compared with standard prenatal care [22].

Here are some aspects and vital signs that can be remotely monitored in pre-eclampsia:

1. **Blood Pressure:** Regular monitoring of blood pressure is a fundamental part of managing pre-eclampsia. Remote monitoring devices can be used to measure blood pressure accurately at home. The data can then be transmitted securely to the healthcare provider's system for review. Any significant fluctuations or sustained high blood pressure levels can be identified promptly, and appropriate actions can be taken.
2. **Heart Rate (Pulse):** Monitoring the heart rate can provide additional insights into the mother's cardiovascular health and help identify any irregularities or signs of distress.
3. **Fetal Heart Rate Monitoring:** For women with pre-eclampsia, it is crucial to monitor the baby's heart rate regularly to ensure the fetus is not experiencing any distress [23].
4. **Weight Monitoring:** Sudden weight gain can be an indicator of fluid retention, a common symptom of pre-eclampsia. Remote monitoring can track weight changes and alert healthcare providers to any concerning trends.
5. **Urine Protein Monitoring:** The presence of protein in the urine (proteinuria) is a significant sign of pre-eclampsia. Remote monitoring systems can be used to assess urine samples regularly.

6. **Symptom Reporting:** Remote monitoring may involve the use of digital platforms or mobile applications where the pregnant woman can report any concerning symptoms she experiences, such as severe headaches, visual disturbances, or abdominal pain. These reports can trigger immediate follow-up actions [24].
7. **Medication Adherence:** If medications are prescribed to manage pre-eclampsia, remote monitoring can include tools to track medication adherence.
8. **Secure Communication with Healthcare Providers:** Remote monitoring systems should allow secure communication between the patient and healthcare providers. This way, any concerns or questions can be addressed promptly.

It is important to ensure that the remote monitoring devices are accurate, reliable, and easy for the patient to use. To ensure that the patient uses the devices correctly and understands the value of routine monitoring, the patient should get the proper training and instruction.

Remote monitoring of vital signs and blood pressure trends in pre-eclampsia can help detect potential complications early, enabling timely interventions and improving pregnancy outcomes. However, it should not replace regular in-person prenatal visits but rather complement them to provide comprehensive care throughout the pregnancy. [25]

12.6 CHALLENGES AND LIMITATIONS

Computer science researchers contribute significantly to disease study and problem-solving by applying computational techniques. They analyze vast datasets, such as genomic information or medical records, to identify patterns and biomarkers associated with diseases. Machine learning algorithms can predict disease outcomes and optimize treatment plans. Researchers also develop healthcare informatics systems to improve patient care and streamline healthcare operations. Additionally, they design simulation models for drug discovery and testing. Through interdisciplinary collaboration with domain experts, computer scientists leverage technology to tackle complex healthcare challenges, enhance diagnosis and treatment, and ultimately contribute to advancements in disease understanding and management.

Detecting pre-eclampsia using AI faces several challenges and limitations. Firstly, acquiring large and diverse datasets of high-quality labeled data for training AI models is difficult due to privacy concerns and the rarity of severe cases. Second, bias in the training data can lead to disparities in healthcare outcomes for different populations. Interpretability of AI models is crucial in medical applications, but many AI algorithms lack transparency, making it hard for healthcare providers to trust their recommendations. Generalizing AI models to diverse populations and avoiding overfitting are also significant concerns. Ethical considerations, regulatory compliance, and the need for rigorous clinical trials further complicate AI implementation. Additionally, integrating AI systems into existing healthcare infrastructure requires substantial investment. Despite these challenges, ongoing research and technological advancements offer hope for leveraging AI's potential in improving pre-eclampsia detection and maternal-fetal healthcare [25].

12.6.1 AVAILABILITY AND QUALITY OF PRE-ECLAMPSIA DATASETS

The availability and quality of pre-eclampsia datasets can present challenges in the development and validation of AI models for detecting this condition. Pre-eclampsia is a relatively rare and complex pregnancy complication, and obtaining comprehensive and well-annotated datasets can be difficult due to various reasons:

1. **Data Privacy and Ethics:** Medical data, especially related to pregnancy and sensitive conditions like pre-eclampsia, is subject to strict privacy regulations. Access to patient records must adhere to ethical guidelines, and researchers must protect the identities and personal information of individuals in the dataset.
2. **Limited Sample Size:** Pre-eclampsia affects a relatively small percentage of pregnancies, making it challenging to gather a sufficiently large dataset to train AI models effectively. Smaller datasets may lead to overfitting and reduced generalization to diverse populations.
3. **Data Collection Consistency:** The quality of the data may vary across different sources and healthcare institutions, leading to inconsistencies and potential biases in the dataset. Ensuring standardized data collection protocols is essential to maintain data quality [25].
4. **Lack of Longitudinal Data:** Effective pre-eclampsia detection often requires longitudinal data to identify trends and changes in vital

signs over time. However, obtaining continuous and comprehensive data throughout the pregnancy can be logistically challenging.

5. **Class Imbalance:** Pre-eclampsia datasets may suffer from class imbalance, where there are significantly more instances of normal pregnancies compared with pre-eclamptic cases. This imbalance can lead to biased model training and reduced sensitivity in detecting pre-eclampsia cases [26].
6. **Incomplete Data:** Some vital signs or patient information might be missing or incomplete in the dataset, which can affect the AI model's performance and reliability.
7. **Heterogeneity of Cases:** Pre-eclampsia is a complex condition with various subtypes and severity levels. Building a dataset that captures this heterogeneity can be challenging but is crucial for a robust AI model.

Despite these challenges, efforts are being made to improve the availability and quality of pre-eclampsia datasets. Collaborative research initiatives, data sharing agreements, and advancements in data anonymization techniques are helping researcher's access larger and more diverse datasets while protecting patient privacy. Additionally, ongoing research in the field of pre-eclampsia and advances in remote monitoring technologies may contribute to the accumulation of more comprehensive and real-time datasets, further supporting the development of AI-based detection methods [27].

12.7 CONCLUSION

In conclusion, the utilization of AI in pre-eclampsia has emerged as a transformative approach with significant potential to revolutionize maternal healthcare. Through the application of advanced algorithms and data-driven insights, AI offers invaluable contributions to early detection, personalized risk assessment, and real-time monitoring of pre-eclampsia. This chapter has presented a comprehensive overview of the key findings and implications of AI in pre-eclampsia, highlighting its promise and impact on maternal and fetal health.

One of the most crucial contributions of AI in pre-eclampsia is its ability to enable early detection. By analyzing diverse datasets, including vital signs, medical history, and genetic factors, AI-driven algorithms can identify subtle indicators of pre-eclampsia, even before symptoms become clinically evident. Early detection allows for timely interventions, including close monitoring, lifestyle adjustments, and medical treatments, reducing the risk of severe complications and improving maternal and fetal outcomes. Moreover, AI can facilitate personalized risk assessment by tailoring risk prediction models based on individual characteristics. This personalized approach allows healthcare providers to allocate resources more efficiently, ensuring that high-risk pregnancies receive targeted attention and care [28].

Real-time remote monitoring is another area where AI shines in pre-eclampsia management. AI-powered remote monitoring systems continuously track vital signs and other relevant data, providing instant alerts for potential pre-eclampsia symptoms. This real-time monitoring enables healthcare providers to detect warning signs early and respond promptly, which can be critical in mitigating adverse events. Additionally, AI serves as a valuable decision-support tool, empowering healthcare professionals to make well-informed and timely choices in managing pre-eclampsia. By providing transparent and interpretable recommendations, AI fosters trust between providers and the technology, leading to more effective and collaborative decision-making.

Furthermore, AI's capacity to analyze large-scale omics datasets offers new insights into the pathophysiology of pre-eclampsia. By examining genomics, proteomics, and metabolomics data, AI can potentially unveil novel biomarkers and therapeutic targets for improved pre-eclampsia management. This represents a promising direction for future research and holds the potential to advance precision medicine approaches in obstetrics [29].

Despite the promising applications, the implementation of AI in pre-eclampsia presents certain challenges. Data quality, privacy concerns, and biases in training data are critical considerations that must be addressed to ensure AI models are robust, reliable, and ethical. To establish standard operating procedures, test AI models through rigorous clinical trials, and set ethical standards for the appropriate use of AI in maternal healthcare, collaboration between AI researchers, healthcare practitioners, and legislators is crucial.

Moreover, there is a need to expand the usage of AI to address maternal health disparities. By extending AI research to low-resource settings and regions with higher maternal mortality rates, we can make pre-eclampsia detection and care more accessible and affordable. Ethical considerations must be at the forefront of these efforts to ensure equitable access to AI-powered healthcare solutions, regardless of patients' socio-economic status or racial and ethnic backgrounds [30].

In conclusion, the integration of AI in pre-eclampsia detection and management holds immense promise for improving maternal and fetal outcomes. The advancements in early detection, personalized risk assessment, and real-time monitoring have the potential to transform maternal healthcare, reducing maternal and fetal morbidity and mortality rates worldwide. As AI technologies continue to evolve, it is imperative to foster collaborative research and prioritize ethical considerations to harness the full potential of AI in supporting the health and well-being of pregnant women and their babies. By embracing AI as a valuable ally, we can pave the way for a new era of maternal healthcare, where data-driven insights and precision medicine intersect to improve the lives of millions of expectant mothers and their children [29, 30].

KEYWORDS

- **artificial intelligence**
- **biomakers**
- **healthcare**
- **machine learning algorithms**
- **pre-eclampsia**

REFERENCES

1. Roberts, J. M., & Cooper, D. W. (2001). Pathogenesis and genetics of pre-eclampsia. *The Lancet*, *357*(9249), 53–56.
2. Redman, C. W., & Sargent, I. L. (2005). Latest advances in understanding pre-eclampsia. *Science*, *308*(5728), 1592–1594.
3. Steegers, E. A., von Dadelszen, P., Duvekot, J. J., & Pijnenborg, R. (2010). Pre-eclampsia. *The Lancet*, *376*(9741), 631–644.
4. Huppertz, B. (2008). Placental origins of preeclampsia: Challenging the current hypothesis. *Hypertension*, *51*(4), 970–975.

5. Brown, M. A., & Magee, L. A. (2008). The hypertensive disorders of pregnancy: ISSHP classification, diagnosis and management recommendations for international practice. *Pregnancy Hypertension, 1*(1), 9–36.
6. Tranquilli, A. L., Brown, M. A., Zeeman, G. G., Dekker, G. A., & Sibai, B. M. (2013). The definition of severe and early-onset pre-eclampsia. Statements from the international society for the study of hypertension in pregnancy (ISSHP). *Pregnancy Hypertension, 3*(1), 44–47.
7. Bartsch, E., Medcalf, K. E., Park, A. L., Ray, J. G., & High Risk of Pre-eclampsia Identification Group. (2015). Clinical risk factors for pre-eclampsia determined in early pregnancy: Systematic review and meta-analysis of large cohort studies. *BMJ, 353*, i1753.
8. Redman, C. W., Sargent, I. L., & Staff, A. C. (2014). IFPA senior award lecture: Making sense of pre-eclampsia - two placental causes of preeclampsia? *Placenta, 35*(Suppl), S20–S25.
9. Poon, L. C., Kametas, N. A., Maiz, N. N., Akolekar, R., & Nicolaides, K. H. (2010). First-trimester screening for pre-eclampsia. Prediction of hypertensive disorders in pregnancy: A prospective multicentre study. *Ultrasound in Obstetrics and Gynecology, 35*(6), 672–680.
10. Hromadnikova, I., Kotlabova, K., & Ivankova, K. (2019). Preeclampsia: Prenatal screening and future personalized preventive approaches. *EPMA Journal, 10*(4), 345–364.
11. Li, M., Zeng, W., Li, Q., Deng, Y., Lv, Y., He, L., ... & Zhang, H. (2017). Identification of pre-eclampsia risk factors by data mining. *BioMed Research International, 2017*.
12. Goldstein, R. F., Abell, S. K., Ranasinha, S., Misso, M., Boyle, J. A., Black, M. H., ... & Teede, H. J. (2017). Association of gestational weight gain with maternal and infant outcomes: a systematic review and meta-analysis. *JAMA, 317*(21), 2207–2225.
13. Payne, B. A., Hutcheon, J. A., Ansermino, J. M., Hall, D. R., Bhutta, Z. A., Bhutta, S. Z., ... & Dadelszen, P. V. (2017). A risk prediction model for the assessment and triage of women with hypertensive disorders of pregnancy in low-resourced settings: The miniPIERS (Pre-eclampsia integrated estimate of risk) multi-country prospective cohort study. *PLoS Medicine, 11*(1), e1002301.
14. Oladapo, O. T., Adetoro, O. O., Ekele, B. A., Chama, C., Etuk, S. J., Aboyeji, A. P., ... & Gülmezoglu, A. M. (2018). When getting there is not enough: A nationwide cross-sectional study of 998 maternal deaths and 1451 near-misses in public tertiary hospitals in a low-income country. *BJOG, 125*(6), 928–938.
15. Deng, Y., Chen, S., Zhao, X., Luo, X., Zhang, Y., Wang, Y., ... & Zhang, H. (2019). An intelligent diagnostic system for the detection of preeclampsia based on deep learning. *Frontiers in Genetics, 10*, 269.
16. Bahado-Singh, R. O., Syngelaki, A., Mandal, R., Graham, S. F., Akolekar, R., Han, B., ... & Nicolaides, K. H. (2017). Metabolomic determination of pathogenesis of late-onset preeclampsia. *Journal of Maternal-Fetal and Neonatal Medicine, 30*(6), 658–664.
17. Ravì, D., Wong, C., Deligianni, F., Berthelot, M., Andreu-Perez, J., Lo, B., & Yang, G. Z. (2017). Deep learning for health informatics. *IEEE Journal of Biomedical and Health Informatics, 21*(1), 4–21.

18. Lavecchia, A. (2019). Deep learning in drug discovery: Opportunities, challenges, and future prospects. *Drug Discovery Today*, *24*(10), 2017–2032.
19. Wang, X., Jiang, W., Maggioni, A., Zhao, S., Wang, Y., & Cheng, H. (2020). A deep learning approach for preeclampsia diagnosis from ultrasound images. *IEEE Transactions on Ultrasonics, Ferroelectrics, and Frequency Control*, *67*(1), 164–172.
20. Kazemi, K., Ong, S. H., Yaghmaie, F., & Redman, C. W. (2014). Non-invasive prediction of pre-eclampsia using transabdominal uterine artery Doppler ultrasound and maternally inherited ancestry polymorphisms. *Ultrasound in Obstetrics and Gynecology*, *43*(5), 530–535.
21. Bilenko, N., Wisner, K., DeMaria, A. N., & McCoy, M. C. (2018). AI for improving screening and diagnosis of pre-eclampsia. *Journal of the American College of Cardiology*, *71*(11), A298.
22. Nebot, A., Jiménez-Carretero, D., Rivas-Medina, A., Gómez-Rioja, R., & Gómez, A. (2018). Computer-aided detection system for early diagnosis of preeclampsia from uterine artery Doppler signals. *Journal of Medical Imaging*, *5*(2), 021208.
23. Tura, A., Damiano, G., Martínez-Terrón, A., Kontopodi, E., Jambawalikar, S. R., Hancock, E., ... & Nicolaides, K. H. (2016). Computational analysis of maternal uterine artery Doppler signals in normal and preeclamptic pregnancies. *Medical Image Analysis*, *31*, 1–15.
24. Sornmo, L., & Laguna, P. (2015). A review of methods for beat-to-beat blood pressure estimation in preeclampsia. *Blood Pressure Monitoring*, *20*(2), 58–61.
25. Jiang, W., Wang, X., Zhang, Y., & Zhou, Y. (2019). Classification of preeclampsia based on an improved deep convolutional neural network with clustering. *Computers in Biology and Medicine*, *107*, 26–31.
26. Bertini, A., Salas, R., Chabert, S., Sobrevia, L., & Pardo, F. (2022). Using machine learning to predict complications in pregnancy: A systematic review. *Frontiers in Bioengineering and Biotechnology*, *9*, 780389.
27. Udenze, I. C., Arikawe, A. P., Makwe, C. C., & Olowoselu, O. F. (2017). A prospective cohort study on the clinical utility of second trimester mean arterial blood pressure in the prediction of late-onset preeclampsia among Nigerian women. *Nigerian Journal of Clinical Practice*, *20*(6), 741–745.
28. Bellos, I., Pergialiotis, V., Loutradis, D., Papapanagiotou, A., & Daskalakis, G. (2020). The role of hemoglobin degradation pathway in preeclampsia: A systematic review and meta-analysis. *Placenta*, 92, 9–16.
29. Smith D. D., & Costantine, M. M. (2022). The role of statins in the prevention of preeclampsia. *American Journal of Obstetrics and Gynecology*, *226*(2), S1171–S1181.
30. Ives, C. W., Sinkey, R., Rajapreyar, I., Tita, A. T. N., & Oparil, S. (2020). Preeclampsia-pathophysiology and clinical presentations: JACC state-of-the-art review. *Journal of the American College of Cardiology*, *76*(14), 1690–1702.
31. Datta, S., Barua, R., & Das, J. (2019). Application of Artificial Intelligence in Modern Healthcare System. doi: 10.5772/intechopen.90454.

CHAPTER 13

Incremental Learning

S. SARANYA, M. SUGANTHAPRIYA, P. R. HARIKRISHNAN, and P. KAVIN

Department of Computer Science, KGiSL Institute of Information Management, Coimbatore, Tamil Nadu, India

ABSTRACT

As a special case of machine learning, incremental learning may continuously learn relevant information from new data without access to the old data. It is thought to be one of the final aims of artificial intelligence technology and is anticipated to have the capacity for remembering. Incremental learning is still difficult to achieve over time. On stationary data distributions, modern deep neural network models excel thanks to batch training. Due to the fact that the distribution of the new data is unknown and has a very different probability from the previous data, this constraint causes catastrophic forgetting in incremental learning scenarios. As a result, a model needs to be both flexible for learning new information and sturdy for consolidating already learned information. Many deep learning techniques for continuous learning have been presented in recent years, but comparing their results is challenging because there is not a universal framework. We discuss task-incremental, domain-incremental, and class-incremental learning as three basic sorts, or "scenarios," of continuous learning to help with this. There are unique difficulties in each of these situations. In terms of difficulty and the efficiency of various techniques, we show that there are significant differences between the three scenarios. By laying a solid framework for precisely identifying benchmark problems, the suggested classification intends to organize the field of continuous learning.

Real-Time Artificial Intelligence (AI): Key Motivations, Technologies, Platforms, and Use Cases. Pethuru Raj, B. Sundaravadivazhagan, V. Kavitha, B. Narendra Kumar Rao, & Hannah Vijaykumar (Eds.)

DOI: 10.1201/9781998511358-13

13.1 INTRODUCTION

The goal of incremental learning is to enable the learning model to modify in response to new input without losing sight of its prior understanding. With incremental learning, sometimes referred to as online learning or lifelong learning, a model is continuously updated and improved over time by learning from new data without having to be totally retrained.

A model is trained using a fixed dataset in traditional batch learning, and once the training is over, its parameters are not changed. The model is created to adapt and learn from new data as it becomes available in incremental learning, though.

Incremental learning has uses in robotics, natural language processing (NLP), computer vision, and recommender systems, among other fields. The stability–flexibility conundrum, which involves balancing the stability of outdated knowledge with the plasticity of fresh knowledge, as well as potential biases brought about by the model's ongoing update are difficulties.

On the basis of the methodology and the manner in which the model is updated with new data, incremental learning in machine learning may be divided into three primary categories [1]. These include task incremental learning, incremental learning in a domain, and incremental learning in a class.

A machine learning technique called incremental learning involves gradually adding new data to a model and then retraining the model using the new data. As a result, the model is able to continuously learn and get better over time.

Online learning, also known as incremental learning, is a subfield of machine learning that involves continuously and in real-time processing of data coming in from a data stream with little to no prior knowledge of the distribution of the predictor variables, sample size, characteristics of the prediction or objective function (such as adequate tuning parameter values), and whether the observations have labels.

Algorithms for incremental learning are adaptable, efficient, and versatile. Incremental learning differs from conventional machine learning in the following ways:

1. A quick and effective data fit enables an incremental model to adjust in real time to changes (or drifts) in the data distribution.
2. The algorithm must be able to quickly create predictions from the most recent version of the model and postpone training the model

since observation labels may be lacking when related predictor data is available.

3. Before incremental learning begins, the population may not have much information available. As a result, the algorithm can be run from scratch. For classification issues, for instance, it is possible that class names will not be known until after the model has processed the observations. If there is sufficient knowledge before learning starts (for instance, you have accurate estimates of the coefficients of the linear model), you can express this knowledge to give the model a head start.
4. Data storage is either inefficient or impossible because observations can come in a stream and the sample size is likely unknown and perhaps enormous. In order to process observations, the algorithm must do so as soon as they are available and before the system discards them. It is difficult or impossible to tune hyperparameters because of this incremental learning feature.

In conventional batch learning, the complete data set is used to train the machine learning model all at once. But incremental learning adopts a different strategy. In sharp contrast to batch learning, which updates its model parameters all at once, it learns from new data points as they become available.

Think of a spam email filter as an example. When using batch learning, a huge number of emails are used to train the filter all at once, after which it is applied to new emails. If spam email characteristics change, the filter may start to malfunction unless it is retrained on a fresh batch of emails that includes the new spam characteristics.

An incremental learning-based spam filter, on the other hand, would modify itself as fresh emails came in, gradually upgrading its understanding of what qualifies as spam. This kind of filter might be able to recognize new spam styles without needing a fresh set of training data if spam tactics change.

What Advantages Does Incremental Learning Offer?

1. **Effective utilization of resources:** Less data must be stored at a time in incremental learning models, which can save a lot of memory. An example would be a bank's fraud detection system, which could update its model each time a transaction is made

rather than keeping track of every transaction and processing it afterward.

2. **Instantaneous adaption:** These models provide real-time adaptability. If we use an AI-based news recommendation system as an illustration, it can learn how a user's preferences change over time and suggest articles based on their most recent interests.
3. **Efficient education:** The machine learning model's capacity to quickly and efficiently learn new tasks can be improved by breaking a task into smaller components. Additionally, incremental learning helps the models' accuracy by enhancing it.
4. **Using nonstationary data to learn:** Incremental learning methods are very useful in a world where data can change quickly. For instance, a weather prediction model can continuously modify its projections in light of the most recent climate information.

What Constraints Apply to Incremental Learning?

1. **Forgetting that is catastrophic:** The model's propensity to forget previous knowledge as it learns new data is "catastrophic forgetting," which is one of the key difficulties with incremental learning.
2. **Managing notion drift is difficult:** Although incremental learning is intended to deal with changing data, it can be difficult to handle sudden shifts in data trends or "concept drift."
3. **Overfitting risk:** Since incremental learning depends on a stream of data, it might overadjust its parameters in response to recent data that might not accurately reflect the distribution as a whole. For instance, an extremely sensitive stock prediction algorithm could produce less precise long-term forecasts as a result of short-term market changes.

13.2 APPLICATIONS FOR REAL-WORLD INCREMENTAL LEARNING EXAMPLES

13.2.1 STOCK MARKET FORECASTING

For applications like stock market forecasting or weather forecasting where the data is continually changing, incremental learning is especially well suited. Since only the fresh data needs to be loaded and processed, it is especially helpful for applications when the data is too big to fit in memory.

However, incremental learning is not without its difficulties. The model must be able to manage idea drift that happens when the data evolves over time, which is one issue. Another difficulty is that updating the data can occasionally make the model forget what it has previously learnt.

Despite these difficulties, incremental learning is a potent technique that can be used to create AI programmes that evolve and learn over time [2].

13.2.2 FRAUD DETECTION IN BANKING

To identify fraudulent transactions, banks employ incremental learning algorithms, similar to the real-time fraud detection system employed by MasterCard. In order to determine the likelihood of fraud, MasterCard's technology examines over 100 different factors for each transaction, including the size, location, and kind of merchant. In order to adjust to changing fraudulent transaction patterns, the system uses incremental learning. For example, if the system starts to identify a new kind of fraud pattern, it can learn this pattern and update the model to find subsequent efforts that are similar, increasing the overall accuracy of fraud detection.

13.3 MACHINE LEARNING WITH IMPLEMENTATION: BATCH (OFFLINE) LEARNING AND INCREMENTAL (ONLINE) LEARNING

With the aid of data, machines are now able to learn much like people do and make judgements. This process is known as machine learning (ML). An ML model is trained by feeding it data, and after learning from the training process while receiving similar sorts of data, the model can do its specialized task based on what it has learned. But what if, while it is running on the production server, we need to retrain our ML model with fresh data? This situation involves the ideas of batch learning and incremental learning, commonly referred to as offline and online learning. You will be able to understand and use both after reading this essay. The following subjects will be covered in this article:

1. Batch learning or offline learning concept.
2. When to use incremental learning versus batch learning.

13.4 BATCH LEARNING OR OFFLINE LEARNING CONCEPT

The process of retraining the production model is known as batch learning, often referred to as offline learning. A model that is now operating on a server or in the cloud can only forecast the results of fresh data based on prior knowledge. However, if we need to retrain the model, we should do so by first replacing the existing model with a new one that is learned using both new and old data. The batch learning procedure is shown in Figure 13.1.

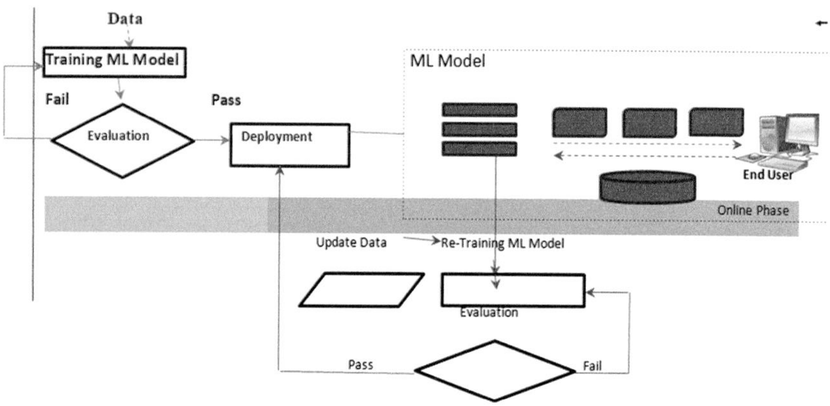

FIGURE 13.1 The batch learning procedure.

13.5 WHEN TO APPLY BATCH LEARNING AND WHEN INCREMENTAL LEARNING

For issues where new data do not arrive often and continuously, batch learning is appropriate. While the model requires quick and ongoing adjustment, incremental learning is suitable. When there is a medium to large volume of new incoming data, batch learning is appropriate. But if there is a lot of fresh data coming in, use incremental learning.

A model that engineers have access to at all times should be the subject of batch learning. Apply incremental learning, however, if the model might be unreachable for a while.

Batch learning is preferable if the model does not demand a lot of time or computational resources. Incremental learning might be preferable, nevertheless, if time and computer resources are limited and the model needs to be retrained.

13.5.1 *TASK INCREMENTAL LEARNING*

The goal of task incremental learning is to get around this limitation by allowing a model to gradually pick up new tasks without losing track of ones it has already learnt. It is particularly important in circumstances where the model must effectively adapt to new tasks that gradually appear over time.

The model in classical machine learning is trained on a predetermined set of tasks, and any new task necessitates retraining the model from scratch on both old and new data. This can be computationally expensive and impractical when dealing with a large number of tasks or continuous data streams.

Some examples:

Initially trained to discriminate between cats and dogs (Task 1), an image classification model gradually picks up the ability to recognize birds (Task 2) without losing sight of Task 1.

It updates its neural network and includes fresh bird photos via regularization. The model consistently gets better at correctly recognizing both cats and dogs as well as birds when it comes across more photographs of birds.

Later, it can keep its competence in earlier tasks while learning a new one, like telling vehicles from bicycles. This gradual approach enables the model to develop flexibility over time and adapt to new visual recognition issues.

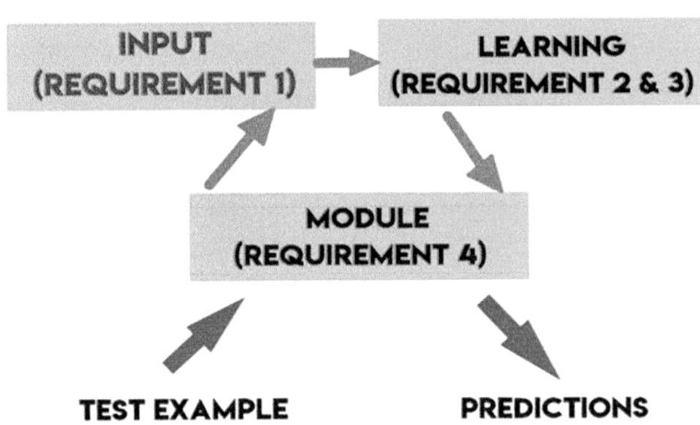

13.5.2 DOMAIN INCREMENTAL LEARNING

Adapting a model to handle new domains, which may have different statistical properties, while keeping knowledge of earlier domains is a challenge that is addressed through domain incremental learning. When the data distribution is nonstationary or the model must manage a constant stream of data from several sources or contexts, it is highly pertinent.

Although it is frequently assumed in classical machine learning that the training and test data originate from the same distribution, this assumption may not hold true in real-world applications due to changes in the data source or environment.

Some examples:

Initially taught to recognize credit card fraud (Domain 1), a fraud detection model gradually picks up the ability to recognize insurance fraud (Domain 2) without losing sight of its prior understanding of Domain 1.

It adjusts its fraud detection features using domain adaptation approaches to address the various data distributions in insurance claims. The model's accuracy in identifying fraudulent activity in both the credit card and insurance domains continues to increase as it comes across incidents of insurance fraud.

Later, it can maintain its competence in the earlier domains while adjusting to a new area, such identifying healthcare fraud. The model gains the ability to detect fraud across a variety of domains thanks to this incremental learning, enabling real-time fraud prevention.

13.5.3 CLASS INCREMENTAL LEARNING

The phrase "class incremental learning" refers to a machine learning paradigm in which a model is trained to continuously and gradually pick up new classes or categories without losing the knowledge of the previously acquired ones [4]. In other words, the model maintains its capacity to recognize the old classes while adapting and updating its parameters to account for the new information as new classes are presented.

Avoiding catastrophic forgetting, which is a tendency for some models to forget previously learned information when presented with fresh data, is the main objective. In situations where the data distribution can change over time and the model needs to dynamically adjust to new and evolving classes, class incremental learning is particularly crucial.

Some examples:

A virtual assistant that has been originally trained to identify fruits (Classes: apple, orange, banana) gradually picks up the ability to identify vegetables (Classes: carrot, turkey berry, tomato), without losing its understanding of fruits. In order to avoid catastrophic forgetting, it repeats and stores fruit samples while training on new vegetable data. The model continuously increases the accuracy of both fruit and vegetable recognition as it encounters new vegetables. Later, it can maintain its competence in the previous classes while learning a new class, like identifying plants. This gradual learning enables the virtual assistant to recognize a wider variety of food products in real time, enhancing its ability to support users with various dietary preferences and recipe recommendations.

13.5.4 TECHNIQUES IN INCREMENTAL LEARNING

1. Replay-based Methods:

This approach will periodically train the model using both new data and a subset of the stored historical data. By doing this, catastrophic forgetting is lessened. Use generative models to create artificial samples from historical data to supplement the present dataset while training, such as generative adversarial networks. Keep a reservoir of previous samples handy for occasional retraining or creating fictitious labels for fresh data.

2. Regularization Techniques:

The learning of significant parameters associated with previous tasks is constrained by the use of elastic weight consolidation (EWC) [3], which is used to regularize terms in the loss function. Synaptic intelligence uses importance-based regularization, like EWC, to safeguard crucial parameters.

3. Dynamic Architectures:

In order to handle new tasks while keeping the parameters discovered for earlier jobs, progressive neural networks (PNN) will enlarge the model architecture. When new information cannot be accommodated in the current design, gradually expand the model architecture.

4. Memory-Augmented Networks:

The model may retrieve previous information quickly with the use of neural networks with external memory components (like neural Turing machines), which facilitates incremental learning.

5. Herding and Clustering:

Employ historical data to create clusters or prototypes that can be used to train a model on new data. k-Means or other clustering algorithms can be used to define task boundaries and reduce forgetting during incremental learning.

6. Learning Without Forgetting (LwF):

Reduce the difference between the predictions on new data and the predictions on old data while training the model on new data. Through distillation or other regularization approaches, keep track of previous tasks.

7. Parallel Networks and Knowledge Distillation:

Train numerous networks individually on various tasks, then aggregate their expertise by distilling it into one main network. This lessens interference during training by allowing separate networks to concentrate on particular tasks.

These methods can be coupled to produce more potent and flexible incremental learning systems since they are not mutually exclusive. The method used will rely on the particular needs of the application and the difficulty of the tasks at hand. A lot of research is being done on incremental learning, and new methods are always being developed to boost its effectiveness and usefulness.

13.5.5 CHALLENGES OF INCREMENTAL LEARNING

The "catastrophic forgetting" phenomenon, in which previously learned information is forgotten when new information is gained, is one of the main obstacles to incremental learning. Since neural networks are notoriously difficult to retrain after being trained on a specific dataset, this is a special issue with them. Creating training data with care is necessary to prevent bias and make sure the AI system can generalize from the data it is given, which is another problem. Last, but not the least, there is the problem of handling nonstationary in the data, which can result in the AI system "chasing its tail" and never arriving to a conclusion.

13.5.6 INCREMENTAL LEARNING USED IN AI

1. A machine learning technique called incremental learning involves gradually adding new data to a model and then retraining the model using the new data. This can be applied to AI applications to enhance the model's performance over time when new data is added.
2. Since the model is continually being retrained on fresh data, incremental learning has the benefit of preventing over fitting. This can be especially helpful in situations when the data is continually changing, like when forecasting the stock market or the weather.
3. The ability to assist the model performs better over time as more data is added is another benefit of incremental learning. This can be especially helpful in situations when the data is continually changing, like when forecasting the stock market or the weather.
4. The effectiveness of AI applications can be increased by using the potent tool of incremental learning. Over time, it can help the model perform better and prevent over fitting.

13.5.7 HOW DOES THE ALGORITHM WORK?

Deep neural networks are primarily utilized for machine perception in numerous applications. Once a network has been trained to perform a particular activity, like classifying birds, it is difficult to educate it to perform new tasks [5]. The recognition of additional bird species or other datasets is then done via incremental learning.

Deep neural networks are here likely to recognize the new network when new data is supplied. That means it is important to remember the previous network after adding new tasks. It will be more efficient to use networks that can gradually assimilate new knowledge, similar to how people learn new memories over time, rather than having to start the model from scratch each time a new task needs to be learnt. There have been numerous attempts to create strategies that reduce catastrophic forgetting; however, these methods have not been directly compared, the tests used to assess them differ greatly, and they have only been tested on small-scale issues (like MNIST).

A scenario example for incremental learning

Here, we use the dataset for autonomous vehicles to teach this technique.

Think about two equations,

$$D_A = \{(\mathrm{xi},\mathrm{yi})\}$$

$$D_B = \{(\mathrm{xi},\mathrm{yi})\}$$

First,

$$\mathrm{Min}\ \theta \sum_{xi,yi \in DA} \mathrm{L}\{y\theta\,(xi), yi\}$$

This equation A will initialize the dataset randomly.

Then,

$$\mathrm{Min}\ \theta \sum_{xi,yi \in DB} \mathrm{L}\{y\theta\,(xi), yi\}$$

The previously stated assignment will be initialized and solved using this equation B one at a time.

The sample used above was made up of two datasets, A and B, which were trained separately without being integrated. There will be a chance of forgetting while training such data.

This is seen as disastrous forgetting or meddling.

A typically yields Worse Performance, while B typically yields Good Performance.

LEARNINGWITHOUTFORGETTING:

Start with:

θ_s: shared parameters

θ_o: task specific parameters for each old task

X_n, Y_n: training data and ground truth on the new task

Initialize:

$Y_o \leftarrow \mathrm{CNN}(X_n, \theta_s, \theta_o)$

$\theta_n \leftarrow \mathrm{RANDINIT}(|\theta_n|)$

Train:

Define $\hat{Y}_o \equiv \mathrm{CNN}(X_n, \hat{\theta}_s, \hat{\theta}_o)$

Define $\hat{Y}_n \equiv \mathrm{CNN}(X_n, \hat{\theta}_s, \hat{\theta}_n)$

$\theta_s^*, \theta_o^*, \theta_n^* \leftarrow \underset{\hat{\theta}_s, \hat{\theta}_o, \hat{\theta}_n}{\mathrm{argmin}} \left(\lambda_o \mathcal{L}_{old}(Y_o, \hat{Y}_o) + \mathcal{L}_{new}(Y_n, \hat{Y}_n) + \mathcal{R}(\hat{\theta}_s, \hat{\theta}_o, \hat{\theta}_n) \right)$

13.5.8 HOW TO MITIGATE FORGETTING?

The following list of downsides results from training again from scratch while comparing both datasets while combining datasets A and B.

1. Many industry networks require days or weeks to train;
2. Power and computer waste
3. Original Training data could not be accessed.
4. Requires cloud or consistent internet access.

Since humans may pick up new information gradually, it is possible to play back previous training data with fresh data.

Old data must be stored, and storage costs increase linearly with tasks.

Evaluating Incremental Learning Algorithms:

1. Data Permutation tasks
2. Incremental class learning
3. Multimodel Learning

Data Permutation Tasks:

Here, we are going to train using images (xi, yi) which consist of class images of numbers 0 to 9.

After training the model, we can predict and classify image using previous training data.

Task 1: Training data – MNIST
Given xi predict yi.
Task 2: Pick an image permutation P2
Training data – MNIST (P2xi, yi)
Given P2xi will predict yi.

13.6 INCREMENTAL CLASS LEARNING

This results in learning a foundational task set, followed by learning new classes, which entails learning through shared features with other classes. A particular application of incremental learning called incremental class learning (ICL) concentrates on the situation where a machine learning model must gradually pick up and recognize additional classes [3].

In other words, the model must be able to accommodate new courses while still remembering information from earlier classes.

Approaches for Continuous Learning:

1. Train whole net with regularization
2. Elastic weight consolidation (EwC)

13.7 REGULARIZATION APPROACHES

In machine learning, regularization strategies are used to reduce over fitting and enhance the generalization capabilities of models. Over fitting, which causes a model to perform poorly on unobserved data, happens when it becomes overly complicated and fits the training data too closely. To limit the complexity of the model and lessen over fitting, regularization techniques impose extra constraints or penalties during model training.

13.8 ELASTIC WEIGHT CONSOLIDATION

In order to solve the issue of catastrophic forgetting, elastic weight consolidation (EWC), a regularization approach, is utilized in incremental or continuous learning [4]. A neural network's performance on older tasks suffers when it is trained on new tasks because it has a tendency to forget the information it learned from earlier tasks. EWC is intended to address this problem by selectively maintaining key weights in the neural network that are essential for the completed tasks while enabling the model to successfully learn new tasks.

13.9 MULTIMODEL LEARNING

When discussing incremental learning, the term "multimodel learning" describes the process of gradually assimilating knowledge from various models. In order to make judgements or complete activities more successfully, it includes merging the information and forecasts from many models. Models are continuously updated and adjusted to new data or tasks as they become available in multimodel learning for incremental learning.

13.10 BENEFITS OF INCREMENTAL LEARNING

Incremental learning sometimes referred to as lifetime learning or online learning has a number of advantageous features that make it an appealing machine learning strategy. The model can continuously learn from new data while preserving information from previous encounters, which is one of its main advantages.

The model can remain accurate and useful in dynamic contexts where data distributions may vary over time thanks to its adaptability. Additionally, incremental learning is very effective since it only needs to process the fresh data, which significantly reduces computational costs and makes it scalable for big datasets.

Incremental learning in AI has several advantages. The ability to learn more effectively is one of its most significant advantages. By splitting the work down into smaller components, incremental learning enables the AI system to learn new tasks more quickly and efficiently. Additionally, by enabling the AI system to concentrate on one task at a time, incremental learning can help to increase the accuracy of the system.

The ability to decrease the amount of data needed for training is another advantage of incremental learning. This is so that the AI system can gain knowledge from a smaller dataset and apply it to a larger dataset later. When there are few training data available, this can be particularly helpful.

Finally, incremental learning can aid to increase the AI system's interpretability. This is due to the fact that by segmenting the work, the system may offer justifications for its predictions. This can be useful for troubleshooting and understanding how the AI system operates.

Research directions for incremental learning in AI:

When it comes to incremental learning in AI, there is still a lot of research to be done. Some potential areas for future investigation include:

1. Examining the best ways to include various data sources in an incremental learning environment
2. Creating innovative concepts for concept drift enhancing current methods for determining when fresh data is available for learning, investigating the best ways to include human feedback into incremental learning systems, and creating incremental learning algorithms that are flexible enough to accommodate changing computational resources.

13.11 CLASS INCREMENTAL

Different continual learning scenarios can be used to empirically evaluate a model's capacity for continuous learning. The limitations and potential of the learning environment are described in each scenario. Here, we argue against the prevalent practise in the literature on continuous learning to experiment primarily with class-incremental scenarios, where classes present in one experience are never brought up again. In light of the fact that class-incremental situations artificially aggravate catastrophic forgetting at the expense of other crucial goals like forward transmission and computational efficiency, we propose that an overemphasis on this setting may be restricting for future research on continuous learning. In fact, repeating of previously encountered notions happens organically in many real-world settings, which helps to mitigate the disruption of prior knowledge. We call for a further investigation of alternate continual learning situations, in which repetition is purposefully incorporated into the flow of incoming knowledge. We outline the benefits that such class-incremental with repetition scenarios could provide for a more thorough evaluation of continuous learning models, starting from already-made suggestions.

13.12 CLASS-INCREMENTAL CONSEQUENCES

The class-incremental learning scenario is currently highly well-liked in the community of continuous learners. Its simplicity and use have encouraged new research and efforts to reduce catastrophic forgetting, the key issue that models encounter when learning in this environment. Task- and domain-incremental learning are further examples of ongoing learning that are discussed in the literature. Learning is task-free, data-incremental, and on-going online [7]. These scenarios suggest other environments for learning, but none of them fully address or analyze the assumptions and limitations that define the class-incremental scenario. We will highlight a few noteworthy exceptions that apply to classification based on repetition of previously seen classes in circumstances of ongoing learning [6]. We now focus on some of the restrictions brought on by the class-incremental algorithm's no repetition restriction.

13.13 REPETITION OCCURS NATURALLY IN MANY REAL-WORLD ENVIRONMENTS

There are few cases where class-incremental learning is coordinated when repetition is derived from the environment. Examples include manipulating several things robotically, forecasting financial trends, and performing autonomous driving duties, among others. A learning agent exposed to a constant flow of information should be able to gradually pick up new information while also being able to drop irrelevant ideas and organize learning according to importance. Not all observed information should be handled equally: if a particular pattern never happens again, pretending to forecast it accurately may be pointless. In fact, it may be possible to view the statistical recurrence of concepts and the interaction between them through time as significant sources of information for assessing the significance of what the agent experiences. If all the information is treated equally, it is exceedingly difficult to decide which concepts to reinforce and which to forget.

Learning in silos inhibits many of the potential insights an agent can gain from the complexity of the environment, ultimately limiting its ability to build a world model that is appropriate for the ever-changing task it must handle.

13.14 DISCUSSION

One of the main problems with artificial intelligence and machine learning is, of course, ongoing learning. For many real-world applications, where the environment is so diverse and complex that an offline training phase may never supply the agent with enough information to succeed, the ability to continually learn from a progressive exposure to new concepts is essential. Research on ongoing education is still in its early stages. As a result, it is only reasonable to rely on skewed experimental setups and to create solutions that are incapable of adapting to any scenario. In this regard, we emphasized how class-incremental scenarios offered a rapid experimental setup that might take advantage of the enormous number of resources already available, particularly in the area of computer vision, to the study of continuous learning. Class-incremental scenarios will still be helpful for the community of lifelong learners because of the rapid prototyping opportunities and the substantial datasets and benchmarks involved. It is crucial to understand that the problems are largely caused by the assumptions underlying class-incremental scenarios, specifically the no repetition restriction. We provided a few different situations. Additionally, we offered a method to create unique data streams for CIR scenarios that carefully balances the model's exposure to new and existing information. The configuration inherent in many real-world applications, which cannot be modeled with a class-incremental scenario, makes the approach particularly suitable for streams with a significant number of little experiences (experiences containing datasets with few patterns).

KEYWORDS

- **batch learning**
- **class-incremental scenarios**
- **data storage**
- **filter**
- **incremental learning**

REFERENCES

1. Stojanov, S., Mishra, S., Thai, N. A., Dhanda, N., Humayun, A., Yu, C., et al. (2019). Incremental object learning from contiguous views. In *Proceedings of the IEEE/CVF Conference on Computer Vision and Pattern Recognition* (pp. 8777–8786). DOI: 10.1109/CVPR.2019.00898
2. Aljundi, R., Kelchtermans, K., & Tuytelaars, T. (2019). Task-free continual learning. In *The IEEE Conference on Computer Vision and Pattern Recognition (CVPR)*. DOI: 10.1109/CVPR.2019.01151
3. Géron, A. Hands-*on machine learning with scikit-learn, keras, and tensorflow: Concepts, tools, and techniques to build intelligent systems*, 2nd edn. (pp. 15–18). Retrieved from https://www.amazon.com/Hands-Machine-Learning-Scikit-Learn-TensorFlow/dp/1492032646
4. Online vs offline machine learning—What's the difference? Retrieved 12 Dec 2022, from https://www.qwak.com/post/online-vs-offline-machine-learning-whats-the-difference

5. Incremental (Online) Learning with Scikit-Multiflow. Retrieved 13 Dec 2022, from https://towardsdatascience.com/incremental-online-learning-with-scikit-multiflow-6b846913a50b
6. Train and test a stream classification model in scikit-multiflow. Retrieved 13 Dec 2022, from https://scikit-multiflow.readthedocs.io/en/stable/user-guide/quick-start.html
7. Machine learning model retraining approach-incremental and continuous model training. Retrieved 13 Dec 2022, from https://www.youtube.com/watch?v=T4y25jc5NyM&list=RDCMUCNU_lfiiWBdtULKOw6X0Dig&start_radio=1&rv=T4y25jc5NyM&t=197

CHAPTER 14

Understanding the Role of Artificial Intelligence in the Industry 4.0 Ecosystem a Digital transformation: Case Studies

P. SUMITRA

PG & Research Department of Computer Science and Application, Vivekanandha College of Arts and Sciences for Women (Autonomous), Elayampalayam, Namakkal (DT), Tiruchengode, India

ABSTRACT

The rate of human evolution and productivity increased exponentially with each third Industrial Revolution, as well as the previous two (Industry 4.0) currently well under way. The way things are created, produced, and delivered is experiencing a significant technological transformation in the manufacturing sector. All physical assets must be quickly and completely digitalized in order to be integrated into digital ecosystems as part of Industry 4.0. It provides significant opportunities for future innovation and business growth. Modern technology, such as artificial intelligence (AI), the internet of things (IoT), big data, machine learning (ML), and other developing technologies, is being used to achieve Industry 4.0. Artificial intelligence has lately seen significant advancements in Industry 4.0 (AI). Industries are focusing on increasing product consistency, productivity, and reducing operating costs in the hopes of achieving these goals by combining human and robotic work. Smart industries' hyperconnected production

Real-Time Artificial Intelligence (AI): Key Motivations, Technologies, Platforms, and Use Cases.
Pethuru Raj, B. Sundaravadivazhagan, V. Kavitha, B. Narendra Kumar Rao, & Hannah Vijaykumar (Eds.)

DOI: 10.1201/9781998511358-14

processes rely on numerous devices that interact via AI automation systems by gathering and comprehending all data kinds. Modern manufacturing can be transformed in significant ways by the use of intelligent automation technologies. Industries will use AI to process data supplied from those devices based on their goal to include IoT devices and linked machines into their equipment. In this chapter, we will discuss about the introduction to Industry 4.0 which will be covered in the first section. The study's second part, which is based on a review of the literature, attempts to summarize how significant AI is to the implementation of Industry 4.0. The third section of the article lists the important developments and numerous difficulties that have made it possible to apply AI. The chapter concludes by outlining and discussing significant AI for Industry 4.0 applications.

14.1 INTRODUCTION

The Germans developed the idea of the fourth industrial revolution, or Industry 4.0, which combine three technical trends: connectivity, intelligence, and flexible automation. The businesses are responsible for fueling the revolution brought on by the exponential expansion of emerging technologies and the transformations they are bringing to the workplace, markets, and the labor force's availability. Quality 4.0 describe how quality will be demonstrated in businesses and how it will be applied to disruptive, new, and digital technologies. In this particular section we are mentioning about the following subdivisions [1–4].

14.1.1 CYBER-PHYSICAL SYSTEMS

Industry 4.0's development phase implements simple, iterative actions and computerized systems that enable integration of physical reality activities with technology and communication infrastructure. System's control components and sensor are interconnected with machines, devices, installations, vessels, networks, and people. Physical and computational capabilities enable interaction through a wide range of different modes [5].

When embedded systems are regulated in contemporary life, things like automobiles, smartphone, and home appliances become a part of your everyday experience. For instance, in many application fields like pharmacy, driving safety measures and the method for the driver to assist cars, automated systems, and machines produced process control, and assistance the process for managing the power supply [6].

1. CPS
2. General
3. Expansion
4. Updates
5. Reliability
6. Data management
7. Privacy
8. Real time
9. Data heterogeneity
10. Security challenge
11. Compromised key attack
12. Man-in-the-middle attack
13. Denial of service attack

14.1.2 IoT

It has been a primary study area for many years. The CPS responds to the internet of things. This protocol aids in making sure that the tactics and components are properly connected to the TCP/IP or Internet-based protocols. Table 14.1 lists the security issues that the Internet of Things is now facing.

14.1.3 INTERNET-OF-SERVICES

The only requirement for using software is an internet connection, together with the necessary communication tools and software development tools [11]. It has been utilized to provide a variety of services, mostly for the medical care of diabetic patients. These services include managing linked devices, particularly mobile phones, and keeping machinery operational whenever a malfunction occurs.

14.2 LITERATURE ASSESSMENT

Digital technology has possibilities increasing the impact of trade actions in the environmental sector; expectations are high for Industry 4.0.

TABLE 14.1 Internet of Things Security Challenges.

S. No	Security Risk Factor	Description
1	Software and firmware vulnerabilities	When applied to real-world situations, many devices are resource-constrained due to a lack of processing power, insufficient access control, funding sources that are not sufficient, inability to update their hardware, and weak defence against physical assaults.
2	Insecure communications	IoT devices with limited resources lack the security mechanisms that were originally built for desktop computers. A dangerous threat is the man-in-the-middle attack, which is one kind of attack. This attack's primary goals are to exchange data with other devices and install malware.
3	Data leaks from IoT systems	When a transmission is not encrypted, hackers can easily access our information, such as our location data, bank details, and family health information. Further, because our data is hosted in the cloud, external attacks potentially affect cloud-hosted services. Most likely, when information is not protected, attackers will be enthusiastic to collect useful data.
4	Malware risks	Set-top boxes, smart watches, and smart TVs were targets of malware assaults that posed a high risk of harm. This specific assault is effective in altering functionality, obtaining user information, and launching other attacks.
5	Cyberattacks	There are some other attacks that are not limited on IoT devices. **1. Denial-of-service attacks (DoS):** Where computing power is scarce, this kind of assault will be especially vulnerable. **2. Denial-of-sleep attacks:** The two basic modes that must be controlled in communication are awake and sleeping modes. This specific attack most likely depletes the battery and disables the sensor mode [7]. **3. Device Spoofing:** It is a widely used trick by affiliate scammers to make it appear on many types of devices [8]. **4. Physical Intrusion:** It is a form of attack phase that directly stole the company's information [9]. **5. Application-based attacks:** It is a form of attack that secretly gets the information on a person's personal and professional life [10].

This section of the chapter deals with certain studies on the environmental sustainability of Industry 4.0.

14.2.1 INDUSTRY 4.0 AND ENVIRONMENT SUSTAINABLE

O'Rielly et al. [12] discussed about ways to help Canadian businesses reduce energy loss. Simo Ford et al. [13] provided a summary of the benefits and drawbacks of additive manufacturing [AM] based on consumption and production. Tim Stock [14] addressed the role of Industry 4.0 technology and sustainability in the environment. Jay [15] discussed about the present state of artificial intelligent technologies and the environmental system needed to use in industry. Sang M. Lee [16] made several recommendations for forecasting the management of ecosystem maintenance and enhancing quality in the 4.0 Industry.

Productivity, policy support, methods for determining cause and effect, and the usage of blockchain technology may all help in this process. Husham [17] addressed about the difficulties with IoT from the aspect of security concerns and data protection, which is based on crypto. Senthil Kumar [18] analyzed and reviewed about the relevance of big data analytics and artificial intelligence (AI), which will serve as a foundation for upcoming research. Maksim [19] analyzed his study on how AI is being integrated into the 4.0 industry, as well as its applications and advantages over programming languages. The studies were conducted using various databases, and they ultimately came to the conclusion that personal interaction is more ideal.

David Mhlanga [20] learned about AI in the financial and educational sectors. When more investment in AI is required, as well as more AI research, we must focus more on new, creative technologies, infrastructure development, and poverty reduction in order to achieve the sustainable development goals. Bakshi [21] focused on the barriers and challenges to improve sustainable in Industry 4.0. Mohammad Seraj [22] investigated a unique method for prediction maintenance in the 4.0 Industry with the aid of the hybrid prediction (HR) approach. The environment's state is provided as input parameters for the Industry 4.0 framework, and the resulting machine productivity is the desired output.

Sameen Khan [23] examined a number of facets of customer behavior in connection to AI and 4.0 Industry items. This specific study was expanded upon using a number of customer behavior criteria, including demand for recognition, purchase intention, and acceptability. Sascha [24] conducted an evaluation of almost 2,365 publications in the Cyber-Physical-System sector, covering around 32 distinct domains, 10 divisions, and 246 field visits. He came to the conclusion that many studies in the 4.0 Industry needed to be improved in the years to come. Sanjay [25] looked at contemporary innovations in technology, focusing on hotels, to make the environment more sustainable.

14.3 BENEFITS OF INDUSTRY 4.0

Literature on AI's "Industry 4.0" that is connected to environmental sustainability from the most recent years, between 2010 and 2023, was reviewed. The researchers thoroughly considered many viewpoints from both the sides. Industry 4.0 technology aids producers in making better, more useful products. They may enter the next production and transporting age with economical, modular automation strengthened by input influenced by data, providing total command over the movement of materials and supplies. It has several advantages, including enhanced performance and competitiveness, increased adaptation and resilience, and increased profitability. Manufacturers are now ahead of the competition thanks to their innovation initiatives. It would also enhance customer service. Despite how fascinating and thrilling smart factories ought to be always the major topic of discussion.

The most crucial action is to make investments in cultural change and capacity building. As part of the "Industry 4.0" concept, the devices will be networked, creating a connection outside the boundaries of the industrial site. The importance of enormously organized data from sensors and equipment cannot be overstated. The workforce might adapt to changing conditions, be ready for further study, and remain relevant by developing their analytics and new technology abilities. Data has become a crucial tool for many businesses.

14.4 ENVIRONMENTAL SUSTAINABILITY REQUIREMENTS

Industry 4.0 is growing, and industries all across the world are changing. The accomplishments of companies that have integrated aspects of Industry 4.0

into their processes. To enable businesses to correctly implement Industry 4.0, a policy must be created. Distinct markets in geographical locations may see different effects of Industry 4.0. New technical capabilities and data resources are the foundation for reevaluating practically every element of traditional industry, including production systems, company floors, and supply chain.

The concept is to combine hardware and software such that computers are employed as part of a machine learning loop rather than as the main task. The IoT and industrial IoT technologies are necessary for any Industry 4.0 endeavor. The physical systems react, analyze processes using software, and record results. As part of the Industry 4.0 philosophy, data is acquired from a range of gadgets and sensors, including robots in manufacturing companies, greenhouse sensor dispensers, and systems in healthcare. IoT seeks to gather data. Computers and networks coordinate and keep an eye on physical processes through feedback loops.

14.4.1 SUSTAINABLE MANUFACTURING

A socio-technical paradigm is known as "Industry 4.0." that unifies big data, CPS, virtual world, using the cloud and digitalization technologies to address how organizational, social, and technological factors interact. Sustainable manufacturing offers a significant opportunity to expand output through Industry 4.0's process simplification, reduction of lead times, and organizational effectiveness. The goal of sustainable manufacturing is to apply the fundamental ideas of sustainable development to industry. It aids in enhancing efficiency on the social, economic, and environmental fronts.

There are still many difficulties and risks that the manufacturing sector must overcome. Hardware at the same time producers and software companies began to specialize the creation of corporate software for large-scale commercial operations with an emphasis on internal reliability and environmental protection. These advancements in the creation of the smart grids, the green energy management, and the distributed output are all part of the environmentally friendly digital transformation. Increasing productivity, improving service to the customer, automating the creation, and integrating the supply are the goals for implementing it.

14.4.2 INDUSTRY 4.0 IN SUSTAINABILITY ENVIRONMENT STUDY

Industry 4.0 is upending the long-standing order of things, altering the rules of the game, and offering the knowledge and direction necessary to boost sustainable efficiency, notably in the manufacturing sector. Performance criteria are evolving nowadays to emphasize reduced emissions and potential gains in efficiency from technology. Upon saving the energy, reducing in carbon, and optimizing the value chain, digital technology has really long helped organizations excel in sustainability. Today's industries are working harder to improve their environmental performance.

Currently, many businesses are focusing their efforts on finding a balance between sustainability aims while still taking profit, and people into consideration. Today, many businesses use a variety of digital technologies to cut down on energy use and production waste. It makes it possible to construct more extensive functions, such as improving social welfare and reducing energy destructive emissions [26–28]. Manufacturing has been a major force behind progress for a long time.

Despite the fact that innovations have been around for a while, implementations have frequently only been prototypes. Additive manufacturing has greatly accelerated progress for many years. Thanks to Industry 4.0 improvements, a significant amount of the work in factories and industrial facilities is now performed by intelligent machinery and other technical equipment. It easy to predict how technology will permeate the global manufacturing industry. In order to satisfy immediate needs and ensure sufficient resources for future generations, a healthy society makes use of its financial, human, and environmental resources.

1. Sustainability Dimension
2. Natural Resource
3. Energy
4. Material
5. Waste and Emission

Utilizing technology effectively reduces waste, saves time, and uses less energy, which also aids in the creation of a smart production system. In a sustainable society, a wide range of individual interests are considered and effectively met. Virtual worlds created by computers help advance industry management methods. The use of resources like tools and tasks is made easier through various immersion technologies.

Growing assurance makes it easier for businesses to maintain their inventory. In a large warehouse, a worker will utilize a smartphone to find a needed component. Industry 4.0 technologies like robotics and modular electronics have aided in the development of plastic production techniques that use the least amount of water and promptly wash away effluent, significantly reducing the impact of plastics on water systems.

14.5 ADOPTION AND TOOLS

14.5.1 ADOPTION IN ENVIRONMENTAL SUSTAINABILITY

Industry 4.0 technology can help the sector become more sustainable. All modern digital businesses select environmental sustainability because they aim to change the world. The primary problems we face today are climate, excessive reliance on nonrenewable energy, loss of diversity, substantial erosion, severe likely disasters, significant carbon dioxide emissions, and poor air and water quality [29–30]. Big data is a critical component of any organization's operations, and it should be taken into account to develop and validate procedures and improve the efficiency of their shop floor. Through industry 4.0, businesses can create a variety of applications and business cases that can be used to improve their operations. Device integrators are in charge of pledging their commitment to change, while technology vendors provide knowledge of real-world, sectoral integration to Industry 4.0 manufacturing scenarios.

One of the most effective ways to improve the efficiency of a factory is by connecting its floor to a digital twin. This can be done through the use of a secure and adaptable digital environment. Having the necessary knowledge about the various risks associated with the digital transformation of an organization is very important for businesses to adopt industry 4.0 [31, 32]. Using data, IoT, and automation, every equipment in a connected factory or industrial IoT plant collaborates to track and monitor every part of production.

The difficulties cleared the way for digitalization, and the founding of new businesses was centered on the creation of an altogether new, contemporary environment that had nothing to do with any activity or transaction. The significance of digital technology in the possible protection of the environment has also been highlighted by the corona virus epidemic. Many countries have created smartphone apps that connect crucial health services and notify users of risk concerns, best practices, and other COVID-19 warnings. Since more people are staying at home and using the internet for daily transportation, shopping, and travel, the amount of traffic on the roads and in the air has dramatically decreased, lowering greenhouse gas (GHG) emissions [33, 34].

14.5.2 TOOLS IN ENVIRONMENTAL SUSTAINABILITY

One of the key elements of industry 4.0 is AI, which provides robots the capacity for thought, learning, and decision-making. Another technological advancement is IoT, which enables machine–human connection. Data-based cloud computing, other Industry 4.0 tenets, and elements of associated products and activities are further important Industry 4.0 instruments for environmental sustainability. In-depth data analysis is a crucial aspect of Industry 4.0.

Its main objective is to acquire customer feedback and data so that businesses may provide them with the products and services they want. Since it provides dependable services on time, resources and time are saved. Without involving any humans, Big Data Analytics is utilized by the control software to give exact commands to these actuators and robots, modifying the actual equipment/machinery settings. Predictive maintenance is a component of big data analysis, which significantly cuts down on response time. Automation of supply chain management is another method that firms use big data analysis.

The way business is done now will change with new scientific advancements. It became a big influence on the global economy and is changing the economy's fundamentals. More information on these specific components may be found under "real-time systems," "big data," "real-time facts," "life cycle analysis," "smart production," "sensor utilization," "digitalization," etc. [35].

14.6 SIGNIFICANT ADVANCEMENTS IN THE CREATION OF A SUSTAINABLE ENVIRONMENT IN INDUSTRY 4.0

Changing the current production, service, and architectural processes is the aim of Industry 4.0. To sustain production and boost overall efficiency, industry 4.0 is used. It combines the most recent developments in "manufacturing technology" with "cyber-physical frameworks" like "Internet of Things" and "Internet of Services." It may be used to create efficient maintenance systems that communicate with one another and transfer data to other systems. By capturing carbon dioxide from an undersea storage facility and maintaining it in the atmosphere throughout the mining process at its source, technologies like carbon capture and storage can significantly lessen the ecological consequences of mining operations.

Modern production complicated, mechanized, and increasingly digital industry is heavily dependent on computers and cutting-edge tools. The production of businesses reliant on natural resources is continually under danger due to a lack of raw materials, deteriorating environmental conditions, and climate change. Their long-term goal is to find viable options that provide economic expansion without increasing damage of the environment [36, 37]. The following are major in Industry 4.0 developments made possible by building a sustainable environment as shown in Table 14.2

14.7 OBJECTIVES OF THE RESEARCH

Industry 4.0 is a revolution, and one of its objectives is to motivate businesses to meet the necessary success requirements. This study meets a variety of research goals, which helps address difficulties related to Industry 4.0's efforts to build healthy environment. The paper's primary research goals are as follows:

- RO1: What can be the outlook of Fifth Industrial Revolution in Smart Factory?
- RO2: How can we reduce the carbon emission in Industry 4.0?
- RO3: How can we reduce the negative impact on environment and socioeconomic aspect?
- RO4: What is required for service management in Industry 5.0?
- RO5: What are the steps that are needed in Product Optimization?
- RO6: Which country is initiating Industry 4.0 with skills and optimization to be flexible?
- RO7: State the condition which will be compatible with change in climatic.
- RO8: What is the role of Industry 4.0 in handling Industry in India?

TABLE 14.2 Developments in Sustainable Environments.

S. No	Name of the Developments	Description
1	Smart Productions	People, processes, and technology should be seamlessly integrated, according to proponents of industry 4.0. To keep track of stocks, reduce unscheduled downtime, and maximize asset usage, chips are increasingly widely utilized in a variety of industries. After the chips were put to use, following design innovations boosted their dependability to the point where they could now be employed in production. Smart production might be defined as the integration of IT technologies functioning in a coordinated manner in real time. Digital technology uses electric fields to automatically find and record tags. Services must routinely update outdated technology and software in order to stay operational. Drones can collect information during various project phases for instrumentation, automatic reporting, inspections, higher-level decision-making, and performance analysis. There are no exceptions when it comes to controlling and monitoring systems that offer crucial features including alarm control, process and water operator control, data gathering, and historical records [38–40].
2	Water Industries	The industry sector is presently going through the same transformation process that financial institutions and retail stores endured for a long time. Particularly with Industry 4.0, new opportunities for better asset management are made possible, including real-time remote tracking, intelligent water monitoring, and alarm-driven preventative maintenance. It suggests strategies to enhance community response and interaction networks, hasten climate response times, and ensure that both rural and urban communities always have access to water. These abilities assist the water business in meeting new needs, effectively allocating infrastructure and resources, and reducing costs while rapidly providing consumers with better services [41].
3	Reducing of Energy Consumption	Industry 4.0 technology may be utilized to assist people grasp their immediate goals as well as make decisions at every level of the innovation process. Together, businesses will make sure that cities play a crucial role in revolution of energy. Businesses and communities provide projects a lot of support, and they also create creative business structures. IoT technology enables energy-efficient improvements to urban quality of life [42].

TABLE 14.2 *(Continued)*

S. No	Name of the Developments	Description
4	Simplicity of information	Industry 4.0 makes use of big data analytics in smart factories, which analyze sensor data from manufacturing unit to foresee the need for maintenance and repair work. To build smart factories, optimize the supply chain, increase pricing, foresee problems, and create new products, big data analytics is necessary. They will be able to find undiscovered output variables thanks to big data analytics. The self-service platform's real-time data deconstruction shows trends, mistakes, and important decision-makers. Large amounts of data from industrial activities can be combined with the use of self-service engineering analysis. The capacity to make operator decisions both within and outside of production units is made possible by connection and information availability [43].
5	To improve air quality	To achieve the national sustainable goals, platforms that maximize the reuse, recycling, and refresh culture in various enterprises on a waste-free basis are required. By developing intelligence, it will significantly cut our energy use, waste generation, and carbon emissions in our cities. Virtual power plants and renewable energy infrastructure, such as solar panels, may be connected to the network of reliable power supply. Road congestion can be reduced and traffic may be managed. The use of automated processes might significantly lower both city emissions and the air quality index. The air becomes cleaner and no pollution is produced [44, 45].

RO9: What will be the outcome of the two companies in Industry?
RO10: What can be merits and forthcoming challenges in shifting Industry 4.0?
RO11: What can be the 4.0 Industry Market size by Application on Industrial Automation and by vertical energy and utilities?
RO12: What are the developments in this era in Market?
RO13: Mention the leading company in 4.0 Industry market.

14.7.1 RESEARCH OBJECTIVES TOWARD APPLICATIONS OF INDUSTRY 4.0 IN ENVIRONMENTAL SUSTAINABILITY

Concurrent data gathering together with machine learning (ML) is utilized for further development in the areas of highly effective and streamlined processes. The merits of sustainability include an increase in resource availability, increased energy production, infrastructure that is deployable in a sustainable environment, worker security and well-being, and ultimately an improvement in living quality.

Smart Factory

According to the report highlighted [46] from ABI research dated on Jan 27, 2021, the third industrial revolution products are to be considered as common technologies [47].

According to ABI research the cellular enabled Industry 4.0 generate in cost savings of the operations nearly 8.50% which is approximately per square meter $200.00 to $600.00 for a year on factory or industries site. As we move on to the Fifth Industrial Revolution, factory has the power to alter how commodities are made and supplied. In this modern era revolution, the manufacturers can hold all the merits of technology.

RO1: What Can be the Outlook of Fifth Industrial Revolution in Smart Factory?

The main target of 5.0 Industry is on both the sides between the customer and the manufacturer in factories side. With the development of new technologies, businesses can now create comfortable working environments. They can solve the problem in an easy manner and have a creative mind [48].

According to oxford economics in Industry 5.0 the innovation must meet the technology. During the pandemic period the rise of robotics has increased in Industry 5.0 [49].

Decreasing Carbon Footprint and Pollution Control in Environmental Sustainability

In order to protect the environment through the production system such as material, sources of energy and knowledge required to convert input to output to produce things, as well as waste and products at the end of their useful lives and GHG emissions, we should reduce the usage of plastic materials, carbon footprint, and water pollution [50].

RO2: How Can We Reduce the Carbon Emission in Industry 4.0?

According to the report [51] dated 25.11.2021 when industrial sector goes to a significant shift then greater the sustainable measures are required. New technologies connect important devices and procedures. The final result is a challenging one in how the production is conducted which lowers their carbon footprint simultaneously achieving their sustainable targets.

1. Industrial Internet of Things (IIOT) and data analytics (DA) can improve the maintenance of asset and durability.
2. Digital twin called the virtual model designed accurately reflecting a physical object removes the wastage and consumption of energy from key development.
3. Automation of robotics which optimizes the energy utilization as well as machine utilization.

Industry 4.0 in Sustainable Buildings

There are several ways that Industry 4.0 is producing sustainable buildings, from architecture to ongoing maintenance. For sustainability, a lot of architects employ software for building information modeling. By utilizing innovative technologies, industrial and commercial projects may coexist with one another in a way that is environmentally benign. After installation, heating and cooling systems will be automatically adjusted for energy efficiency. Whenever people leave the location by using sensors and other IoT devices, with proper instruction it automatically switches off the lights and air conditioning.

RO3: How Can We Reduce the Negative Impact on Environment and Socioeconomic Aspect?

According to the reporting status of Procure [52] dated on 13.05.2021, natural wealth is being used up by us at an extraordinary pace. The building industry must be accountable for the effects of the built environment and how it is interconnected with the environment. Adopting sustainable construction practices will lessen environmental harm while improving adverse effects. As a result, the capital cost and the long-term asset value will be equal. Building design has a vital part in biodiversity because it helps people live in a healthy, ergonomic environment.

Industry 4.0 in Service Management

By connecting measuring devices and manufacturing companies' whole information and computerization infrastructure in order to manage energy and services, Industry 4.0 is made possible. It helps to increase the capacity for gathering, sending, and storing massive amounts of data about the use, production, and transition of energy sources. Growth of data and easier access to computing power make it possible for specialized AI algorithms to more easily recognize and analyze interest patterns across a range of sectors.

RO4: What Is Required for Service Management in Industry 5.0?

According to Saeid [53] the upcoming Industry 5.0 revolution is the network sensor. Smart houses, cobots, and intelligence systems are a few examples. This enables more rapid operations for assessment and change. In the fifth industrial revolution, process efficiency is improved by fusing human intellect with machine inventiveness.

Solution for the Given RO4

1. Sensor Network Interoperability—A sensor network with rudimentary intelligence and processing power that would enable some local data preprocessing might reduce the communication of high bandwidth. This lessens network congestion and delay. Using the sensor network, a standard framework for information transfer is needed [54].

Industry 4.0 in Product Optimization

Cloud computing is a single entity. It may be used to forecast process failure, improve the product, and so forth. It is effective in transmitting knowledge about machines, sensors, gadgets, robots, and living things.

RO5: What Are the Steps That Are Needed in Product Optimization?

The four-step smart production management system (SPMS) suggested in several brainstorming meetings held at various departmental levels and with the aid of prior research.

Step 1: Observation—visit the shop floor and data acquisition system
Step 2: Analysis the program and data analytics
Step 3: Modification in shop floor management system—lean and smart manufacturing
Step 4: Validation of smart production system
Step 5: Production enhancement in a case study of industry 4.0.

Industry 4.0 in Skill and Flexible Optimization

Over the past several hundred years, much technological advancement has revolutionized business and the industrial sector. Systems for training and preparation are required to aid in the development of skills necessary for success in industry 4.0 vocations. Robots are a crucial component of terms of technological advancements and are utilized in industrial production starting plasma cutting, welding, assembling, and finally completing the entire one.

Additionally, robots may offer various benefits to suppliers of all sizes. If used effectively, automaton systems may improve value control, save working environment, reduce bottlenecks, boost productivity, raise worker happiness, and more. Industry 4.0 already has a lot of technological components; thus, the revolution combines products, systems, and computers

RO6: Which Country is Initiating Industry 4.0 with Skills and Optimization to be Flexible?

According to the report of ISA China, Japan, and India play a major role in skill and flexible optimization.

China: It includes "Made in China 2025"

China started a game of catch-up that has never been played before. China 2015 hold the idea of the fourth industrial revolution made in China 2025 initiative, a 10-year plan to completely prop up China trade aim of catching up in production powerhouses like Germany and fending off competition from other developing nations with lower labor costs.

According to IFR, the number of robots purchased in China increased by 56% to roughly 57,000 units in 2014. China is expected to have deployed more than three-quarters of all industrial robots worldwide by 2018. The strategy focuses on 10 industries, including advanced robotics and computerized machinery, aerospace technology, hybrid and electric vehicles, and biological medicine.

Country of Japan: It includes "JAPAN"

The target of Japan includes

1. Establishing a supply/engineering chain between small- and medium-sized enterprises, resulting in excellent quality and productivity
2. Providing service models to the manufacturing sector that use their goods as essential competitive elements.
3. Lowering costs and enhancing platform-based manufacturing operation management software flexibility.

India: "Make in India"

India plans to apply Industry 4.0 ideas and cutting-edge technology in initiatives for production, smart buildings, and general infrastructure as part of the country's Digital India agenda. Expanding all over India's manufacturing sector, Prime Minister Narendra Modi introduced Make in India on September 25, 2014.

The Government of India launched the "Make in India" campaign both household and foreign companies to manufacture their products in India. At 2015 HANNOVER MESSE, India made this declaration on a worldwide scale as the partner nation [55].

Industry 4.0 in Climatic Change

The world economy is on the verge of a turning moment. It may also significantly affect the long-term health of economies by introducing new business models and enhancing protection, inclusivity, accessibility, and protracted growth for all stakeholders. The exponential roadmap enables collaboration between scientists, tech entrepreneurs, big business, and nonprofit organizations.

The quick development of digital technology is changing corporate practices and opening up new customer prospects. This offers useful tactics and resources for tackling climate change and other sustainability-related issues, having a beneficial effect on industries.

RO7: State the Condition Which Will be Compatible with Change in Climatic

Energy efficiency is a growing problem for many industrial businesses as the use of industry 4.0 technology develops. Energy conservation through increased and more effective usage of renewable energy sources is crucial to sustainability. Technology is enabling improved management of complex systems and delivering solutions for saving the energy, efficient in manufacturing, to control the process, to control the emissions, safety improvement, and maintain the machine.

Industry 4.0 in Handling Industry Functions

The integration of 3D printing and nanotechnology into supply chains and manufacturing processes has increased production flexibility and improved energy efficiency. It may give the product's complete assembly instructions so that production engineers can show the maintenance of the problems on the shop floor in real time. When a company connects to cloud storage, all relevant data is accessible wherever it is required. By combining cameras and sensors, engineers may gather more data to aid in more efficiently completing their duties. The importance of having access to the most recent data must be recognized by industries considering automation to assist them manage whole industrial activities.

RO8: What Is the Role of Industry 4.0 in Handling Industry in India?

According to Times of India [56], large enterprises and SMEs may accomplish the aforementioned aims quickly and affordably by adopting Industry 4.0 principles like automation, adaptability, and interoperability. This change is being made possible by sensors, cognitive systems, networks, communication, compute, storage, applications, and competitive business models since they are more affordable and widely available.

A product must have high usefulness, established dependability, desired safety, and alluring elegance in order for a buyer to purchase it at a price that is also reasonable. According to recent advancements in a few key sectors, the advantages of such a transformation would outweigh those that India has already reaped through decades of devoted and continuous innovation in the realms of communication and roadway infrastructure. Conventional manufacturing and production methods are changing.

Due to new Industry 4.0 technologies, traditional manufacturing and production procedures are being transformed, which makes it possible to significantly enhance corporate operations, processes, productivity, quality, and customer service. A "manufacturing & production cloud," the industrial sector's "cloud" counterpart of IT, is required for the Make-In-India project to be a smashing success. To strengthen the Indian industrial sector and advance manufacturing and output, a comparable environment is needed. Small enterprises and startups may create and adopt innovation thanks to it.

Small enterprises and startups may create and adopt innovation thanks to it. To strengthen the Indian industrial sector and advance manufacturing and output, a comparable environment is needed. By adopting the Industry 4.0 framework, major businesses and mini, medium, and moderate enterprises may create the "manufacturing & production cloud," which is fueled by seamless communication and cooperation between systems and stakeholders.

Industry 4.0 in Minimum Wastage

A new form of waste in 4.0 Industry is tabulated in Table 14.3 [57]. The organization has daily improvement meetings, strives to cut lead times, gathers data from various sources, and transforms the data into information in order to anticipate and eliminate waste.

RO9: What Will be the Outcome of the Two Companies in Industry?

According to Table 14.3, data analysis seeks to make long-term predictions using trial-and-error methods and efforts that take failure into account. Forecasting based on ineffective analysis should not merely be written off as a failure. It is also a significant contributor to the learning curve for the business. Time and value are two factors from the Table 14.3 states that

TABLE 14.3 Industry 4.0: a New Form of Waste.

Types of Division	Serial No.	Time	Method	Name of the Company	Contact Details
Strategic	1	One hour	Face-to-face interview	ABT	Head, Production Digital Transformation
	2	One hour	Face-to-face interview	ABT	Head, Digital Machining
	3	Two hours	Face-to-face interview	Bloom Berk	IT Strategy & Innovation Manager
Production Area	4	30 minutes	Interview through phone call	ABT	Business Process, Expert Manufacturing Execution
	5	30 minutes	Interview through Phone call	Bloom Berk	Plant Manager
HR Manager	6	1 hour	Interview through phone call	ABT	Manager Recruitment
	7	30 minutes	Interview through phone call	Bloom Berk	Human Resource

Company ABT uses to gauge its effectiveness. After manufacturing staff get used to the changes, investments in digital technology will produce positive results.

Additionally, the existing manufacturing clients buy products, extend their relationships, and provide limitless potential for further product enhancement. The major way that the company learns is through data analysis, which also provides an opportunity to make better judgements. Time and client feedback are two important factors from the Table 14.3 states that Company Bloom Berk uses to gauge performance. Novelty and resourcefulness are crucial in working process, and benefit greatly from flexibility with possibilities for error.

Industry 4.0 in Automobile

According to the research, 49.0% of automakers have already invested more than $250 million in smart manufacturing. This data makes it very clear that the automotive industry is keen to make the switch to Industry 4.0. In order to go beyond optimization and automation, Industry 4.0 in the automotive industry is changing the whole manufacturing process to be more customer-centric and successful at managing logistics.

A market research study claims that 24.0% of automakers' facilities will be "smart factories" by the end of 2022. It is clear from the evidence that the automotive industry is keen to make the switch to Industry 4.0 [58].

RO10: What Can be Merits and Forthcoming Challenges in Shifting Industry 4.0?

The majority of OEMs and suppliers find it difficult to make the shift to Industry 4.0. They require the right equipment and personnel to put the procedures and systems in place before they can begin using Industry 4.0 automotive technology. They also require the necessary knowledge to address the numerous logistical issues that they may encounter, such as:

1. Removing the worry that robots will take the place of people in the workforce
2. Addressing the difficulties and expenses associated with implementing new technology
3. Reducing dangers to the cybersecurity of their manufacturing data systems

Merits

The automotive industry can benefit in many ways from the adoption of Industry 4.0 as shown here:

1. Lower total operating and maintenance expenses
2. Lower rates of product failure and rework
3. Encourage new possibilities and ideas
4. Better use their supplies and resources
5. Increase throughput while reducing downtime

Industry 4.0 in Efficiency of Market

Progressive firms use renewable energy sources in the manufacturing, marketing, and the supply chain for the power sector and emerging technology. Additionally, firms utilize innovative strategies and technologies to maximize efficiency, minimize duplication, and make better use of energy. The connections and issues with sustainable production include overlap areas, aligning, and classification qualities of various processes via efficient and cost-effective waste management to improve market modeling and efficiency.

RO11: What Can be the 4.0 Industry Market Size by Application on Industrial Automation and by Vertical Energy and Utilities?

At a predicted CAGR of 16.0% during the projection period of 2022–2029, the worldwide industry 4.0 markets increase from Rs. 10,853,278,474.00 in 2022 to Rs. 31,234,244,158.00 by 2029. According to fortune business insight reporting status proved the market trend from the years 2017 to 2029 [59].

RO12: What Are the Developments in This Era in Market?

As of November 2021, the specific firm Siemens is expanding the Internet of Things with a stand-alone digital and vendor, distinct application store for customers, and cross-industry partnerships. Sentaca is one of the leading consulting and solution providers as of February 2022. Hybrid cloud consulting companies support communication service providers across multiple platforms to transform their own businesses.

RO13: Mention the Leading Company in 4.0 Industry Market

General Electric is ranked #1 in the market, followed by Siemens, Honeywell, ABB, and Emerson Electric.

14.8 CONCLUSION

The "Industry 4.0" revolution contributes production sectors. The use of much less resources is made possible by Industry 4.0 technologies, which enable more quick and efficient innovation. A data analyst, a machine learning professional, and a big data specialist are required to manage the data, extract insights, optimize processes and equipment, and maintain the factory since smart factories produce enormous volumes of data. Conclusions can almost always be drawn from data. It is now up to us to make the most of this period, improve the environment, and make it more meaningful, purposeful, and sustainable.

However, businesses may easily obtain information on all of these problems and work toward building a factory that can handle power, material, and energy limits thanks to the usage of technology. In the upcoming technological revolution, the rules controlling how we interact with one another and with robots must be established. By using the technology, manufacturers can be able to manage in all aspects and meet the expanding demand. 5G networks provide the flexibility that businesses require for present and upcoming systems, which is essential for any type of activity and applications. Additionally, they provide the delay that manufacturers require. The emerging trend known as "industry 5.0" has enhanced collaboration and human–machine interaction. The time has come.

KEYWORDS

- **artificial intelligence**
- **digital transformation**
- **Industry 4.0**
- **innovation ecosystem**
- **IoT**

REFERENCES

1. https://www.shutterstock.com/search/cyber-physical-systems.
2. https://sovrin.org/library-iot/.
3. https://ostechnix.com/cloud-computing-basics/.
4. https://blog.westerndigital.com/cognitive-computing-data-driven-world/.
5. Muataz, S. H., Husham, I., & Salman, A.L. (2019). A review of cyber of industry 4.0: Components and security challenges. In *2nd International Conference on Advance and Scientific Innovation, Journal of Physics, Conference Series*. IOP Publishing (pp. 1–6). DOI: 10.1088/1742-6596/1424/1/012029.
6. Broy, M. (2013). Engineering cyber-physical systems: challenges and foundations. In: M. Aiguier, Y. Caseau, D. Krob, and A. Rauzy, A. (Eds.), *Complex systems design & management*. Springer.
7. https://www.geeksforgeeks.org/denial-of-service-ddos-attack/.
8. https://networksimulationtools.com/spoofing-attack-network-projects/.
9. https://link.springer.com/chapter/10.1007/978-1-4302-6383-8_14.
10. https://www.acunetix.com/websitesecurity/web-application-attack/.
11. Syed, I. S., Cesar, S., Edward, S., & Carlos, T. (2015) Virtural engineering object/virtual engineering process: A specialized form of Cyber Physical System for Industries 4.0. *19th International Conference on Knowledge Based and Intelligent Information and Engineering Systems, Procedia Computer Science, 60*, 1146–1155.
12. Rielly, K. O., & Jeswiet, J. (2014) Strategies to improve industrial energy efficiency. *21st CIRP Conference on Life Cycle Engineering, Procedia CIRP, 15*, 325–330.
13. Simon, F., & Melanie, D. (2016) Additive manufacturing and sustainability: an exploratory study of the advantages and challenges. *Journal of Cleaner Production, 137*, 1573–1587.
14. Tim, S., Michael, O., Sascha, K., & Holger, K. (2018). Industry 4.0 as enabler for a sustainable development: A qualitative assessment of its ecological and social potential. *Process Safety and Environmental Protection, 118*, 254–267.
15. Jay, L., Hossein, D., Jaskaran, S., & Vibhore, P. (2018). Industrial artificial intelligence for Industry 4.0-based manufacturing systems. *Manufacturing Letters, 18*, 20–23.
16. Sang, M. L., Don Hee, L., & Youn, S. K. (2019). The quality management ecosystem for predictive maintenance in the Industry 4.0 era. *International Journal of Quality Innovation, 5*, 1–11.
17. Husham, I., AL-Salman, & Muataz, H. S. (2019). A review cyber of Industry 4.0 (Cyber-Physical Systems (CPS), the Internet of Things (IoT) and the Internet of Services (IOS)) components and security challenges, *Journal of Physics, Conference Series, 1427*, 1–7.
18. Senthil, K. J., Mohamed, R., Kashif, A., Ala, A. F., & Mohsen, G. (2020). The duo of artificial intelligence and big data for Industry 4.0 review of applications, techniques, challenges, and future research directions. *IEEE Internet of Things Journal,* 1–33.
19. Maksim, S., & Georgi, T. (2020). The use of artificial intelligence in Industry 4.0. *Bulgarian Academy of Sciences, Problems of Engineering Cybernetics and Robotics, 73*, 17–29.
20. David, M. (2021) Artificial Intelligence in the Industry 4.0, and its impact on poverty, innovation, infrastructure development, and the sustainable development goals: Lessons from emerging economics? *Sustainability and Industry 4.0 in Civil and Infrastructure Engineering, 13*(11), 5788.

21. Bakshi, B., & Paulson, J. (2022) Sustainability and Industry 4.0: Obstacles and opportunities. *IEEE American Control Conference (ACC).*
22. Mohammad, S., Osama, K., Mohd Zaheen, K., Mohd Pavez, Bhupendra, B., Amaan, U., & Md Toufique Alam (2022). Analytical research of artificial intelligent models for machining industry under varying environmental strategies: An industry 4.0 approach. *Sustainable Operations and Computers, 3*, 176–187, 2022.
23. Sameern, K., Sarika, T., Maryam, F., & Mohd Zaheen, K. (2022). Impact of artificial intelligent and industry 4.0 based products on consumer behaviour characteristics: A meta-analysis-based review. *Sustainable Operations and Computers, 3*, 218–225.
24. Saecha. (2022). Cyber-physical systems in the context of Industry 4.0: A review. *Categorization and Outlook, Information Systems Frontiers,* 1–42.
25. Sanjay, N., Florian, K., & Christodoulidou, N. (2023). Industry 4.0 applications towards sustainability in hospitality: First waves in the guest room. *Journal of Global Business Insights, 8*(1), 31–48.
26. Ghobakhloo, M. (2022). Industry 4.0, digitization, and opportunities for sustainability. *Journal of Cleaner Production, 252*, 119869.
27. Olah, J., Aburumman, N., Popp, J., Khan, M. A., Haddad, H., & Kitukutha, N. (2020). Impact of Industry 4.0 on environmental sustainability. *Sustainability, 12* (11), 4674.
28. Tiwari, K., Khan, M. S. (2020). Sustainability accounting and reporting in the Industry 4.0. *Journal of Cleaner Production, 258*, 120783.
29. Micieta, B., Binasova, V., Lieskovsky, R., Kracovic, M., & Dulina, L. (2019). Product segmentation and sustainability in customized assembly with respect to the basic elements of industry 4.0. *Sustainability, 11*(21), 6057.
30. Vuksanovc, I., Herceg, V. K., Mijuskovic, V. M., & Herceg, T. (2020). Challenges and driving forces for Industry 4.0 implementation. *Sustainability, 12*(10), 4208.
31. Sima, V., Gheorghe, I. G., Subic, G., & Nancu, D. (2020). Influences of the Industry 4.0 revolution on the human capital development and consumer behaviour: A systematic review. *Sustainability, 12*(10), 4035.
32. Man, J. C. De., & Strandhagen, J. O. (2017). An Industry 4.0 research agenda for sustainable business models. *Procedia CIRP, 63*, 721–726.
33. Narayanamurthy, G., & Tortorella, G. (2021). Impact of COVID-19 outbreak on employee performance moderating role of industry 4.0 base technologies. *International Journal of Production Economics,* 108075.
34. Acioli, C., Scavarda, A., & Reis, A. (2021). Applying Industry 4.0 technologies in the COVID-19 sustainable chains. *International Journal Productivity and Performance Management.*
35. Xu, L. D., Xu, E. L., & Li, L. (2018). Industry 4.0: State of the art and future trends. *International Journal of Production Research, 56*(8), 2941–2962.
36. Azizi, A. (2020). Applications of artificial intelligence techniques to enhance sustainability of industry 4.0: Design of an artificial neural network model as dynamic behaviour optimizer of robotic arms. *Complexity.*
37. Liu, Y., Zim, Q., & Seuring, S. (2020). New technologies in operations and supply chains: Implications for sustainability. *International Journal of Production Economics, 229*, 107889.

38. Fatorachian, H., Kazemi, H. (2018). A critical investigation of Industry 4.0 in manufacturing: Theoretical operationalisation framework. *Production, Planning & Control,* 633–644.
39. Zambon, I., Cecchini, M., Egidi, G., Saporito, M. G., & Colantoni, A. (2019). Revolution 4.0: Industry vs agriculture in a future development for SMEs. *Processes, 7*(1), 36.
40. Totorella, G. L., & Fettermann, D. (2018). Implementation of Industry 4.0 and lean production in Brazilian manufacturing companies. *International Journal of Production Research, 56*(8), 2975–2987.
41. Lom, M., Pribyl, O., & Svitek, M. (2016). Industry 4.0 as a part of smart cities. In: *2016 Smart Cities Symposium Prague (SCSP)* (pp. 1–6). IEEE.
42. Nunes, M. I., Pereira, A. C., & Alves, A. C. (2017). Smart products development approaches for Industry 4.0. *Procedia Manufacturing, 13*, 1215–1222.
43. Ciffolilli, A., & Muscio, A. (2018). Industry 4.0: National and regional comparative advantages in key enabling technologies. *European Planning Studies, 26*(12), 2323–2343.
44. Xu, L. D., & Duan, L. (2019). Big data for cyber-physical systems in industry 4.0: A survey. *Enterprise Information Systems, 13*(2), 148–169.
45. Rossi, J., Bianchini, A., & Guarnieri, P. (2020). Circular economy model enhanced by intelligent assets from Industry 4.0: The proposition of an innovative tool to analyze case studies. *Sustainability, 12*(17), 7147.
46. https://www.ericsson.com/en/internet-of-things/audience-page/abi-research-industry-40-roi.
47. https://trendlog.dk/the-story-about-the-third-industrial-revolution/.
48. https://www.isa.org/intech-home/2018/march-april/features/welcome-to-industry-5-0.
49. https://blog.oxfordeconomics.com/world-post-covid/in-the-5th-industrial-revolution-creativity-must-meet-technology.
50. Judit, O., Nemer, A., Jozsef, P., Muhammad, A. K., Hossam, H., & Nicodemus, K. (2020). Impact of Industry 4.0 on environment sustainability. *Sustainability, 12*(11).
51. https://www.industrialtransformationnetwork.com/article/articles/how-industry-40-is-helping-manufacturers-reduce-carbon-emissions.
52. https://procurepartnerships.co.uk/how-can-we-improve-the-negative-impact-construction-has-on-the-environment/.
53. Nahavandi, S. (2019). Industry 5.0—A human-centric solution. *Sustainability, 11*(16), 4371.
54. Flora, A., Valentina, C., Andrea, G., & Antonino, M. (2011). Semantic enriched data model for sensor network interoperability. *Simulation Modelling Practice and Theory,* 19, 1745–1757.
55. https://www.isa.org/intech-home/2016/may-june/features/industry-4-0-intelligent-and-flexible-production.
56. https://timesofindia.indiatimes.com/blogs/voices/role-of-industry-4-0-in-accelerating-make-in-india-goals/.
57. Jamila, A., & Robin, H. (2020). Digital Muda—The new form of waste by Industry 4.0.*Operations and Supply Chain Management, 13* (3), 269–278.
58. https://www.fortunebusinessinsights.com/industry-4-0-market-102375.
59. https://www.marketsandmarkets.com/Market-Reports/industry-4-market-102536746.html?gclid=EAIaIQobChMI1qnQt9nk_AIVWX8rCh2H0A7CEAAYASAAEgJqSfD_BwE.

CHAPTER 15

Artificial Intelligence for the Digital Era: Unleashing the Power of Intelligent Technologies

N. A. NATRAJ

Symbiosis Institute of Digital and Telecom Management, Symbiosis International (Deemed University), Pune, India

ABSTRACT

In the digital age, artificial intelligence (AI) has transformed companies, society, and human–machine interaction. The article examines AI's rapid development and impact on the digital world. It then analyzes AI applications in several sectors, highlighting its paradigm-shifting implications on healthcare, banking, manufacturing, and entertainment and a complete description of its effects across industries. Data—the foundation of AI—is stressed. This chapter successfully handles data gathering, storage, and ethical issues that stress the need for responsible AI development. The narrative intelligently pivots to critically examine AI's future in the digital age. The chapter discusses upcoming breakthroughs, trends, and the social impacts of AI, preparing readers for a transformational future. In the conclusion, key ideas are presented, emphasizing ethical AI deployment and practices. This chapter repeatedly urges stakeholders to take AI's future seriously. It will serve as a comprehensive manual that connects theoretical foundations with practical applications, giving readers a complete understanding of AI in the digital age and its implications for intelligent technology.

Real-Time Artificial Intelligence (AI): Key Motivations, Technologies, Platforms, and Use Cases. Pethuru Raj, B. Sundaravadivazhagan, V. Kavitha, B. Narendra Kumar Rao, & Hannah Vijaykumar (Eds.)

DOI: 10.1201/9781998511358-15

15.1 INTRODUCTION

15.1.1 UNVEILING THE HISTORICAL TAPESTRY OF ARTIFICIAL INTELLIGENCE IN THE DIGITAL ERA

Artificial intelligence (AI) emerges as the masterstroke in the intricate fabric of the digital era. AI is a harmonious symphony of technology reverberating throughout every contemporary dimension. The effect of AI is not just a technological trend; rather, it is a transformative force that cuts across industries, molds societies, and intricately weaves itself into the fabric of our day-to-day existence at a time when we are on the edge of developments that have never been seen before. This introduction acts as a portal, asking readers to tour the dynamic terrain of AI and urging them to realize the tremendous depth of its impact on our collective destiny. It beckons readers to traverse the field of AI and serves as a portal. The confluence of numerous technical advances characterizes the advent of the digital era, and at the very heart of this convergence is the symphony of AI, which orchestrates progress with unrivalled accuracy. Artificial intelligence is a technology force that, in contrast to any other technological force, possesses the unique ability to learn, adapt, and evolve, thus mimicking the very essence of human intelligence [1]. Because of this potential, AI has been propelled from the realm of theoretical possibility into the forefront of practical application, drastically altering how we live, work, and interact with one another.

At the core of this revolutionary trip is the awareness that AI is not only a tool but rather a dynamic force that is reshaping the landscape of industries. This realization is at the heart of this revolutionary journey. The application of AI is having a ubiquitous impact, as it is streamlining processes, enhancing the quality of decisions, and introducing novel opportunities across a wide variety of business sectors, including healthcare, finance, manufacturing, and the entertainment industry. It functions as an engine of innovation, bringing forth things that were previously impossible even in the world of science fiction. This is all because of the internet. The AI symphony, which concurrently builds the framework for a future defined by intelligent technology, is responsible for the orchestration of precision medicine, the empowerment of financial forecasts, and the transformation of manufacturing processes, all of which are being orchestrated by the AI symphony. Outside of the area of industries, the impact of AI can be felt all the way down to the fundamental components that make up society systems.

It has become a transforming force that influences how we communicate with one another, how we obtain access to information, and how we comprehend the world around us. The impact that AI will have on society will be multifaceted, and as a result, discussions concerning ethics, privacy, and the equitable distribution of rewards have been sparked as a result. The symphony of AI's rich melodies compels us to ask difficult questions about the role that technology plays in shaping the common goals and aspirations of our society. One day, AI will be so ingrained in our daily activities that it will be indistinguishable from a partner who is physically present. AI makes our experiences better and broadens our capabilities in a number of different ways. For example, it enables virtual assistants to anticipate our needs and it enables recommendation systems to customize content to our preferences. Both of these improvements increase our overall quality of life. It is the unseen force that is the driving force behind the luxuries that we now take for granted, and it is gently altering the way that we navigate the digital landscape. This introduction serves as a portal into this dynamic environment and invites readers to contemplate the far-reaching impacts that AI will have on our collective destiny in the future. This is an invitation to explore the symphony of innovation, where the transformative melodies of AI may be heard echoing in the boardrooms of businesses, the halls of healthcare institutions, and the living rooms of households all around the world. As we prepare to embark on this journey, the symphony of AI encourages us to pay close attention to its complexities and fully comprehend the huge influence it will have on the canvas that is our digital existence. This invitation arrives just as we are getting ready to embark on this journey.

15.1.2 BACKGROUND AND CONTEXT

A fascinating journey through AI's history is needed to understand its importance in the digital age. This decades-long story of inventiveness, innovation, and computational ambitions shapes how robots, born of the human mind, have come to emulate intelligence. Visionaries like Alan Turing and John McCarthy seeded AI seeds in the mid-20th century. In the 1950s, Turing developed the Turing Test to evaluate a machine's human-like intellect. McCarthy created "AI" in 1956 and organized the Dartmouth Conference, which launched AI research. These early days were marked by ambitious goals and a belief that machines could fulfil human functions.

However, the moment's technology fell short of the ideal, resulting in the "AI winter," decreasing funding and interest in AI research. The narrative changed in the late 20th century as processing power and algorithms evolved. Once restricted to theory, AI returned. AI applications were used in expert medical diagnosis and industrial automation as expert systems and rule-based techniques grew prominent. After decades of collaboration and technological improvement, intelligent machines emerged in the 21st century. Machine learning (ML), an AI branch, became popular, allowing systems to learn from data and improve performance. Rule-based programming gave way to adaptive and self-learning algorithms. AI's evolution is a story of a persistent desire to create robots that can mimic human intellect. Natural language processing (NLP), computer vision, and neural network advances color the story. These advances allowed machines to grasp language, recognize data patterns, and navigate physical places with unprecedented sophistication. AI in the digital age is built on this historical context. Scholars and scientists' computational aspirations of machines speaking natural language and solving problems have become intelligent systems incorporated into our daily lives. We stand on the shoulders of Turing and McCarthy, and the history of AI provides essential insights. It highlights the hurdles, breakthroughs, and iterative process that has taken AI from concept to digital revolution. The historical tapestry of tenacity, creativity, and the tenacious human spirit pulls us toward an era where intelligent technologies are a reality [2].

This chapter provides clear aims to navigate intelligent technologies' complexity amid the huge influence of AI. The goal is to reveal AI's fundamental principles and the complex mechanisms that make robots live. Beyond that, the chapter examines AI's uses in healthcare, banking, manufacturing, entertainment, and other fields. Readers are also encouraged to evaluate the social impacts of AI by critically examining its ethical issues. Finally, the chapter looks ahead to see how AI will change human–machine interaction in the digital age.

15.2 AI APPLICATIONS ACROSS INDUSTRIES

Artificial intelligence has significantly changed many industries and our way of life, including work and recreation. We will examine the revolutionary uses of AI in four key sectors in this section: manufacturing, entertainment, finance, and healthcare.

15.2.1 IMPACT OF AI IN THE HEALTHCARE SECTOR

Artificial intelligence is a disruptive entity in the dynamic healthcare industry, fundamentally altering established models of patient care and medical decision-making [11]. This technological collaborator has been crucial in enhancing the capabilities of medical practitioners by providing unparalleled levels of accuracy, productivity, and ingenuity. This segment delves into the diverse range of AI's implementations in the healthcare industry, encompassing surgical robotics, predictive analytics, diagnostic tools, and AI-powered telemedicine platforms.

15.2.1.1 AI IN DIAGNOSTIC IMAGING: REVOLUTIONIZING DISEASE DETECTION

An enormous step forward in interpreting medical images has been made possible by the development of AI-powered diagnostic tools supported by cutting-edge ML algorithms [8]. The ability to analyze complex medical images with excellent accuracy, such as X-rays and MRIs, has far-reaching ramifications for diagnosing a medical condition. AI speeds up diagnosis and, more importantly, makes it easier to spot signs of disease in their earliest stages by capitalizing on ML's ability to identify irregularities and patterns in medical images.

The relevance of this technology rests not only in its speed but also in its capacity to improve early intervention. Both of these aspects contribute to the overall picture. A more timely diagnosis of a problem, whether it be a tumor or a neurological disorder, allows medical personnel to create more effective treatment regimens and improves the likelihood of positive patient outcomes. The powerful ally that AI brings to diagnostic imaging facilitates a paradigm shift toward preventive and precision medicine.

15.2.1.2 NATURAL LANGUAGE PROCESSING (NLP): ILLUMINATING INSIGHTS FROM MEDICAL DATA

Natural language processing, which enables AI to sort through large volumes of information with an unparalleled level of efficiency, appears as a guiding beacon in the vast sea of medical literature and patient records. NLP makes it possible for AI to sort through massive volumes of information with an unprecedented level of efficiency. This language intelligence is helpful in the identification of relevant information, which lays the groundwork for the creation of individualized treatment plans, which in turn are beneficial to patients.

The capability of NLP to extract information from unstructured medical texts has the potential to fully revolutionize the way in which medical practitioners access and make use of information. This transformation could be made possible by advances in AI. AI can assist in recognizing trends, correlations, and critical data points by searching through patient records, research papers, and medical literature. Using the facts to your advantage, you should be able to accomplish this goal. This not only makes the processes of decision-making more efficient, but it also adds to the development of personalized medicine, which is the practice of adapting therapies to the particular characteristics of each unique patient. Over the past few years, one of the most prominent trends in the medical field has been the rise of personalized medicine.

15.2.1.3 PREDICTIVE ANALYTICS: FORECASTING DISEASE TRENDS AND OPTIMIZING RESOURCES

Healthcare practitioners can benefit greatly from predictive analytics models that leverage ML. By estimating illness trends using vast amounts of data, these models enable proactive disease prevention and optimal resource allocation. Artificial intelligence has predictive power not just for treating individual patients but also for managing communal health [11]. Based on historical data, these models can forecast disease outbreaks, spot emerging health trends, and help healthcare organizations allocate resources effectively. In addition to improving patient outcomes, this proactive approach strengthens healthcare systems' overall capacity to address public health issues.

15.2.1.4 PRECISION BEYOND HUMAN CAPABILITIES: ROBOTICS IN SURGERY

The area of surgery is changing dramatically with the introduction of robots into surgical procedures. Artificial intelligence-driven robotics surpass human abilities in accuracy and dexterity, enhancing the precision of intricate surgical operations. Under the direction of AI algorithms, these robotic assistants work in tandem with surgeons to achieve optimal outcomes. Robotic technologies provide control, stability, and visualization in a variety of medical operations, from simple to intricate. The increased capabilities of these AI-powered assistants may be used by surgeons to carry out procedures more precisely, thus reducing recuperation times and enhancing patient satisfaction after surgery.

15.2.1.5 AI-POWERED TELEMEDICINE PLATFORMS: TRANSFORMING HEALTHCARE DELIVERY

With the help of AI-driven technology, telemedicine systems have opened up previously unimaginable levels of access to medical care. Chatbots that are run by AI serve as the initial point of contact, streamlining the interactions between patients and providing quick access to medical information. This not only increases the efficiency with which healthcare is delivered, but it also ensures that patients receive information that is current and accurate. The marriage of AI and telemedicine eliminates conventional obstacles by making it possible to conduct consultations, monitoring, and even diagnosis over the internet. Chatbots powered by AI have the ability to evaluate patient problems, schedule appointments, and make preliminary advice by utilizing sophisticated algorithms. This helps to democratize healthcare services.

Ultimately, the use of AI in the healthcare sector is part of a revolutionary wave that is changing the way the sector looks. Together, these technologies—which range from robotic surgery to personalized treatment plans, from accurate diagnostics to AI-driven telemedicine—will create a future in which healthcare is not only more effective but also more patient-centered. The healthcare sector is expected to be significantly impacted by AI as it develops further [11]. This will mark the beginning of a new age in medical practice where enhanced patient outcomes, accessibility, and precision are given top emphasis.

15.2.2 AI IN FINANCE: NAVIGATING THE DIGITAL ERA

Leading the charge in the digital age is the financial industry, where AI is reshaping the world via unprecedented innovation. The financial industry is leading this shift in our period as well. This study delves into the various applications of AI in the financial industry, focusing on ML in financial forecasting, AI's role in cybersecurity and fraud detection, integrating chatbots and virtual assistants, and blockchain technology's revolutionary potential. We pay particular attention to the use of AI in cybersecurity and fraud detection, chatbot and virtual assistant integration, and ML in financial forecasting.

15.2.2.1 MACHINE LEARNING IN FINANCIAL FORECASTING

Machine learning in financial forecasting is a transformational application that employs advanced algorithms to analyze historical and real-time data, discovering insights that drive decision-making in the ever-evolving world of finance. In other words, ML in financial forecasting is an application that will change the way financial forecasting is done. This application does historical and real-time data analysis with the help of sophisticated algorithms. It is notoriously difficult to convey the complexity and nuances of financial markets using the methodologies that have been used traditionally. Because of this, ML now has the opportunity to come in and fill the hole with the data-driven precision it possesses.

Machine learning models are particularly effective in the field of financial forecasting because of their ability to recognize complicated patterns that are hidden inside of enormous datasets. Because of this, it is now possible to generate more precise projections of market patterns and asset prices. Since ML algorithms process such enormous volumes of data as the historical performance of markets, economic indicators, and events occurring in the geopolitical realm, they are able to recognize patterns that human analysts may be unable to due to the fact that they are more likely to miss them. Because they are able to process a diverse range of real-time data sources, financial experts have the potential to get a more in-depth understanding of the current state of the market. This enables them to make more informed investment decisions.

The creation of predictive models that anticipate changes in market conditions is one of the most significant uses of ML in the field of financial forecasting. These models are able to adapt and develop when new data is accumulated, which enables continual elaboration and enhancement of their capabilities. Financial institutions obtain insights about anticipated movements in asset values through the use of regression analysis, time-series forecasting, and other ML techniques. These insights assist the institutions in making educated decisions regarding investment strategies.

Robo-advisors are a manifestation of ML in the financial services industry. They give personalized investment advice based on an individual's risk profile as well as their financial goals. These ML algorithms are utilized by these digital platforms in order to conduct market analysis and optimize investment portfolios. This democratization of sophisticated financial planning makes it possible for a larger audience to acquire specialized investment strategies and reap the benefits of doing so.

In essence, ML in financial forecasting overcomes the limitations of traditional methods by harnessing the power of algorithms to analyze vast datasets, recognize patterns, and make predictions with a level of precision and adaptability that reshapes how financial professionals approach market dynamics. This is accomplished by harnessing the power of algorithms to analyze vast datasets, recognize patterns, and make predictions. Incorporating ML into financial forecasting promises not only increased accuracy in anticipating market trends but also a more flexible and responsive approach to the intricacies of the financial landscape as the digital era unfolds. This is due to the fact that ML is capable of learning from its own past mistakes.

15.2.2.2 AI IN FRAUD DETECTION AND CYBERSECURITY

Artificial intelligence (AI) is of utmost importance in enhancing cybersecurity measures and transforming fraud detection, providing a proactive defence against the ever-changing landscape of cyber threats. With the increasing migration of financial transactions and sensitive data to digital platforms, the protection of financial institutions and their clients becomes more dependent on the integration of AI-driven solutions.

With the aid of sophisticated pattern recognition and analysis, AI excels at identifying anomalies in transactions for the purpose of fraud detection. Algorithms capable of ML are capable of real-time analysis of massive datasets, identifying anomalous patterns that are suggestive of fraudulent activities. The ability to detect fraud in real time is of utmost importance in preventing the reputational and financial harm that can result from fraudulent activities; it enables prompt intervention and reaction.

The potential of AI to continuously adapt to new data and learn from it is a key advantage that it offers in the field of fraud detection. Machine learning models, in contrast to rule-based systems, which are subject to the ever-changing techniques of hackers, are continuously improving to counter emerging threats. The capacity for adaptation helps to reduce the number of false positives, which in turn helps to avoid the mistaken labeling of honest business dealings as fraudulent. The effectiveness of fraud detection systems as a whole is consequently improved as a result of this.

In the field of cybersecurity, threat analysis driven by AI adds an extra layer of defence against potential online dangers. AI systems are able to discover potential security breaches and vulnerabilities because they examine enormous quantities of data, such as user activity, network traffic, and system logs. Because AI is able to identify patterns that represent malicious conduct and react immediately, organizations that utilize this proactive strategy are able to keep a competitive edge over cyber threats. This is because AI is capable of identifying patterns that signify malicious behavior.

Threat intelligence, anomaly detection, and behavioral analysis are all examples of cybersecurity procedures that can be enabled by AI. Artificial intelligence is able to recognize unusual user behaviors that may be an indication of a security breach and can do so by tracking departures from the norm. Moreover, AI systems are able to do threat intelligence data analysis, which enables them to keep abreast of the most recent cybersecurity threats and adapt their defences accordingly.

The constantly shifting characteristics of cyberattacks have caused a sea change in the way that AI is used to the tasks of detecting fraud and ensuring network security. Artificial intelligence (AI) emerges as a powerful partner in the process of protecting the integrity of digital ecosystems, financial transactions, and sensitive data. This is accomplished by employing a fusion of ML algorithms, behavioral analysis, and real-time threat detection. As businesses are forced to contend with the complexities of the digital age, the incorporation of AI into fraud detection and cybersecurity has emerged as a strategic imperative rather than a mere necessity for maintaining confidence, fortification, and adaptability in the face of continuously developing cyber risks.

15.2.2.3 CHATBOTS AND VIRTUAL ASSISTANTS IN FINANCE

Chatbots and virtual assistants have become significant tools in the financial industry, revolutionizing the manner in which clients engage with financial institutions and improving overall user experiences. In the ever-evolving field of finance, the incorporation of AI has led to the emergence of intelligent conversational agents, who function as virtual assistants, offering a diverse range of services and support to individuals.

One of the primary utilizations of chatbots and virtual assistants within the realm of banking pertains to client engagements. AI-powered virtual assistants provide smooth and customized interactions through the utilization of NLP and ML techniques. Intelligent agents possess the capability to comprehend user inquiries, furnish details regarding account balances, transaction histories, and moreover proffer discernment into financial patterns. The constant availability of virtual assistants, operating 24/7, facilitates client access to support at any given time, hence enhancing customer service and overall happiness.

The utilization of chatbots contributes to the improvement of inquiry resolution efficiency and speed. These AI-powered solutions have the capability to rapidly analyze client inquiries, delivering timely and precise responses. Chatbots play a significant role in enhancing customer service efficiency within the banking and insurance sectors. They accomplish this by addressing commonly asked questions, providing guidance on account procedures, and offering assistance with fundamental financial queries.

In addition to facilitating consumer interactions, chatbots serve a vital function in the automation of mundane chores. Chatbots play a significant role in enhancing operational efficiency by doing repetitive duties such as assisting users in account setup and guiding them through transaction processes. This enables human agents to allocate their time and efforts toward more intricate and value-added operations.

Within the realm of financial institutions, chatbots play a significant role in enhancing the onboarding procedure for prospective consumers. The individuals in question possess the ability to provide guidance to users during the process of establishing an account, elucidate the intricacies of various financial goods and services, and even offer aid in the successful completion of application forms. This not only improves the overall client experience but also guarantees that users are provided with a seamless and user-friendly initiation to financial services.

The utilization of chatbots and virtual assistants transcends conventional banking services. Within the insurance industry, virtual assistants have the capability to effectively navigate users through policy details, facilitate the processing of claims, and furnish comprehensive information regarding coverage alternatives. The inherent adaptability of these AI-driven solutions renders them indispensable in diverse fields encompassing the wider financial industry.

In summary, the incorporation of chatbots and virtual assistants inside the financial sector represents a significant transformation in the manner in which customers engage and receive services. With the ongoing progress of AI, it is anticipated that these intelligent agents will undergo additional development, resulting in the provision of increasingly personalized and complex services [3]. The current era of finance is characterized by the smooth incorporation of AI-powered conversational agents, which provide individuals with a financial experience that is more accessible, efficient, and user-friendly.

15.2.2.4 BLOCKCHAIN TECHNOLOGY IN FINANCE

The introduction of blockchain technology has been acknowledged as a significant catalyst in the financial industry. This is due to the fact that it provides mechanisms that are both safe and transparent for the conduct of financial transactions. Blockchain is essentially a decentralized and distributed ledger system that efficiently documents transactions over a network of interconnected computers. This fundamentally describes blockchain as an open-source technology. In the world of finance, blockchain technology provides a system for documenting and authenticating transactions that is strong and unchangeable, hence reducing the risk of fraudulent activities and unauthorized entrance.

The potential of blockchain technology to ensure the safety of financial transactions is one of the key advantages that may be gained by applying this technology in the context of the financial sector. The blockchain technology establishes a secure and transparent ledger of transactions by utilizing cryptographic principles. This enhances the ledger's resistance to unauthorized adjustments or fraudulent activities by making the ledger more difficult to falsify. The improvement of the accuracy of financial data assists to instil users with a greater sense of confidence.

The blockchain technology has proven useful in a variety of different financial processes, in addition to its application in digital currencies like Bitcoin. Because the terms of a smart contract are directly written into computer code, this sort of contractual arrangement is capable of automatically carrying out the terms of the agreement [8]. Smart contracts are also known as executable smart contracts. The use of middlemen is made less necessary because of the implementation of smart contracts, which can automate and enforce contractual agreements. Additionally, because the structure of blockchain technology is decentralized, there is less of a need to depend on a single governing authority. This helps to develop a framework for financial transactions that is both more resilient and efficient. A great amount of interest has been generated in blockchain's ability to revolutionize traditional financial systems, making it a central focus in the era of digital technology. This interest has been generated by the continued research of blockchain applications by the financial industry.

15.2.3 AI IN MANUFACTURING SECTOR

Manufacturing is undergoing a major revolution due to the seamless integration of AI technologies. This revolution in manufacturing introduces a new era of efficiency and innovation. AI is revolutionizing the industry with predictive maintenance, quality control, collaborative robotics, and optimization of supply chain [14]. Implementing AI in the manufacturing industry is a paradigm change that will usher in a new era characterized by increased productivity, innovation, and competitiveness. AI technologies are revolutionizing many different elements of manufacturing processes, delivering solutions that streamline operations, increase product quality, and optimize the overall production efficiency. The following are some of the most important sectors in which AI is having a big impact on the industrial sector:

15.2.3.1 AI-DRIVEN PREDICTIVE MAINTENANCE

Predictive maintenance is revolutionizing the way manufacturers manage their equipment. For the purpose of predicting probable faults in equipment, ML systems examine both historical and real-time data. This preventative method makes it possible to arrange maintenance actions in a smart manner, hence reducing unplanned downtime and improving production schedules.

15.2.3.2 COMPUTER VISION FOR QUALITY CONTROL

Artificial intelligence-enhanced computer vision systems improve quality control procedures by allowing for the identification of defects in real time while manufacturing is in progress. These technologies detect flaws with the use of sophisticated algorithms for picture identification, guaranteeing that only products of the highest possible quality are shipped out of the manufacturing facility. This not only enhances the quality of the product but also decreases the amount of manual inspections that are required.

15.2.3.3 COLLABORATIVE ROBOTS (COBOTS)

Collaborative Robots, sometimes referred to as Cobots, are robots that integrate AI in order to work alongside human operators in production facilities. These robots are also known as Cobots. The utilization of these intelligent robots contributes to an increase in overall productivity by providing assistance with tasks that are either physically demanding or involve repeated motions. Cobots are a type of robot that place an emphasis on the safety of the workplace by utilizing a variety of smart sensors and adaptive ML. Because they possess these characteristics, Cobots are able to handle complicated environments and work in seamless collaboration with human workers.

15.2.3.4 OPTIMIZATION OF THE SUPPLY CHAIN

Supply chain operations may be made more efficient by employing AI-driven algorithms that analyze data in real time and in the past. AI has the capability to achieve this. This entails making the best use of shipping routes, reducing transportation costs, and generally increasing the overall effectiveness of logistical operations. Furthermore, the application of AI results in significant enhancements to the processes of inventory management and demand forecasting. As a result, manufacturers find it much easier to keep their stock levels at optimal levels and respond quickly and effectively to changes in the market.

15.2.3.5 AUTOMATION OF PROCESSES AND PROCESS OPTIMIZATION

Artificial intelligence is employed to enhance the manufacturing processes, ranging from resource allocation to production scheduling. This covers both process optimization and automation. By employing AI-driven technologies to increase operational efficiency, regular jobs may be automated. This process increases operational efficiency. This includes the application of AI technology to evaluate production data in order to constantly improve operations, the use of intelligent scheduling, and resource optimization.

15.2.3.6 AI IN OTHER INDUSTRIAL SECTORS

Artificial intelligence (AI) is transforming manufacturing in a number of ways, including supply chain resilience, product design, and energy management. In the field of energy management, ML algorithms are used in industrial facilities to analyze patterns of energy usage through the use of AI. The knowledge acquired makes it possible to pinpoint inefficient procedures and create suggestions for improving energy efficiency. This lowers operating expenses while also meeting environmental goals. Artificial intelligence (AI) is a key component in product design and innovation since it helps engineers create more creative and effective products [6]. Using AI algorithms for generative design—a process that studies several possible design outcomes—has become a standard procedure, encouraging innovation and productivity in production processes. AI also helps to improve the resilience and visibility of the supply chain. Manufacturers use AI to track and assess every phase of the production process, obtaining a complete picture of the supply chain. Businesses are able to recognize possible risks and react quickly to interruptions because of this increased visibility, which eventually strengthens the supply chain's overall resilience. Efficiency, sustainability, and flexibility are major forces driving growth in manufacturing, and the integration of AI across these dimensions heralds a transformational age in this field.

The application of AI in manufacturing is not only an increase in the capacity of existing technologies; rather, it signifies a profound paradigm shift in the manner in which manufacturing processes are conceived of and carried out. Manufacturers are in a position to benefit from increased productivity, greater product quality, and the potential to keep their competitive advantage in an economy that is constantly altering as a result of the continued development of technology related to AI.

15.2.4 AI IN THE ENTERTAINMENT INDUSTRY

The rapid advancement of AI is directly responsible for a number of paradigm shifts that are currently taking place within the entertainment industry. Whether it is through the generation of tailored suggestions or the construction of realistic special effects, AI is assisting in the creation of experiences that are more engaging and immersive for viewers.

15.2.4.1 SUGGESTIONS OF CUSTOMIZED CONTENT BASED ON YOUR PREFERENCES

A user's viewing history, interests, and activity are analyzed by systems powered by AI in order to generate personalized recommendations for material to that user. Streaming services use AI to better understand individual tastes, hence enhancing consumer happiness and engagement by generating recommendations for films, television series, music, and other content that is uniquely suited to each viewer. This can be done by making suggestions for films, television series, music, and other content.

15.2.4.2 CONTENT GENERATION AND SCRIPTWRITING

The use of AI to assist in the generation of content as well as the writing of scripts is becoming increasingly commonplace. In order to provide insights, algorithms for NLP analyze patterns in social media, comments from audiences, and historical data. These insights can serve as a guide for the creation of material that is both interesting and relevant, and they can be leveraged to do so. Some AI systems are even able to contribute to the writing of screenplays, which they do according to the specific genres, topics, or preferences of the audience they are writing for.

15.2.4.3 DEEPFAKE TECHNOLOGY

Deepfake technology, which is powered by AI, makes it possible to create very realistic special effects for use in motion pictures and television programmes. These effects can be seen in movies and on television [4]. With the use of this technology, filmmakers are able to recreate settings that would either be challenging or prohibitively expensive to shoot in real life. The seamless integration of computer-generated graphics with live-action video paves the way for the realization of this possibility. Deepfakes, despite the fact that they raise ethical problems, also offer new chances for creative narrative telling. This is despite the fact that deepfakes give rise to the questions in the first place.

15.2.4.4 ENHANCEMENT OF VISUAL EFFECTS AND ANIMATION

The use of AI is presently playing a key role in the enhancement of visual effects and animation within the entertainment industry. Rendering procedures can be improved with the help of algorithms that learn from their previous iterations, animation workflows can be streamlined, and the resulting images can even have a more realistic aspect. The production process moves along more quickly as a result of this, and it also makes it feasible to create content that is more visually appealing and immersive.

In the field of music production and composition, AI is making significant progress. It can now compose whole new works and do in-depth analysis of large music archives. It is essential to the music business, helping with anything from producing background music for movies to supporting artists during their creative processes. AI also plays a key role in creating immersive augmented and virtual reality experiences. By utilizing AI, these technologies may react and change in response to user activities, resulting in dynamic virtual worlds that can be tailored to a person's tastes. This makes a variety of settings, such as virtual tours and gaming, more realistic and engaging.

Artificial intelligence (AI) technology is being used in the entertainment industry to analyze social media data, gauge audience attitude, spot trends, and improve marketing tactics. Social media data and audience engagement become essential elements that help the industry better understand target market preferences, improve audience interaction, and personalize advertising campaigns. AI is also used in digital platforms for automatic content moderation, which makes sure user-generated content complies with social standards. This automatic method is especially helpful on sites where user-generated content exchanges are the major source of safety and civility online. The way AI is influencing music, VR experiences, and content moderation is just one example of the many ways it is being used to change the entertainment business.

The use of dynamic pricing models and tailored monetization strategies is made possible by AI.

15.2.4.5 STRATEGIES OF INDIVIDUALIZED COMMERCIALIZATION

The use of AI allows for dynamic pricing models. AI might be used by streaming services, for example, to vary subscription rates in response to changes in user behavior, consumption patterns, and the dynamics of the market. Streaming services could also use AI to identify new users and recommend content to them.

15.2.5 CROSS-INDUSTRY IMPACT OF AI IN DIGITAL ERA

The term "artificial intelligence" (AI) refers to a phenomenon that has grown into a widespread force that is crossing traditional industry boundaries and profoundly transforming the way organizations are run. Advanced AI algorithms are currently undergoing a diagnostic process revolution, which is one of the most notable examples of how AI is having an effect across industries [6]. AI supports medical personnel in early disease diagnosis and accurate interpretation of medical imaging data, considerably improving patient outcomes. This is accomplished through the analysis of medical imaging data. In addition, apps driven by AI that are used in drug discovery help to speed up the process of identifying possible candidates, which streamlines the research and development process for new medications. The disruptive power of AI in the healthcare industry has extended to telemedicine, which uses AI-powered chatbots to improve first patient contacts, with the goals of enhancing accessibility and optimizing the delivery of treatment.

The cross-industry influence of AI can be seen in many different sectors of the financial sector, including ML models that drive financial forecasts and algorithmic improvements that enhance fraud detection and cyber-security. Because of the role that AI plays in enhancing decision-making processes and providing individualized services, financial institutions are increasingly turning to AI-driven virtual assistants to provide personalized investment advice and improve their interactions with customers. The retail industry is undergoing a paradigm shift as a result of AI's analysis of enormous datasets to deliver personalized shopping recommendations, improve overall customer pleasure, and optimize inventory management. Deepfake is a form of AI that is being embraced by the entertainment industry. This has led to the creation of special effects and computer-generated visuals that are extremely lifelike. The adaptability of AI in reshaping the entertainment scene is further demonstrated by its role in content recommendations and music production.

Applications of AI are undergoing a transition in the manufacturing industry. One example of this is predictive maintenance, in which algorithms forecast when equipment will fail and optimize production schedules to minimize downtime. Computer vision that is powered by AI can improve quality control by locating flaws in the production process in real time. This helps to ensure that goods adhere to rigorous quality requirements. Machine learning is used in supply chain optimization, another use of AI that may be used across industries. This helps to optimize logistics, inventory management, and demand forecasting, which in turn promotes efficiency and responsiveness. AI has the potential to improve the education industry through the development of adaptive learning platforms that can personalize educational experiences for students depending on their particular requirements and success. The transportation sector is moving in the direction of enhanced safety and efficiency by adopting autonomous vehicles that are guided and controlled by AI. This represents a fundamental shift in the landscape of the business. AI's influence can be seen in other industries, such as the energy sector, where it is used in smart grids to improve the efficiency of energy distribution; and in the telecoms industry, where it is used to improve network performance and enable more effective customer support through the use of chatbots.

The agricultural industry is undergoing a digital transformation thanks to the implementation of AI-driven precision farming techniques. These approaches analyze sensor and satellite data in order to make educated decisions regarding crop management. The ability of AI to drive efficiency, innovation, and adaptation across a spectrum of sectors exemplifies the revolutionary potential of this technology in the larger landscape of business and industry. The cross-industry impact of AI is characterized by its ability to drive efficiency, innovation, and adaptability across a spectrum of sectors. The cross-industry impact is anticipated to intensify as organizations continue to integrate and enhance AI technologies. This will unleash new opportunities for efficiency, intelligence, and strategic decision-making in the digital era.

15.3 DATA: THE FUEL OF AI

Data is the uncontested fuel that accelerates innovation, shapes intelligence, and defines the capabilities of intelligent systems in the field of AI. Data serves as the fundamental building element upon which AI systems learn, adapt, and make informed judgements; hence, the relationship between AI and data can be described as symbiotic. This symbiosis lies at the basis of AI's ability to mimic human intelligence and give significant insights across a wide range of fields, and it is essential to the field's further development.

15.3.1 THE IMPORTANCE OF DATA IN ARTIFICIAL INTELLIGENCE

Data is the fuel that keeps AI running, and it plays a critical part in the creation, training, and operation of AI systems. In other words, data is the AI's "lifeblood." The ability to learn and make intelligent decisions is at the heart of AI's capabilities. This is a process that significantly relies on varied datasets that have been carefully curated [7]. The efficiency and precision of AI algorithms are directly influenced, both directly and indirectly, by the quantity and quality of the data. When AI models are trained on labeled datasets, a process known as supervised learning takes place. This allows the system to understand patterns and make predictions based on historical examples. The depth and breadth of available data gives AI the ability to generalize and apply learned skills to the previously unexplored scenarios.

In addition, the iterative nature of ML is fueled by the data that is collected. Artificial intelligence models have the ability to learn, improve their predictions, and expand their skills over time when they are continuously exposed to fresh data. Artificial intelligence systems may have difficulty understanding complicated patterns if they do not have access to a strong and diverse dataset, which might result in inferior performance and limited applicability. Because of this, the success of AI applications is inextricably linked to the availability, quality, and diversity of data. As a result, data is an essential component in the development and efficiency of AI technologies.

15.3.2 OBSTACLES ENCOUNTERED DURING THE DATA COLLECTION AND STORAGE PROCESS

The process of collecting and storing enormous volumes of data presents its own unique set of obstacles, despite the fact that AI relies heavily on data in order to function properly. Maintaining the integrity and precision of the data constitutes one of the most significant obstacles. Data that is inaccurate or biased can result in AI models that are skewed, which can either continue or even exacerbate existing prejudices. When attempting to find solutions to these problems, it is essential to take into account the dataset's representativeness, as well as the data's source and the possibility of inaccuracies in the data.

The protection of sensitive information is still another important problem. There is an increasing requirement for organizations to protect the privacy of individuals and comply with legislation controlling the use of data as more and more massive databases are accumulated. A continuing difficulty that calls for careful considerations and stringent privacy policies is finding a way to strike a balance between utilizing data for the growth of AI and protecting the privacy of individual users.

The enormous amount of data that is generated on a worldwide scale presents issues for both storage and processing. For training purposes, AI models, particularly those based on deep learning, frequently call for enormous datasets. This, in turn, calls for a robust computational infrastructure as well as storage solutions that are particularly effective. Because the costs of storing and maintaining massive datasets can be rather high, it is absolutely necessary for businesses to make investments in storage solutions that are both scalable and cost-effective.

In addition, the process of collecting data and integrating it is made more complicated by the fact that there are a wide variety of data sources and formats. Data may need to be reconciled and integrated from a variety of sources in order for AI systems to function properly. These sources may include structured databases, unstructured text, pictures, and sensor data. This integration challenge brings to light the significance of data pretreatment and normalization for the purpose of ensuring that AI models are able to make good use of a variety of datasets.

15.3.3 ETHICAL CONSIDERATIONS TO KEEP IN MIND WHEN WORKING WITH AI DATA

Ethical considerations in data handling and usage have become of the utmost importance as AI gets more thoroughly interwoven into various sectors of society. The use of skewed or prejudiced data can contribute to the maintenance of the existing social inequalities and provide biased results from AI. The creation of ethical AI demands a dedication to the principles of justice, openness, and accountability in the manner in which data is gathered, vetted, and utilized.

The importance of ensuring diversity and representativeness in training data is one of the most important ethical considerations. When training datasets largely represent certain demographics or opinions, the potential for biases to emerge increases. AI developers have a responsibility to work toward inclusiveness by searching for varied datasets that cover a wide range of ethnic, cultural, and socioeconomic aspects. In addition, continual monitoring and auditing of AI systems can assist in identifying and addressing biases that may manifest themselves during the lifecycle of the system.

When dealing with AI data, protecting users' privacy should be a top priority. In order to protect the private information of individuals, businesses and other organizations need to implement strong privacy policies and practices. Techniques such as anonymization and encryption can be utilized to safeguard personal information while still retaining the capacity to conduct insightful research. Consent and openness are essential components because they ensure that individuals are informed about how their data will be used and that they have the choice to opt out of having their data used in a certain way.

The ethical implications extend to the openness of the decision-making processes as well as the algorithms used by AI. To ensure that judgements created by AI can be comprehended and evaluated appropriately, it is essential that these decisions be explainable and interpretable. This transparency not only helps to build confidence but also makes it possible to hold AI systems accountable in situations where they have an effect on the lives of individuals, such as in the fields of healthcare, finance, or criminal justice.

In summary, data is unquestionably the gasoline that powers AI ahead, developing its capabilities and determining the influence it will have on society. The power of data to facilitate learning, generalization, and adaptation is a reflection of the critical role that data plays in AI. However, the difficulties involved with the collecting of data, storage of data, and ethical considerations in managing data highlight the necessity for responsible and careful practices in the creation of AI. To ensure that AI technologies contribute positively to society while still respecting individual rights and values, it is vital to strike a balance between leveraging the power of data for AI developments and keeping ethical principles. This will ensure that AI technologies will be able to meet these goals.

15.4 THE FUTURE OF AI IN THE DIGITAL ERA

Artificial intelligence is developing as a revolutionary force that is transforming industries, society, and the very fabric of our day-to-day lives. This spike in technological developments is taking place during the digital era, which is witnessing an unparalleled rise in technological advancements. As we make our way over the terrain of the digital future, several important dimensions begin to emerge into view. These dimensions provide a glimpse into the path that AI may take and the possible influence it may have.

15.4.1 NEW DEVELOPMENTS IN THE FIELD OF AI

15.4.1.1 EXPLAINABLE ARTIFICIAL INTELLIGENCE (XAI)

Explainable AI (XAI) is becoming a greater focus in the field of AI, which is one of the developing trends in this field. The need for transparency in the decision-making processes of AI systems is developing in proportion to the increasing complexity of these systems. The goal of XAI is to improve the interpretability and understandability of AI systems, particularly in fields such as healthcare, finance, and legal systems where accountability and interpretability are of the utmost importance.

15.4.1.2 ARTIFICIAL INTELLIGENCE AT THE PERIPHERY

The future of AI will have both centralized and decentralized computing at the periphery. Centralized processing will still play an important role. Edge AI is a trend that brings intelligence closer to the source of the data, hence reducing latency and enhancing real-time processing capabilities. This trend is part of a larger movement. This transition is particularly critical in applications such as those that demand quick decision-making, such as the Internet of Things (IoT), driverless vehicles, and smart cities [12, 15].

15.4.1.3 THE ROLE OF AI IN CREATIVE DOMAINS

Artificial intelligence is moving beyond its analytical roots to become a cocreator in creative domains. AI is being increasingly seen as a potential partner in the creative process due to its ability to generate musical compositions, create visual works of art, and assist in the production of material. This movement challenges perceptions about the limits of computer and human creativity, which opens up new avenues for the creative expression of new ideas.

15.4.2 PROSPECTIVE CHANGES AND ADVANCEMENTS

15.4.2.1 ARTIFICIAL INTELLIGENCE AND QUANTUM COMPUTING

The convergence of AI and quantum computing carries with it a vast amount of untapped potential. The processing power made available by quantum computing has the ability to completely change the landscape of AI by delivering answers to challenging issues on a scale that has never been seen before [10, 19]. This has substantial ramifications for a wide range of fields, including cryptography and optimization, as well as the creation of methods for ML that are more trustworthy.

15.4.2.2 HUMAN ENHANCEMENT

The purpose of AI is gradually shifting away from automating tasks and more toward improving the lives of humans. Because of this, a symbiotic relationship between humans and technology is fostered as a result of the utilization of technologies that enhance both cognitive and physical functions. In the not too distant future, innovations such as brain–machine interfaces and exoskeletons hold the promise of redefining the connection between humans and AI and opening up new paths for collaboration. These developments are also expected to open up new pathways for research and development.

15.4.2.3 THE ROLE OF AI IN DRUG DISCOVERY AND HEALTHCARE

The healthcare industry is on the edge of revolutionary advances, with AI speeding up the processes of drug discovery, personalizing treatment programmes, and facilitating more accurate diagnostics. The convergence of AI, genomics, and medical research carries the promise of introducing a new era of precision medicine by making it possible to develop groundbreaking treatments and therapies.

15.4.3 THE REPERCUSSIONS FOR SOCIETY AND THE WORLD AT LARGE

15.4.3.1 WORKFORCE TRANSFORMATION

The implementation of AI across industries will inevitably result in a shift in the nature of labor performed by humans. The nature of work is shifting, with more importance being placed on occupations that complement the capabilities of AI, such as data science, AI ethics, and human–AI collaboration. In order to successfully navigate this workforce shift, efforts aimed at upskilling and reskilling will play a critical role [5, 16].

15.4.3.2 DIGITAL INCLUSION AND ACCESSIBILITY

As AI grows more pervasive, it is absolutely necessary to ensure that digital inclusion and accessibility are met. It is vital, in order to cater to various populations, to bridge the digital gap and remove biases in AI systems. Fostering innovation and preventing the worsening of societal disparities can both be accomplished by working toward more equitable access to AI technologies.

15.4.3.3 THE ROLE OF AI IN GOVERNANCE AND REGULATION

Governance structures and laws are attempting to keep up with the rapid improvements being made in AI. It is expected that in the future, ethical norms, data privacy legislation, and multinational cooperation will be refined in order to develop a responsible AI ecosystem. Establishing public trust in AI and guaranteeing its responsible application require striking a balance between technological advancement and ethical concerns.

15.4.4 THE IMPORTANCE OF AI IN CONFRONTING WORLDWIDE PROBLEMS

15.4.4.1 CLIMATE CHANGE AND SUSTAINABILITY

Artificial intelligence is positioned to play a catalytic role in the effort to address environmental concerns. Applications of AI have the potential to make major contributions to sustainable practices, including the optimization of energy usage, assistance in climate modeling, and management of resources. The promise of AI to promote environmental stewardship is illustrated by recent advancements in precision agriculture and smart energy grids.

15.4.4.2 HEALTHCARE AND PANDEMIC PREPAREDNESS

Recent occurrences on a global scale highlight the crucial role that AI may play in providing healthcare and providing a response to pandemics. Predictive analytics powered by AI, epidemiological modeling, and the discovery of new drugs are all essential components of fortifying global healthcare systems against new dangers. It is expected that in the future there will be more investments in technologies that utilize AI to improve public health preparedness.

15.4.4.3 CYBERSECURITY CHALLENGES

The proliferation of digital landscapes has led to an increase in the number of cybersecurity risks. In order to strengthen cybersecurity defences, AI is essential because of its capacity to examine enormous databases and identify anomalies. In the future, there will be an emphasis placed on AI-driven threat intelligence, autonomous security systems, and preventative measures to combat growing cyber threats.

15.4.5 THE IMPORTANCE OF IMPLEMENTING ETHICAL AI PRACTICES

15.4.5.1 FAIRNESS AND BIAS MITIGATION

There is a compelling need to address bias in AI algorithms and make certain that their results are fair. This is an important ethical aspect. Artificial intelligence systems need to be built and trained to eliminate biases, promote equity, and prevent discriminatory repercussions on a wide variety of user groups.

15.4.5.2 TRANSPARENCY AND ACCOUNTABILITY

Ethical AI practices require that transparency and accountability be maintained throughout the creation and deployment of AI systems [9, 20]. Users and other stakeholders should have a solid grasp of the inner workings of AI systems, and procedures for accountability should be put in place to deal with any unexpected effects that may arise.

15.4.5.3 PRIVACY AND DATA PROTECTION

It is essential to uphold ethical standards by protecting the personal information of users and respecting their right to privacy. Applications that use AI need to comply with severe laws and regulations around data protection. Finding a middle ground between conserving individual privacy and fostering creative endeavors can be facilitated by putting into practice methods that protect privacy, such as federated learning.

In final analysis, the future of AI in the digital era will be defined by the confluence of a number of different factors, including rising trends, inventive developments, societal repercussions, global effect, and an unyielding demand for ethical practices. As AI continues to advance, its revolutionary potential holds the promise of addressing complex challenges and ushering in a new era of human–machine collaboration, innovation, and responsible technological advancement [16]. This promise is supported by the fact that AI is currently undergoing rapid development. To successfully navigate this future, researchers, governments, corporate leaders, and society as a whole will need to make a determined effort to guarantee that the benefits of AI are tapped in an ethical and inclusive manner [17, 18].

15.5 CONCLUSION

Artificial intelligence (AI) is emerging as a transformational force that is transforming industries, societies, and people's day-to-day lives as we stand on the precipice of the digital era. The voyage into this future will require cautious navigation, taking into account developing trends such as explainable AI, edge AI, and the role that AI plays in creative expression. The paradigm shift toward human and AI collaboration, in which AI serves as a cocreator and enhancer of human potential, highlights the adaptability and versatility of AI in fulfilling the ever-changing demands of the society. Nevertheless, this period comes with it a number of problems, including the demand for effective governance as well as workforce reforms, digital inclusion efforts, and more. The worldwide significance of AI extends beyond the technological achievements that have been made, which calls for a comprehensive knowledge of its capabilities, limitations, and ethical implications. In the modern day, AI is not only a technological tool but also a catalyst for the development of society. As such, it is necessary to strike a balance between utilizing its power for the benefit of society and minimizing any potential downsides. Collective effort is required to achieve this delicate equilibrium. This action should include investments in education, international collaboration on governance, promotion of ethical practices, assuring digital inclusiveness, advancing research and innovation, fostering public involvement, and continuously monitoring and adapting to changing conditions. We can create a digital age in which technology serves as a force for positive transformation, innovation, and the improvement of humanity if we accept the potential of AI and do so in a responsible and inclusive manner.

KEYWORDS

- **5G**
- **AI**
- **blockchain**
- **data analysis**
- **digitalization**
- **finance**
- **healthcare**
- **Industry 4.0**

REFERENCES

1. Borges, A. F. S., Laurindo, F. J. B., Spínola, M. M., Gonçalves, R. F., & Mattos, C. A. (2021). The strategic use of artificial intelligence in the digital era: Systematic literature review and future research directions. *International Journal of Information Management, 57*, 102225. https://doi.org/10.1016/j.ijinfomgt.2020.102225.
2. Perifanis, N.-A., & Kitsios, F. Investigating the influence of artificial intelligence on business value in the digital era of strategy: A literature review. *Information, 14* (2), 85. https://doi.org/10.3390/info14020085.
3. Balsalobre-Lorente, D., Abbas, J., He, C., Pilař, L., & Shah, S. A. R. (2023). Tourism, urbanization and natural resources rents matter for environmental sustainability: The leading role of AI and ICT on sustainable development goals in the digital era. *Resources Policy, 82*, 103445. https://doi.org/10.1016/j.resourpol.2023.103445.
4. Karnouskos, S. (2020). Artificial intelligence in digital media: The era of Deepfakes. *IEEE Transactions on Technology and Society, 1* (3), 138–147. https://doi.org/10.1109/tts.2020.3001312.
5. Zhang, Z., Wen, F., Sun, Z., Guo, X., He, T., Lee, C. (2022). Artificial intelligence-enabled sensing technologies in the 5G/internet of things era: From virtual reality/augmented reality to the digital twin. *Advanced Intelligent Systems, 4* (7). https://doi.org/10.1002/aisy.202100228.
6. Polak, P. (2021). Welcome to the digital era—The impact of AI on business and society. *Society, 58* (3), 177–178. https://doi.org/10.1007/s12115-021-00588-6.
7. Sewpersadh, N. S. (2023). Disruptive business value models in the digital era. *Journal of Innovation and Entrepreneurship, 12* (1). https://doi.org/10.1186/s13731-022-00252-1.
8. Farzana, A., Hallur, G. G. & Natraj, N. A. (2023). The study of techno-commercial factors influencing the adoption of hyper-automation across select industry verticals. In *Somaiya International Conference on Technology and Information Management (SICTIM) IEEE*. https://doi.org/10.1109/SICTIM56495.2023.10105014.

9. Almawgani, A. H. M., Reddy, B. R., Alsuwian, T., Ashok, P., Rathish, C. R., & Madhan, M. G. (2023). Impact of optical device parameters on the performance characteristics of temperature dependent quantum cascade lasers. *Optical and Quantum Electronics, 55*(10). https://doi.org/10.1007/s11082-023-05197-7.
10. Pardeshi, K., Pathak, P., & Alsadoon, Z. (2023). Applications of artificial intelligence and machine learning in e-commerce. In *Proceedings of the 15th International Conference on X-ray Microscopy—XRM2022*. https://doi.org/10.1063/5.0170750.
11. Gautam, A., Chirputkar, A., & Pathak, P. (2022). Opportunities and challenges in the application of artificial intelligence-based technologies in the healthcare industry. In *2022 International Interdisciplinary Humanitarian Conference for Sustainability (IIHC)*. https://doi.org/10.1109/iihc55949.2022.10059767.
12. Mohan Das, R., Arun Kumar, U., Gopinath, S., Gomathy, V., Natraj, N. A., Anushkannan, N. K., & Balashanmugham, A. (2023). A novel deep learning-based approach for detecting attacks in social IoT. *Soft Computing*. https://doi.org/10.1007/s00500-023-08389-1.
13. Mohamed, D., Al-Dossari, M., Savarimuthu, N., Natraj, N. A., El-Gawaad, N. S. A., & Aly, A. H. (2023). Study of three types of photonic crystal materials: A comparative and computational study. *Optical and Quantum Electronics, 55* (6). https://doi.org/10.1007/s11082-023-04714-y.
14. Singh, A., & Prabhu, S. (2023). Agile blockchain-based risk management framework with integrated artificial intelligence in a supply chain industry. In *2023 Second International Conference on Augmented Intelligence and Sustainable Systems (ICAISS)*. https://doi.org/10.1109/icaiss58487.2023.10250525.
15. Akash, & Kulkarni, P. (2021). Transformation of mobility industry by advanced digital technologies. *Journal of Physics: Conference Series, 1964* (4), 042020. https://doi.org/10.1088/1742-6596/1964/4/042020.
16. Mantey, E. A., Zhou, C., Srividhya, S. R., Jain, S. K., & Sundaravadivazhagan, B. (2022). Integrated blockchain-deep learning approach for analyzing the electronic health records recommender system. *Frontiers in Public Health, 10*. https://doi.org/10.3389/fpubh.2022.905265.
17. Vaitheeswaran, G., Sundaravadivazhagan, B., & Karthikeyan. (2023). Some studies of the latest artificial intelligence applications of drones are explored in detail with application phenomena. *Drone Technology* (pp. 99–116). https://doi.org/10.1002/9781394168002.ch5.
18. Karthikeyan, R., Balamurugan, V., Cyriac, R., & Sundaravadivazhagan, B. (2022). COSCO2: AI-augmented evolutionary algorithm based workload prediction framework for sustainable cloud data centers. *Transactions on Emerging Telecommunications Technologies, 34* (1). https://doi.org/10.1002/ett.4652.
19. Mohemed Ashik M., S., B., Subashini, B. (2021). A review on Internet of Things (IoT): Security challenges, issues and the countermeasures approaches. *Psychology and Education Journal, 58* (2),–6560. https://doi.org/10.17762/pae.v58i2.3188.
20. Natraj, N. A., Kamatchi Sundari, V., Ananthi, K., Rathika, S., Indira, G., & Rathish, C. R. (2022). Security enhancement of fog nodes in IoT networks using the IBF scheme. In *Third International Conference on Image Processing and Capsule Networks* (pp. 119–129). https://doi.org/10.1007/978-3-031-12413-6_10.

CHAPTER 16

An Enhanced Hidden Markov Model for Predicting Two-Tier Web Page and Improving Accuracy

OMOSEBI PAUL ADEOYE[1], ADEWOLE ADETUNJI PHILIP[2], and SENNAIKE OLADIPUPO AYODEJI[2]

[1]*Department of Computer and Information Sciences, Trinity University, Yaba, Lagos, Nigeria*

[2]*Department of Computer Sciences, University of Lagos, Akoka, Lagos, Nigeria*

ABSTRACT

The weblog contains a substantial amount of information about user requests. To offer suggestions for websites, minimize latency, and engage in other online activities such as advertising, it is critical to be able to anticipate future requests made by users based on the pages they have seen in the past. These applications compromise on the intricacy of their models and the accuracy of their forecasts. A web page prediction model that builds and provides a classifier using sessions as training examples. The former system, which used 30 min as the default timeout, has been replaced with one that produces sessions by evaluating the average time spent on website access. Because sessions are utilized as training examples, this is done. Hidden Markov models have lately gained popularity for predicting the next page a user will visit on a website based on navigational history saved in a weblog. This usage-based technique has the potential for further

Real-Time Artificial Intelligence (AI): Key Motivations, Technologies, Platforms, and Use Cases.
Pethuru Raj, B. Sundaravadivazhagan, V. Kavitha, B. Narendra Kumar Rao, & Hannah Vijaykumar (Eds.)

DOI: 10.1201/9781998511358-16

improvement to improve forecast accuracy. The hidden Markov model with an autoregressive model is suggested as a way of improving web page prediction accuracy by 83% compared with the conventional hidden Markov model.

16.1 INTRODUCTION

The utilization of web applications as a platform for information engagement with users is becoming an increasingly popular practice, hence the issue of attempting to anticipate a user's actions while visiting a website has become increasingly urgent. These applications need to be able to quickly process big data and present users promptly with accurate information that is gleaned from online repositories. Crawling the web and processing massive amounts of data both need a large amount of computational power. To simplify the process in terms of both space and time, new methods need to be developed, but these methods cannot be developed without first being constructed. It is considered that data mining, which refers to the process of extracting certain information based on defined criteria, is a solution that can assist in solving problems [23].

We have witnessed a rise in the number of applications that are moving away from ways that are based on desktop programs and instead converting to methods that are based on web pages. In particular, services, such as shopping, billing, communication, and transportation may all be accomplished through the use of an Internet browser or an application that is hosted on a web page as opposed to a sizable application that must be downloaded and installed into the device. This is because Internet browsers and web-based applications do not require as much storage space as traditional applications do. Web page applications are widely used as a medium through which individuals can communicate with one another and share information. However, there is a typical difficulty with this method, and that problem is the processing of massive amounts of information, which is commonly referred to as big data [10].

Websites that provide online news are faced with this challenge because these websites are required to serve thousands of people in a short period of time and have a lot of content that needs to be arranged for simple filtering. The challenge lies in the fact that these websites have a lot of content that need to be arranged. Even though web apps have to deal with massive volumes of data, they continue to be the easiest option for customers to receive the most recent information. It is tempting to a huge number of users to be able to load an application in a matter of seconds without the need for installation. This leads to the demand for massive quantities of processing power to offer users the results they were anticipating in a matter of seconds. By categorizing the data in a more effective manner, which may be accomplished with the help of classifiers, it may be possible to lessen the quantity of traffic that is generated on the network. The procedure is referred to as data mining, which is an industry term [6].

The method of data mining is performed when a user is filtering through a huge quantity of data to expose characteristics within the records of a large amount of data to minimize the amount of data that need to be evaluated. This is accomplished via the usage of a large number of data. These factors vary depending on the application; for instance, with online news, the features may be the news items, the behavior of visitors, and the frequency with which they visit, and so forth. The discipline of data mining may be subdivided into a variety of subfields; one of these subfields is known as the predictive branch, and it addresses issues relating to the classification, regression, time series analysis, and prediction, among other things. There is also the descriptive branch, which includes processes such as categorizing, summarizing, discovering rules of the association, and developing sequences [21].

One of the uses of data mining is the enhancement of search engines through the classification of online pages. Another application of data mining is the identification of usage patterns, which makes it possible to comprehend the activities that people engage in when using a web application. You will be able to retain customers using the program for longer periods and more frequently if you have a good understanding of what it is that customers desire to see included in the program. It is possible to improve the accuracy of predictions made using this usage-based method by combining it with the structural qualities of the web pages [24].

Web page classification is essential for getting web page information and it falls under network content download. It is vital for directory websites, subject browsers, and search engine user experience. Online consumers often believe that technology and financial site homepages provide news from many fields. Misconception: Large-scale news requires automated classification. Real-time systems that categorize news websites effectively will be important. Grab a certain amount of data from web pages and combine it with existing web page classification technologies (which refers to constructing the text classification system) to automatically classify web pages [16, 22].

Recently, Markov models have seen a lot of usage for predicting the next page a user will visit on a website based on the navigational history that is stored in the weblog. In this study, they provide an improved hidden Markov model (HMM) method that takes into account the features of the Web by relying not only on HMM but also on autoregression to analyze data. This method was developed as part of this research. The phrase "prediction accuracy" will be used throughout this text to refer to a projected web page that has a greater overall prediction accuracy. This phrase describes a web page that has a higher level of prediction accuracy.

16.2 RELATED WORKS

Machine learning algorithms were proposed to classify web news documents effectively [13]. Web applications are a popular way to communicate with people. These apps must quickly evaluate big data and give consumers credible news portal information, which might be tough. Web crawling and data analysis need a lot of computer power, thus innovative techniques must be found to reduce the time and space needed. Data mining, the process of retrieving information based on given features, is thought to mitigate the concerns stated above. In this work, they provide a web model that can extract news articles and organize them into four categories: business, technology and science, health, and entertainment. Four machine learning classifiers are compared SVM, kNN, DT, and LSM (LSTM). The accuracy and receiver operating characteristic (ROC) curves of these classifiers are compared. kNN's accuracy was 89%, whereas SVM's was 95%.

The optimization algorithm automatically adjusts the size of the learning rate, making the research of Web Page Classification Algorithm (WPCA) based on Deep Learning (DL) more efficient [25]. WPCA DL is the article's main source. This article uses keyword weight calculation to reduce the effect of high-frequency phrases. WPCA is improved by reducing low-frequency word weights. Optimizer affects learning rate. DL-based WPCA research is simplified. DL WPCA takes 354 s against 2436 s for standard. DL WPCA uses 186.25 s less memory than the traditional algorithm (6.35 s). DL-based WPCAs are faster, more efficient, and use less memory. This study recommends expressing website text as a weighted directed graph to construct a semantic network. Chart making. First, assign language package chart nodes to the bureau. Similarity and co-word mining connect words based on comparable edges. Web page performance is used to assess WPC structure. Network file feature extraction and tf-idf show web page features. Vector space commonly represents web pages. Experiments show that these tactics can satisfy customers.

Machine learning may be used to forecast gift industry website rankings: SEO variables. Machine learning is used to predict a web page' search engine position [18]. Using a list of keywords for 30 gift-related blogs, we get 733 first-page Google results. We rank these sites using 30 criteria. It was observed that the XGBoost model excels at predicting genuine search ranks with an average accuracy of 0.86. According to the featured study, a website's internal and external linkages, web domain security, and H3 heading length have the most influence. The inclusion of the term in the domain URL, H1 headers, and content has the least influence on a website. The results show the importance of SEO linkages. Internet marketers and content providers need actionable information.

For phishing website prediction, machine learning was suggested [2]. Phishing takes personal and sensitive data like Internet passwords. With phishing attacks on the rise and their detrimental impact on businesses and consumers, organizations and individuals must be able to differentiate phishing websites from lawful ones. This study predicts phishing websites to protect customers. This study predicts ML. Businesses, the government, and others mine data for insights. Enterprises worldwide utilize machine learning to assess, categorize, and forecast findings for estimating and planning. Logical regression, decision trees, and random forest classification are used to predict fake websites. Results are plotted. Data came from UCI's machine learning library. Modifying the seed value improves accuracy to 95.93, 97.96, and 98.78%. Future research and best practices may offer a more reliable way to anticipate a phishing website's URL and attributes.

A suggestion and navigation prediction system was suggested [20]. Weblog contains plenty of user request data. Predicting users' future requests based on visited pages can improve page suggestions, latency, and online advertising. Apps compromise accuracy and complexity. They propose a web navigation prediction framework for web page recommendation. Instead of 30 min, training examples use the average page load time. This design is compared with a two-tier prediction system using benchmark datasets. WNPWR is more accurate and faster than two-tier prediction.

Markov model predicted a user's web link and route [3]. In the study, web pages and connections abound where users get more data daily, and about one million web pages are added. Weblogs track readership, web mining employs logs. Log file sizes are growing due to increased usage. This study compares Markov and Kth Markov models for web prediction, improves route scalability with a modified Markov model, and presents a two-tier prediction architecture that creates a classifier EC using training instances and classifiers. The study also shows how this can improve forecast time and accuracy, using benchmark data sets, contrasting the Markov model and ARM methods. Improved Markov model lowers paths without reducing accuracy and results confirm that all-Kth model accuracy increases with the order.

A hybrid approach was used to predict the user's web navigation behavior [14]. Diverse websites and increased traffic complicate the Internet. it is difficult to predict user needs. Long training, increasing prediction time, low accuracy, and memory limits are obstacles to prediction. When multiple models are available, the system enhances forecast accuracy. The system reduces prediction challenges, gives accurate results, simplifies the approach, and reduces false predictions. The new hybrid model combines the Markov model with HMM to provide appropriate website results. We tested the hybrid model's accuracy, precision, and miss-prediction using varied datasets.

Deep learning was presented for online ad interest and response prediction [7]. Digital advertising models user interest and behavior: Interests influence ad reaction and activity (Ad). An ad viewer's interests can influence their buying decisions. Existing ad-click prediction methods use a static feature set to train machine learning classifiers. These algorithms disregard temporal variance and user behavior changes to learn. LSTMcp and LSTMip are proposed for click prediction and interest modeling. They predict a user's ad-click likelihood and ad campaign click probability. They also collect page information as a temporal sequence and apply LSTM to understand user interests. Addressing sequences and temporal variety of user inquiries increases ad response and click prediction.

16.3 MATERIALS AND METHODS

This study proposes a machine learning-based web prediction methodology. This chapter presents HMM algorithms for classifying web page labels.

This approach uses an updated autoregressive algorithm to extract line segment data from input data images and assign a unique value to each segment. Support vector machine and k-nearest neighbor categorize chosen features. Figure 16.1 shows the proposed workflow of the study.

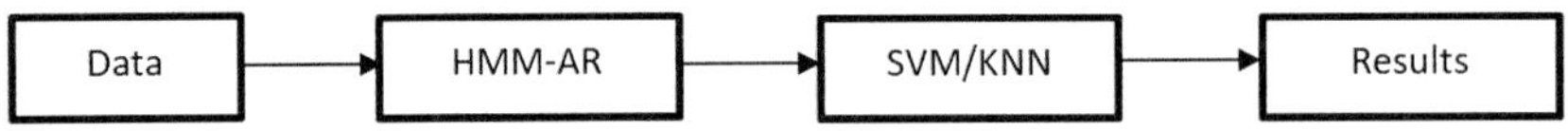

FIGURE 16.1 Proposed workflow.

16.3.1 DATASET

11,430 URLs and 87 extracted attributes are in the proposed dataset. The dataset tests machine learning-based web page prediction methods. About 56 are acquired from URL structure and syntax, 24 from related page content, and 7 from external service inquiries. Half of the URLs are phishing and half are legitimate. Along with the dataset, we give Python scripts for feature extraction. May 2020 datasets are currently on Mendeley: dataset A: a list of URLs and related DOM tree objects for duplicating and testing innovative URL and content-based features that replace phishing websites' short-term existence. https://data.mendeley.com/datasets/c2gw7fy2j4/3.

Dataset B contains feature values for training classifiers. This dataset is indexed with URLs, thus the index must be erased before testing [8]. Python is used for teaching, testing, modeling, and development.

16.3.2 HIDDEN MARKOV MODEL (HMM)

Hidden Markov models predict user behavior without data. utilizing with one or more actions per state to model resident activities. AI is combined with a model state, similar in device, location, and time. Otherwise, AI is generated. Frequency track changes. Discrete modeling has few states and frequent transitions. If all actions were stated, the prediction would be impossible. Hidden Markov models are used for voice recognition, data extraction, and anomaly detection (Lane, 1999). Baum–Welch or hand-crafted HMMs are feasible (Baum, 1972). The new study predicts tasks using HMMs. Clusters are HMM hidden states. State clusters are hidden.

State-hidden tasks. Vi separated N clusters. The training uses symbols. Ms. O shows action. The answer is A. B shows a sign in N states. Each state is unique [9].

Recurring actions require ergodic models. Our model only defines N and M, thus needs A, B, and = (A, B). Baum–welch triple initialization and EM training: The forward technique (Rabiner, 1989) determines sequence probability. The forward technique compares OBS to HMMs. The model best explains sequence. Alternative assessment methods employ one model and several observation sequences. Oi + k + 1 and Oi + 1 ... Oi + k. M sequences have the same first k but different last symbols [5]. Then k + 1. Top N guesses are valid, where N is 3, 5, or custom. This requires k-activities. Keeping the whole observation sequence for forecasting may be expensive; therefore, just store the latest k symbols or activities.

16.3.3 AUTOREGRESSIVE MODEL (AR)

Autoregressive models predict time series values using past inputs. Autoregression uses historical data to forecast future time steps. Simple concepts can predict time series. AR models incorporate forecasting, statistics, and parameter estimation. Parameter estimation estimates future time series points. Time series modeling is evaluated using several criteria.

The ARMA (p, q) model merges AR (p) and MA (q) models for univariate time series modeling. Here is how the ARMA (p, q) model looks:

$$yt = c + AR(p) + MA(q) + \varepsilon t$$

$$yt = c + \phi 1 yt - 1 + \phi 2 yt - 2 + \cdots + \phi pyt - p + \varepsilon t$$

yt is the current value, c is constant, ϕi is the autoregressive coefficients, $yt\text{-}i$ is the previous value and εt is the error term. If the series is not stationary at the first level, remove it d (d = 1, 2, 3...) times. ARIMA (p, d, q) is a time series model. Autoregressive model construction involves identifying, estimating, and diagnosing parameters. After validating the stationarity of the time series, the AR and MA terms are obtained from the ACF plot for model identification. ACF is a statistical metric of correlation used to examine if earlier time series values have a link with the newest data. Then, ARIMA parameters (p, d, q) are calculated using least squares.

The purpose of an autoregressive model is to make predictions about the variable of interest by employing a linear combination of the previous values of the variable denoted as "self" in the autoregression (AR) notation. This notation shows that the variable is being regressed against itself. AR is comparable to multiple regression, with the difference being that the AR model uses the lagging values of the time series *yt* as its predictors.

Algorithm 1: The Procedure for Modeling Using HMM-AR.

Initialize Sample S; set of the visited web by users with the sequence of pages needed to predict pages;

    If HMM is a used prediction;

        then the probability is given;

        if the not probability is given;

        all orders of HMMs are calculated

    Output: Subset samples

Data: Obtain Selected HMM Data

Result: Choose the Optimized AR model;

Starting;

    Fragment data to training and Testing;

    Transform data by HMM selected Data;

    While minimum and partial correlations **do**

        If the sample is accepted **then**

            Evaluate model

        Else

            Obtain values;

    End

    End

16.3.4 KNN ALGORITHM

KNN is a machine learning classification method. It is instance-based or lazy learning. This classification method uses local approximation and defers calculation till categorization. It saves all the instances in the supplied dataset and classifies new cases using Euclidean distance as a similarity metric. The K most similar events, termed neighbors, are then controlled by searching the whole test prep set for another data point. The anticipated outcome is calculated by reducing the yield variable for K situations when neighbors cast a majority vote. In website prediction, researchers employ a KNN classifier [15].

Classification drives data mining. It is employed in business, management, and research. Decision tree, SVM, Bayesian, kNN, ANN, and so forth. Machine learning algorithm KNN: In the KNN algorithm, compute the distance between point A and all other points to eliminate the k-nearest points, and then count the k remaining points. A has the most. KNN employs 'near red, near black' to infer category from neighbors. Distance between the sorting sample and training sample of the known category is obtained. Neighbors' classification determines sample data categories. Calculate the test item's distance from each training object; find the nearest k training objects as neighbors' test objects; categorize test objects using k Kordia's core categories [19].

Algorithm 2: KNN-code [11].

Picks k neighbors.

K neighbors need Euclidean distances.

Take K closest neighbors using Euclidean distance.

Count each category's K neighbors.

The category with the most neighbors obtains new data points.

The k-NN model is ready.

K-data Classifications

Keep training tuples.

Classify unknown tuples

    Use Kth value

    Find nearby k training tuples.

    Assign unseen tuple's training class from neighboring pairs.

Finish

Standardize data before applying the formula kth value to determine Euclidean distance. Math is streamlined. Normalize attribute values [0, 1]. 0 is the lowest value, and 1 is the highest. Min-max normalization is (v − minA)/(max − minA). This normalizes A's v to [0, 1]. maxA and minA are A's maximum and minimum [4].

k controls how many nearest neighbors to be utilized in categorization. In k-neighbors, the unseen tuple class is the majority. The unseen tuple's class attribute is the average of the k-nearest training tuples' class attributes if the data set's class attribute is actual. Regression predicts class characteristics. KNN has k = 1. The unknown tuple class is the nearest training data point, class. Which distance metric should be used, and should all traits be weighted equally? [11]

16.3.5 SVM-ALGORITHM

SVM is another predictive analytics classification approach. In most situations, SVM is used as a binary classifier, that is, to segregate information focused by class, either 0 or 1. To achieve this, a vector machine chooses a variable-space hyperplane. Using this hyperplane, vector machine learning determines the optimal class separation coefficients. For context-aware mobile services, researchers employ SVM classifiers [1].

Vapnik (1995) proposed SVM, a common machine learning method. SVMs are used for text categorization, hand-written digit identification, tone recognition, photo classification and object detection, and gene expression data analysis. The cost and kernel parameter adjustments affect SVM performance for certain datasets. Cross-validation is needed to discover the optimal parameter settings. Model choice: SVM outperforms other data classification algorithms in accuracy, according to a recent study. Classification and regression employ SVMs, a statistical learning-based machine technique. They are linear. SVM optimizes the geometric margin and reduces classification error. SVM utilizes MMCs. SVM reduces risk (SRM). SVM translates the input vector to a higher-dimensional space with a hyperplane. Two parallel hyperplanes flank the data-separating hyperplane. SVM optimizes the distance between hyperplanes. More space between parallel hyperplanes improves classifier generalization error [12].

This study compares the expected reaction with the actual response (i.e., the ground truth) to determine prediction accuracy. Recall/sensitivity, precision, ROC Precision is the percentage of accurately anticipated phone call activities to all expected activities (both correctly and incorrectly). Recall measures completeness by comparing the number of accurately anticipated phone call activities to the total relevant activities. The recall is sensitive. These are determined from prediction findings' true-positive, false-positive, and false-negative rates. ROC includes true-positive rate and false-positive rate to assess classifier output quality [17].

F-measure/kappa F-measure blends accuracy and recall into a single score, the harmonic mean. Kappa contrasts actual and projected accuracy (random chance). F-measure and Kappa definitions:

$$\text{F-measure} = 2 * \text{Precision} * \text{Recall} / \text{Precision} + \text{Recall}$$

$$\text{Kappa} = (\text{observed accuracy} - \text{expected accuracy}) / (1 - \text{expected accuracy})$$

16.4 RESULTS

This study uses web page prediction methods from https://data.mendeley.com/datasets/c2gw7fy2j4/3. Features include: 56 come from URL structure and syntax, 24 from page content, and 7 from other services. The collection comprises 50% phishing and 50% authentic URLs. Along with the dataset, we provide Python scripts for feature extraction. "!pip install" installed dependencies and libraries in Jupyter lab. PyTorch, Numpy, Pandas, Matplotlib, Sklearn, Shuttle, and Pillow were imported. The training and validation datasets were compiled from Mendeley.

Figure 16.2 shows the development environment for interactive data processing where we have the count of status of the URLs which shows whether it is either legitimate or phishing.

A heatmap is a graphical representation of data where the individual values contained in a matrix are represented as colors.

Figure 16.3 shows a heatmap of the dataset, with each matrix value colored differently.

Using the loaded dataset, various experiments were proposed. This implementation utilizes hmm-learn for HMM training and decoding and a model selection to determine the optimal number of parameters. Figure 16.4 displays the dataset with HMM, and Figures 16.5 and 16.6 show the confusion matrix showing the dataset with HMM and SVM and HMM and KNN results, respectively.

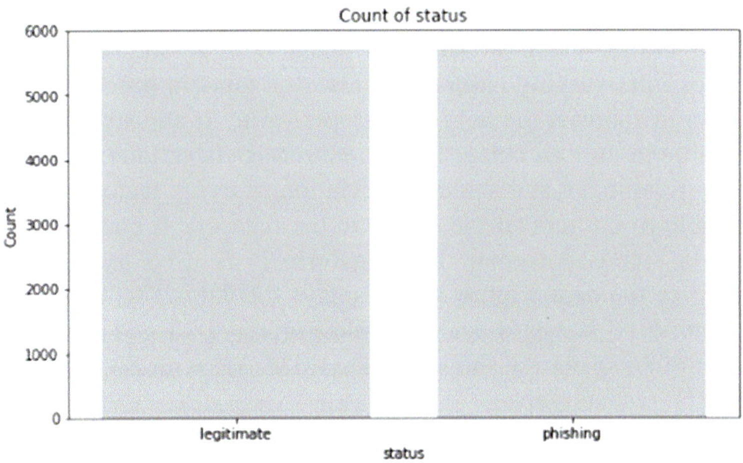

FIGURE 16.2 Loaded data counts.

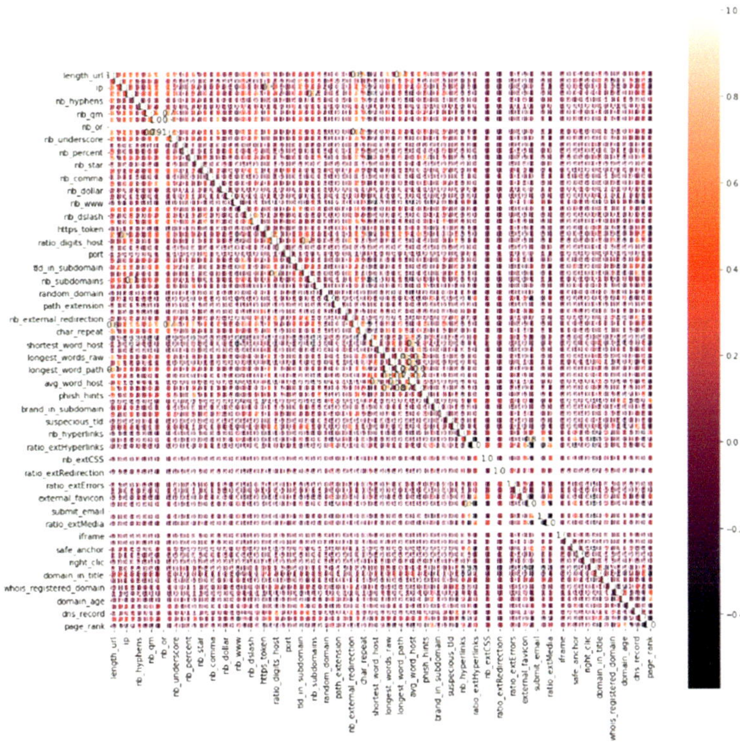

FIGURE 16.3 The heatmap of the dataset.

To boost HMM's performance, this study incorporated an autoregressive model using a standard scalar. An autoregressive model describes time-varying natural processes as random processes. It uses past data to forecast the next time step's value. It adjusts each feature or variable to unit variance. This operation is standalone. It calculates the empirical mean and standard deviation of every feature, making it vulnerable to outliers (if they exist in the dataset). It converts Python data to a standard format. Fit transform () is used to convert and standardize the data. Figure 16.7 displays the dataset with HMM with Autoregressive standard scalar, while Figures 16.8 and 16.9 demonstrate confusion matrix showing the HMM-AR with SVM and KNN classifiers, respectively.

```
data = pd.read_csv(r'/content/dataset_B_05_2020 (1).csv')
x = data.drop(columns=['url', 'status']).values
y = data.iloc[:, -1].values
#y = y.reshape(-1, 1)
#encoder = OrdinalEncoder()
# transform data
#y = encoder.fit_transform(y)
#x = encoder.fit_transform(x)
print(x)

[[ 37.  19.   0. ...   1.   1.   4.]
 [ 77.  23.   1. ...   0.   1.   2.]
 [126.  50.   1. ...   0.   1.   0.]
 ...
 [105.  16.   1. ...   0.   1.  10.]
 [ 38.  30.   0. ...   0.   0.   4.]
 [477.  14.   1. ...   1.   1.   0.]]

# Implementation of the markov model
markov = hmm.GaussianHMM(n_components = 3, covariance_type = "diag", n_iter = 50, random_state = 42)
markov.fit(x)

GaussianHMM(n_components=3, n_iter=50, random_state=42)
```

FIGURE 16.4 Dataset with HMM.

This study included multiple experiments. Table 16.1 displays HMM with and without AR and SVM and KNN classification results.

To improve web page prediction, this study created a model to predict the next website visited. The proposed model includes HMM, SVM, and KNN classifiers. This concept works for all web applications. In Table 16.1, the results of training and testing classifiers using a large data set to enhance prediction accuracy are shown.

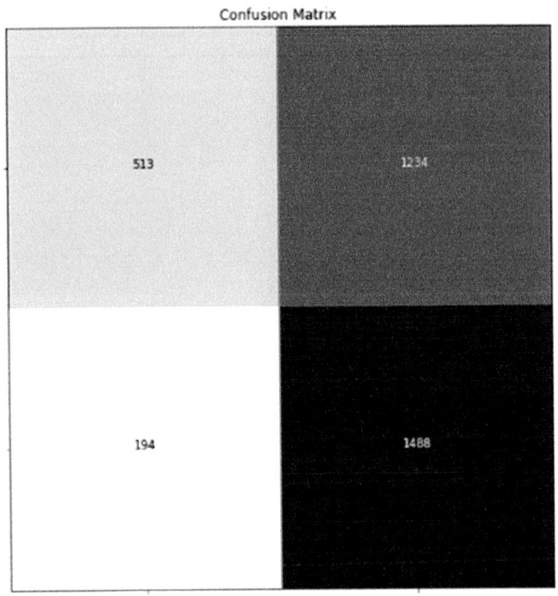

FIGURE 16.5 Confusion matrix showing the dataset with HMM and SVM (TP = 513; TN = 1488; FP = 1234; FN = 194).

FIGURE 16.6 Confusion matrix showing the dataset with HMM and KNN (TP = 1417; TN = 1444; FP = 330; FN = 238).

```
data = pd.read_csv(r'/content/dataset_B_05_2020 (1).csv')
x = data.drop(columns=['url', 'status']).values
y = data.iloc[:, -1].values
y = y.reshape(-1, 1)
encoder = OrdinalEncoder()
# transform data
y = encoder.fit_transform(y)
x = encoder.fit_transform(x)
print(x)
#sc = StandardScaler()
#x = sc.fit_transform(x)
#y = sc.fit(y)

[[ 25.  15.  0. ...  1.  1.  4.]
 [ 65.  19.  1. ...  0.  1.  2.]
 [114.  46.  1. ...  0.  1.  0.]
 ...
 [ 93.  12.  1. ...  0.  1. 10.]
 [ 26.  26.  0. ...  0.  0.  4.]
 [305.  10.  1. ...  1.  1.  0.]]
```

FIGURE 16.7 HMM with a standard scalar.

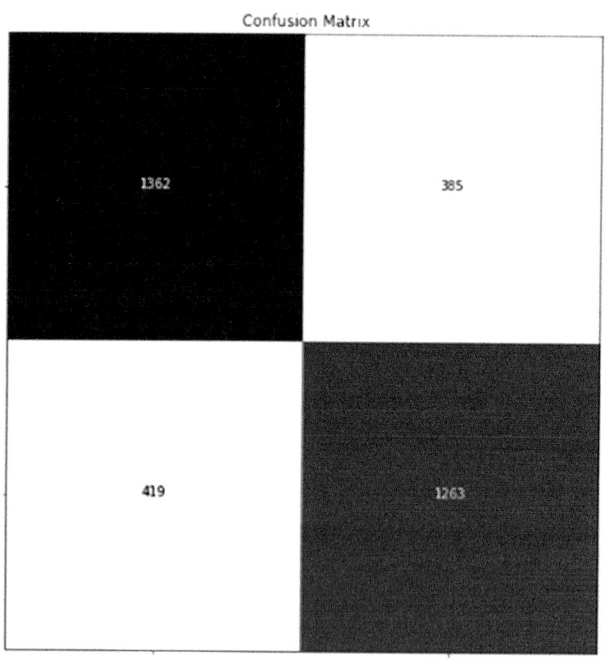

FIGURE 16.8 Confusion matrix of HMM-AR with SVM (TP = 1362; TN = 1263; FP = 385; FN = 419).

FIGURE 16.9 Confusion matrix of HMM-AR with KNN (TP = 1405; TN = 1371; FP = 342; FN = 311).

TABLE 16.1 HMM with and without AR and SVM and KNN Classification Results.

Measures	HMM-SVM	HMM-KNN	HMM-AR-SVM	HMM-AR-KNN	Derivations
Accuracy	58.36	83.44	77.21	81.14	ACC = (TP + TN) / (P + N)
Sensitivity	72.56	85.62	76.47	81.88	TPR = TP / (TP + FN)
Specificity	54.67	81.40	76.64	80.04	SPC = TN / (FP + TN)
Precision	29.36	81.11	77.96	80.42	PPV = TP / (TP + FP)
F1-Score	41.81	83.30	77.21	81.14	F1 = 2TP / (2TP + FP + FN)
Matthews correlation coefficient	22.03	66.99	53.08	61.92	TP * TN – FP * FN / sqrt ((TP + FP) * (TP + FN) * (TN + FP) * (TN+FN))

16.5 CONCLUSIONS

Web mining studies session data and finds a user's prior session websites to answer the prediction question. This predicts the pages the visitor will visit next. This research covers web data mining and theory. Web mining is defined, taxonomized, described, and predicted. Different writers' strengths and weaknesses are analyzed. Dynamic online data allow for more research, such as developing better prediction systems. The author wants to enhance web page prediction and next-user access. He wants it to be accurate, speedy, and error-free. Predicting website use is helpful. Internet ads are more specific. Web page access forecasts will inform advertising. Access prediction increases architecture, personalization, and market segmentation. Internet content simplifies research. Log record patterns can predict future events. Analysis of input data predicts user behavior. Web servers and search engines enhance predictions. Online predicts. Web search and website results increase. Web prediction makes suggestions. In these systems, we find the top n people who share a target user's interests or hobbies to boost website traffic and provide recommendations.

KEYWORDS

- **auto regressive**
- **hidden Markov model**
- **KNN**
- **SVM**
- **web page**

REFERENCES

1. Awad, M., & Khanna, R. (2015). Support vector machines for classification. In *Efficient learning machines* (pp. 39–66). Berkeley, CA: Apress. https://doi.org/10.1007/978-1-4302-5990-9_3.
2. Awasthi, A., & Goel, N. (2021). Phishing website prediction: A machine learning approach. In C. R. Panigrahi, B. Pati, B. K. Pattanayak, S. Amic, & K. C. Li (Eds.), *Progress in advanced computing and intelligent engineering: Advances in intelligent systems and computing* (Vol. 1299, pp. 143–152). Singapore: Springer. https://doi.org/10.1007/978-981-33-4299-6_12.
3. Charpate, A., Bramhankar, C., Gaikawad, P., & Londhe, A. D. (2015). Prediction of link and path for user's web browsing using Markov model. *International Journal of Computer Science and Mobile Computing*, *4*(2), 144–148.'s

4. Cherif, W. (2018). Optimization of K-NN algorithm by clustering and reliability coefficients: Application to breast cancer diagnosis. *Procedia Computer Science*, *127*, 293–299. https://doi.org/10.1016/j.procs.2018.01.125.
5. Churbanov, A., & Winters-Hilt, S. (2008). Implementing EM and Viterbi algorithms for hidden Markov model in linear memory. *BMC Bioinformatics*, *9*(1), 224. https://doi.org/10.1186/1471-2105-9-224.
6. Dwivedi, Y. K., Ismagilova, E., Hughes, D. L., Carlson, J., Filieri, R., Jacobson, J., Jain, V., Karjaluoto, H., Kefi, H., Krishen, A. S., Kumar, V., Rahman, M. M., Raman, R., Rauschnabel, P. A., Rowley, J., Salo, J., Tran, G. A., & Wang, Y. (2021). Setting the future of digital and social media marketing research: Perspectives and research propositions. *International Journal of Information Management*, *59*, 102168. https://doi.org/10.1016/j.ijinfomgt.2020.102168.
7. Gharibshah, Z., Zhu, X., Hainline, A., & Conway, M. (2020). Deep learning for user interest and response prediction in online display advertising. *Data Science and Engineering*, *5*(1), 12–26. https://doi.org/10.1007/s41019-019-00115-y.
8. Hannousse, A., & Yahiouche, S. (2021). Web page phishing detection. *Mendeley Data*. https://doi.org/10.17632/c2gw7fy2j4.3.
9. Hofmann, P., & Tashman, Z. (2020). Hidden Markov models and their application for predicting failure events. In V. V. Krzhizhanovskaya, et al., (Eds.), *Computational Science – ICCS 2020. ICCS 2020: Lecture Notes in Computer Science* (Vol. 12139, pp. 464–477). Springer, Cham. https://doi.org/10.1007/978-3-030-50420-5_35.
10. Ivan, C., & Popa, R. (2015). A cloud-based mobile dispatching system with built-in social CRM component: Design and implementation. *Computers*, *4*(3), 176–214. https://doi.org/10.3390/computers4030176.
11. Lamba, A., & Kumar, D. (2016). Survey on KNN and its variants. *International Journal of Advanced Research in Computer and Communication Engineering*, *5*(5), 430–435.
12. Lanza, B., & Parashar, D. (2021). Do support vector machines play a role in stratifying patient population based on cancer biomarkers? *Archives of Proteomics and Bioinformatics*, *2*(1), 20–38.
13. Mulahuwaish, A., Gyorick, K., Ghafoor, K. Z., Maghdid, H. S., & Rawat, D. B. (2020). Efficient classification model of web news documents using machine learning algorithms for accurate information. *Computers and Security*, *98*, 102006. https://doi.org/10.1016/j.cose.2020.102006.
14. Narvekar, M., & Banu, S. S. (2015). Predicting user's web navigation behavior using hybrid approach. *Procedia Computer Science*, *45*, 3–12. https://doi.org/10.1016/j.procs.2015.03.073.
15. Prasath, V. B. S., Haneen Arafat, A. A., Hassanat, A. B. A., Lasassmeh, O., Tarawneh, A. S., Alhasanat, M. B., & Salman, H. S. E. (2019). Effects of distance measure choice on KNN classifier performance - A review. *ArXiv:1708.04321v3*, 1–39.
16. Qi, X., & Davison, B. D. (2009). Web page classification: Features and algorithms. *ACM Computing Surveys*, *41*(2), 1–31. https://doi.org/10.1145/1459352.1459357.
17. Saito, T., & Rehmsmeier, M. (2015). The precision-recall plot is more informative than the ROC plot when evaluating binary classifiers on imbalanced datasets. *PLoS One*, *10*(3), e0118432. https://doi.org/10.1371/journal.pone.0118432.
18. Salminen, J., Corporan, J., Marttila, R., Salenius, T., & Jansen, B. J. (2019). Using machine learning to predict ranking of webpages in the gift industry. *Proceedings of*

the 9th International Conference on Information Systems and Technologies (pp. 1–8). New York, NY, USA: Association for Computing Machinery. https://doi.org/10.1145/3361570.3361578.

19. Sarker, I. H. (2021). Machine learning: Algorithms, real-world applications, and research directions. *SN Computer Science*, *2*(3), 160. https://doi.org/10.1007/s42979-021-00592-x.
20. Sejal, D., Kamalakant, T., Anvekar, D., Venugopal, K. R., Iyengar, S. S., & Patnaik, L. M. (2016). Webpage recommendation with web navigation prediction framework. *International Journal of Knowledge and Web Intelligence*, *5*(3), 230. https://doi.org/10.1504/IJKWI.2016.078733.
21. Sivarajah, U., Kamal, M. M., Irani, Z., & Weerakkody, V. (2017). Critical analysis of big data challenges and analytical methods. *Journal of Business Research*, *70*, 263–286. https://doi.org/10.1016/j.jbusres.2016.08.001.
22. Husain, T., Sani, A., Ardhiansyah, M., & Wiliani, N. (2020). Online Shop is an interactive media information society based on search engine optimization (SEO). *International Journal of Computer Trends and Technology*, *68*(3), 53–57. https://doi.org/10.14445/22312803/IJCTT-V68I3P110.
23. Vassakis, K., Petrakis, E., & Kopanakis, I. (2018). Big data analytics: Applications, prospects, and challenges. In G. Skourletopoulos, G. Mastorakis, C. Mavromoustakis, C. Dobre, & E. Pallis (Eds.), *Mobile big data: Lecture notes on data engineering and communications technologies* (Vol. 10, pp. 3–20). Springer, Cham. https://doi.org/10.1007/978-3-319-67925-9_1.
24. Ventola, C. L. (2018). Big data and pharmacovigilance: Data mining for adverse drug events and interactions. *P & T: A Peer-Reviewed Journal for Formulary Management*, *43*(6), 340–351.
25. Yu, Y. (2022). Web page classification algorithm based on deep learning. *Computational Intelligence and Neuroscience*, *2022*, 1–10. https://doi.org/10.1155/2022/9534918.

CHAPTER 17

Operationalizing Machine Learning: The Path to MLOps Excellence

GURRAM SUNITHA

Department of CSE, School of Computing, Mohan Babu University, Tirupati, Andhra Pradesh, India

ABSTRACT

Successful deployment and management of machine learning (ML) models has become paramount in the ever-evolving landscape of artificial intelligence and ML. This chapter explores an emerging discipline "machine learning operations (MLOps)" which combines DevOps practices with ML workflows to streamline the operationalization of ML models. This chapter explores the challenges associated with ML model deployment and the need for an organized and efficient MLOps framework. This chapter outlines the key components of MLOps, including version control, reproducibility, model monitoring, and automated testing. It explores the significance of continuous integration and continuous deployment pipelines in MLOps, highlighting the benefits of automating the ML model development and deployment process. Finally, the chapter concludes with a discussion on the future of MLOps and the emerging trends in the field. It emphasizes the need for organizations to invest in robust MLOps practices to ensure the scalability, reliability, and maintainability of their ML systems. By adopting MLOps principles, organizations can accelerate the deployment process, enhance collaboration, and maximize the value derived from their AI/ML initiatives.

Real-Time Artificial Intelligence (AI): Key Motivations, Technologies, Platforms, and Use Cases.
Pethuru Raj, B. Sundaravadivazhagan, V. Kavitha, B. Narendra Kumar Rao, & Hannah Vijaykumar (Eds.)

DOI: 10.1201/9781998511358-17

17.1 INTRODUCTION

Machine learning operations (MLOps) is a critical part of the artificial intelligence (AI) and machine learning (ML) world [35]. It is about making ML work in the real world, where it can bring meaningful changes. This chapter explores MLOps to understand why it is so important. Also, this chapter explores the challenges it helps us overcome.

17.1.1 UNDERSTANDING THE IMPORTANCE OF MLOPS

MLOps is like the engine that drives a car. Without it, ML models might work well in a controlled environment, but can stumble when faced with real-world data and demands [1].

Consider the following scenario. An efficient ML model is trained and tested. It identifies animals from images with appreciable accuracy. Let us say that the need arose to deploy ML model in an e-commerce site. The purpose is to use the model to suggest pet-animals-related products to users. This is where MLOps helps. It ensures that the learning model works in the live environment; that it is updated when new data come in; and that it does not suddenly start suggesting bird-related products instead of pet-animal-related products.

17.1.2 KEY CHALLENGES IN DEPLOYING MACHINE LEARNING MODELS

Bringing ML models into the real world is critically challenging [33]. Some of the key challenges are discussed below:

1. **Data Variability:** Data can be manipulated and controlled in a controlled environment. But real-world data will be messy. It might be incomplete, noisy, and different from the data on which the ML model was trained. MLOps helps you adapt to this data variability.
2. **Model Updates:** As time flows, new data are gathered. Hence, ML models need regular updates. MLOps automates model updation so that the model does not become outdated.
3. **Scaling Up:** MLOps ensures that the ML model scales smoothly.
4. **Monitoring:** MLOps sets up monitoring on the deployed ML models.

5. **Security and Compliance:** User data are private. To maintain its sensitivity, certain regulations must be followed. MLOps helps maintain security and compliance.
6. **Team Collaboration:** MLOps is for everyone who is involved in deploying and maintaining ML models. MLOps provides a structured way for achieving team collaboration.
7. **Reproducibility:** Reproducibility is the capability of being able to recreate the process. MLOps makes it easier [23].

MLOps takes the magic of ML and makes it practical. It is a useful tool to deploy ML models in the real-world. It is a blend of good practices, smart tools, and a dash of automation [32].

17.2 FOUNDATION OF MLOPS

MLOps stands on two pillars: DevOps and ML [16]. It is where these two tangles make ML models work effectively in real-life scenarios. The core aspects that form the foundation of MLOps are discussed in this section.

17.2.1 THE INTERSECTION OF DevOps AND MACHINE LEARNING

DevOps ensures that different processes are pipelined together harmoniously. When DevOps is applied to the ML pipeline, it becomes the bridge that connects the process of developing and deploying software with the complexities of ML [15].

In traditional software development, DevOps focuses on collaboration, automation, and continuous integration and deployment (CI/CD) to deliver software reliably and quickly. With ML, DevOps principles are adapted to manage the unique challenges of building, testing, and deploying ML models [19].

MLOps borrows DevOps practices, such as version control, automated testing, and CI, and applies them specifically to ML workflows [34]. It is about creating a smooth path for ML models from development to deployment. Thus, ensuring they function well in production environments [26].

17.2.2 MLOPS LIFECYCLE: FROM DEVELOPMENT TO DEPLOYMENT

MLOps effectively manages the process of constructing the ML model until its deployment [11]. Figure 17.1 serves as a visual guide to understanding the workflow and the various stages in the MLOps process. Each of these steps represents a phase in the lifecycle of ML operations, emphasizing the iterative nature and continuous improvement involved in managing ML models effectively.

1. **Development:** Data scientists and ML engineers experiment with algorithms, features, and data to create models that perform well [20]. MLOps supports this stage by facilitating collaboration, version control, and reproducibility. It ensures that different team members can work on the same project efficiently and that experiments can be replicated.
2. **Training:** Models are trained using historical data to learn patterns and make predictions. MLOps helps in automating this process and managing resources efficiently [22]. It ensures that training environments are set up correctly and that the training process is efficiently monitored.
3. **Testing:** ML models need rigorous testing. MLOps aids in setting up automated tests to ensure that ML models are accurate and robust before deployment. Such automated tests evaluate ML models on different datasets to check their generalizability.
4. **Deployment:** MLOps automates the deployment process. Such automation makes deployment consistent and repeatable. Automation involves packaging the model, setting up infrastructure, and integrating it into the production environment. Continuous monitoring and feedback loops are established to ensure that the ML model performs as expected.

MLOps lifecycle is an iterative process. It has continuous improvements and refinements based on real-world performance. It is about maintaining models over time and adapting them to changing data and environments. Thus, ensuring they remain effective and reliable.

17.3 COLLABORATION IN MLOps

Collaboration is fundamental for the successful deployment and management of ML models using MLOps [29]. This section explores the roles of different teams and how cross-functional collaboration plays a crucial part in MLOps.

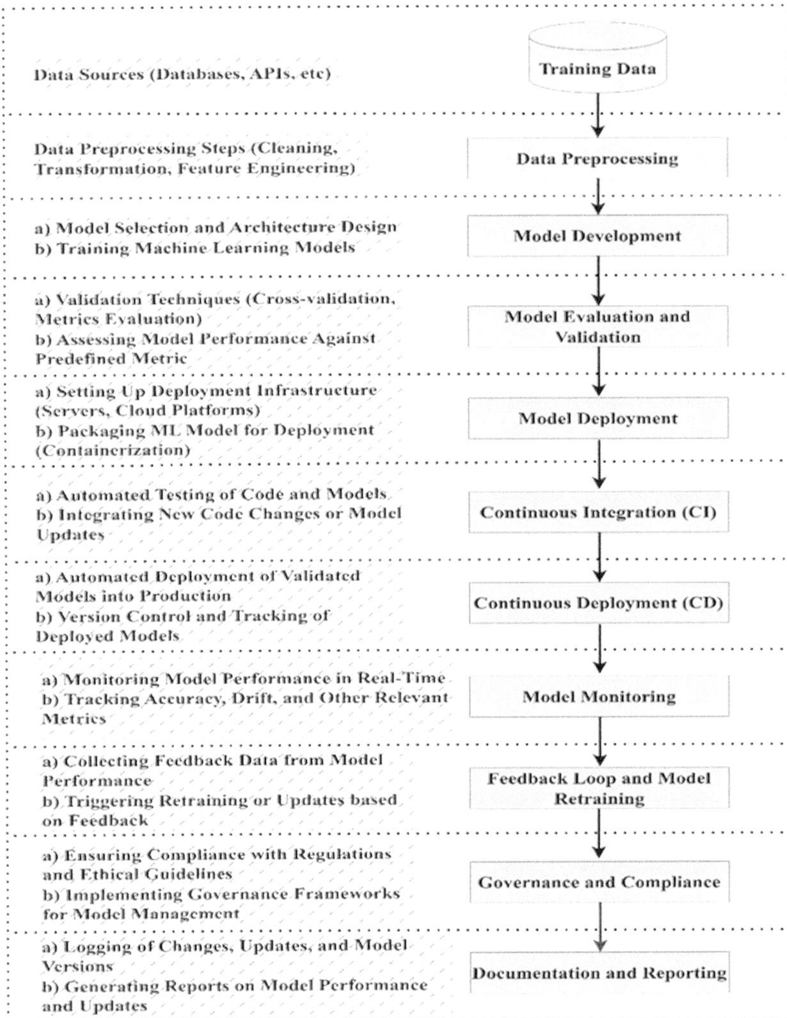

FIGURE 17.1 MLOps workflow diagram.

17.3.1 THE ROLE OF DATA SCIENTISTS, ML ENGINEERS, AND OPERATIONS TEAMS

Data Scientists gain insights into data by exploring patterns and designing ML models [20]. They analyze data to create ML models. These models can make predictions based on extracted patterns.

ML Engineers work on implementing these models into production environments. They focus on ensuring the scalability, reliability, and efficiency of these models. Their expertise lies in optimizing the model deployment process. Also, integrating the model seamlessly with existing systems is important.

Operations teams handle the infrastructure and deployment aspects. They manage the servers, networks, and systems necessary for deploying models. They ensure that the models are integrated into the production environment without disruptions.

17.3.2 FOSTERING CROSS-FUNCTIONAL COLLABORATION

Collaboration among MLOps teams is crucial for success [12, 13]. It is about breaking down silos and working together effectively toward a common goal. The common goal shall be to deploy accurate and reliable ML models. To foster collaboration (Figure 17.2):

1. **Clear Communication:** Teams need to communicate effectively. Thus, ensuring that everyone understands the goals, challenges, and requirements of the project. Clear communication channels help in avoiding misunderstandings and delays.
2. **Shared Goals:** All team's objectives shall be aligned for achieving successful deployment of ML models. Each team needs to understand how their role contributes to the overall objective.
3. **Collaborative Tools and Processes:** It is essential to implement tools to allow different teams to work together seamlessly. This includes tools that facilitate collaboration. For example, version control systems, project management tools, communication platforms, etc.
4. **Cross-Training and Knowledge Sharing:** Knowledge-sharing sessions and cross-training among teams shall be encouraged. This helps in better understanding each other's roles and challenges. This promotes a culture of learning and collaboration.
5. **Feedback Loops:** Feedback mechanisms shall be established for the purpose of ensuring information flows smoothly between teams. This helps in identifying issues early and making necessary improvements.

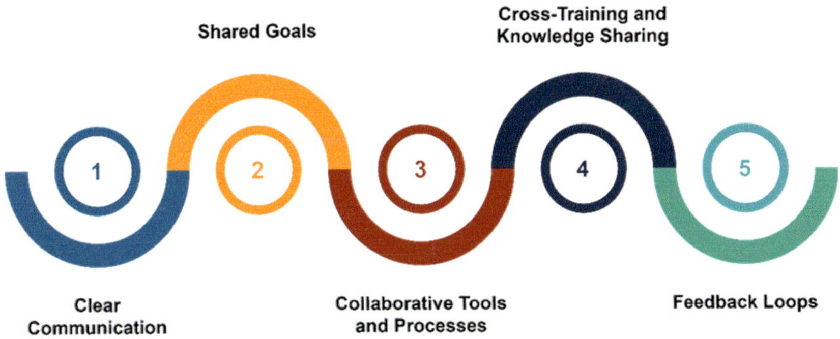

FIGURE 17.2 Fostering cross-functional collaboration for MLOps.

Effective collaboration ensures that ML models function optimally in real-world scenarios. It is about leveraging the strengths of each team to achieve a common objective – "deploying and managing ML models efficiently and effectively in production environments."

17.4 VERSION CONTROL AND REPRODUCIBILITY

In MLOps, version control and reproducibility are fundamental aspects [37]. They ensure consistency and reliability in managing ML models. The essential practices in MLOps are leveraging Git for version control and prioritizing reproducibility [23]. They provide the foundation for transparent collaboration, reliable model development, and deployment in real-world settings. Ultimately contributing to the success and credibility of ML initiatives. The section discussed the significance of the importance of reproducibility in MLOps and using Git for version control.

17.4.1 GIT AND VERSION CONTROL FOR MACHINE LEARNING

Git is a widely used version control system. It plays a key role in managing changes to ML models. Changes to the ML model is defined by the changes to its associated code, datasets, and configurations [4]. Git helps to track such modifications to the model. It also enables reverting to previous versions if needed. Git also facilitates collaboration among team members.

In MLOps, Git enables data scientists and ML engineers to collaborate effectively. It helps in managing different versions of models, experiments, and datasets. Thus, providing a transparent record of changes made throughout the model development pipeline. Teams can work concurrently on different aspects of the model without interfering with each other's work. This is made possible by Git by using branches and merge requests. Git's branching system allows for experimentation with different approaches and features. Teams can create separate branches to explore new ideas without affecting the main codebase. Once these changes are tested and validated, they can be merged back into the main branch.

17.4.2 REPRODUCIBILITY IN MLOPS: ENSURING CONSISTENCY

In the context of MLOps, reproducibility refers to the ability to recreate the exact environment and conditions under which an ML model was developed, trained, and evaluated. It is crucial for ensuring consistency, transparency, and reliability in ML workflows [23]. It is essential to document and manage all the elements involved in model development to achieve reproducibility,. such as code, dependencies, hyperparameters, and datasets. This documentation is often encapsulated in scripts or configuration files. It allows others to replicate the same results or build upon existing work. Reproducibility in MLOps helps mitigate issues related to changes in software versions, data discrepancies, and undocumented modifications. Thus, ensuring that the results obtained during development can be reliably reproduced when deploying the model in a production environment.

Reproducibility enables teams to verify and validate models effectively. It maintains consistency in environments and dependencies to do so. It also facilitates model auditing, regulatory compliance, and collaboration among different teams working on similar projects.

17.5 MODEL PACKAGING AND CONTAINERIZATION

Packaging models for portability and utilizing containers play a vital role in the context of MLOps [24]. Thus, ensuring seamless deployment and efficient management of ML models across different environments. This section expresses an opinion on the significance of packaging models in MLOps workflows in terms of portability and the utilization of containers.

17.5.1 PACKAGING MODELS FOR PORTABILITY

Packaging models involve encapsulating all the necessary components into a format. So to ensure easy transportation and deployment across various environments. Components, such as code, dependencies, and configurations. This portability is crucial as it allows models developed in one environment to be run consistently in others. This is possible irrespective of differences in infrastructure and tools.

Teams can avoid compatibility issues and streamline deployment by packaging models into standalone units. Commonly used formats for packaging models include serialized files (like pickle files in Python) or standardized formats like ONNX. These allow models to be used across different frameworks.

Portability simplifies the process of sharing ML models between team members. Portability enables the deployment of ML models in different computing environments. Thus, ensuring that the model performs consistently regardless of where it is used.

17.5.2 UTILIZING CONTAINERS IN MLOPS WORKFLOWS

Containers have emerged as a powerful tool in MLOps. It allows the packaging of applications and their dependencies into isolated and lightweight units. Docker is a widely used containerization platform [2]. It facilitates the creation and deployment of containers housing ML models. Containers offer a standardized environment for running applications. Thus, making it easier to ensure consistency between development, testing, and production environments. They encapsulate the model, its dependencies, libraries, and configurations. A container is a self-contained unit that can be easily moved across various platforms without any compatibility issues.

Containers promote scalability and reproducibility by ensuring that the model behaves uniformly, irrespective of the underlying infrastructure. This is particularly beneficial while deploying models in cloud environments, on-premises servers, and edge devices. Utilizing containers in MLOps workflows allows teams to streamline the deployment process. Thus, reducing deployment time, and enhancing the overall reliability of ML models in production environments. It enables efficient scaling, easy management, and consistent performance of models across diverse computing environments.

In MLOps workflows, packaging models and leveraging containers are critical strategies for the consistent and efficient deployment of ML models. They provide flexibility, scalability, and reliability for successful model deployment and management in various operational settings.

17.6 CI/CD PIPELINES FOR MLOPS

CI/CD pipelines form the backbone of MLOps [9]. They enable automated workflows to streamline the development, testing, and deployment of ML models. The significance of building CI pipelines and enabling CD of ML models within the MLOps framework is presented below.

CI/CD pipelines are integral to MLOps. They provide automation and standardization throughout the ML model development lifecycle. They enable efficient collaboration among teams and maintain code quality. They facilitate rapid and reliable deployment of ML models, ultimately contributing to the agility and effectiveness of MLOps.

CI/CD pipeline in Figure 17.3 emphasizes automation, continuous testing, monitoring, and feedback loops. Thus, ensuring seamless integration and deployment of ML models while maintaining reliability and performance across different environments in an MLOps context.

17.6.1 BUILDING CONTINUOUS INTEGRATION (CI) PIPELINES

CI involves automating the integration of code changes into a shared repository. In the context of MLOps, CI pipelines are designed to merge code changes made to ML models into a central repository. This is done by triggering automated testing and validation processes [17].

CI pipelines help maintain the integrity of the codebase. They ensure that changes made by different teams are integrated smoothly. These pipelines execute a series of predefined steps. For example, code compilation, unit testing, quality checks, etc. They provide immediate feedback to the development team.

CI pipelines facilitate the validation of changes to ML models. They ensure that modifications in code, data, and features do not adversely affect the ML model's performance. They enable data scientists and ML engineers to collaborate seamlessly. While maintaining a high level of code quality and consistency.

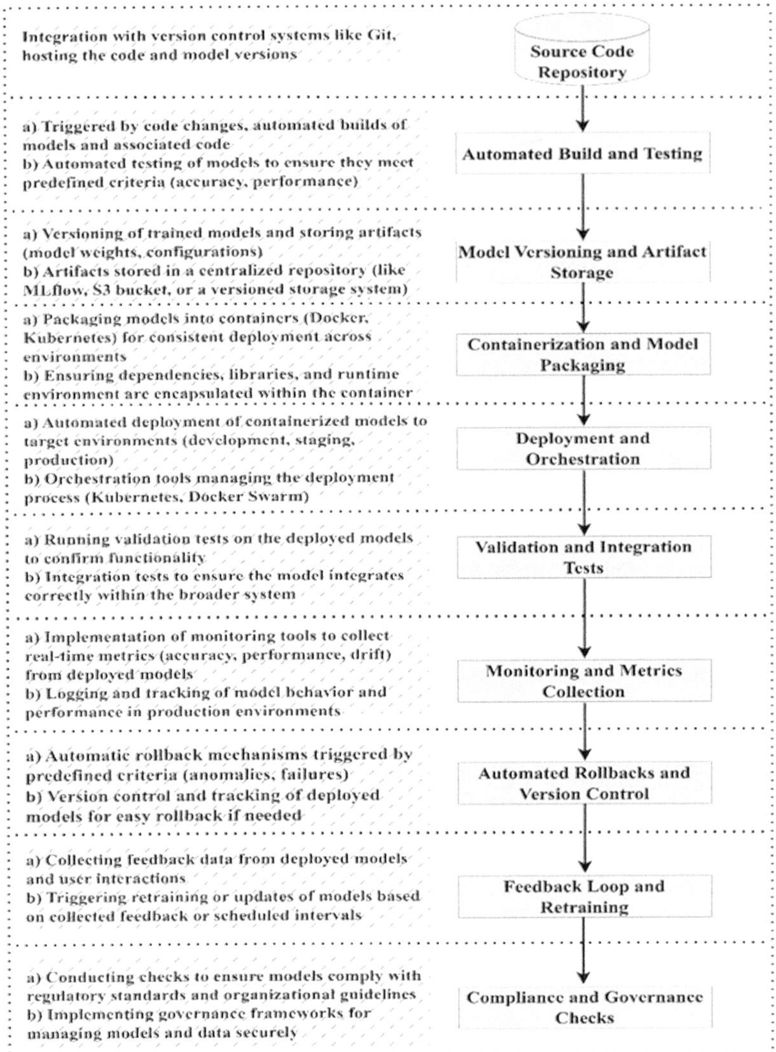

FIGURE 17.3 CI/CD pipeline for MLOps.

17.6.2 ENABLING CONTINUOUS DEPLOYMENT (CD) OF MACHINE LEARNING MODELS

CD in MLOps focuses on automating the deployment process of ML models into production environments [30]. CD pipelines take the validated

ML models from the CI phase and automatically deploy them to the target environment. Thus, ensuring a rapid and consistent deployment cycle. CD pipelines for ML models involve various steps, including model packaging, environment setup, and deployment orchestration. These pipelines ensure that models undergo necessary validation and quality checks before being released into production.

By enabling CD of ML models, organizations can accelerate the time-to-market for new model iterations. Agility can be enhanced in responding to changing requirements, and the manual effort involved in deployment processes can be reduced. This automated approach ensures that models are deployed reliably and consistently. Thus, minimizing the risk of errors and discrepancies in production.

17.7 MODEL MONITORING AND MANAGEMENT

In MLOps, model monitoring and management are essential aspects. They ensure the ongoing performance, reliability, and adaptability of ML models in production. Within the MLOps framework, the significant challenges are – real-time monitoring of model performance and strategies for model retraining and updating.

Real-time monitoring of model performance and strategic approaches to model retraining are integral components of MLOps. They enable proactive maintenance, continual improvement, and adaptation of ML models. Thus, ensuring their sustained performance and relevance in dynamic real-world environments.

17.7.1 REAL-TIME MONITORING OF MODEL PERFORMANCE

Real-time monitoring involves continuously observing and evaluating the performance of deployed ML models in production environments [10]. It aims to detect anomalies, assess model behavior, and ensure that the model operates as intended. Monitoring metrics like accuracy, precision, recall, and other performance indicators allow teams to identify deviations from expected behavior. Anomalies might arise due to changes in input data patterns, shifts in user behavior, or environmental changes. Timely detection of such anomalies helps in taking corrective actions promptly.

Automated monitoring tools and systems enable teams to set thresholds and alerts [32]. These tools trigger notifications when the model's performance falls below predefined thresholds. They enable rapid intervention and prevent potential issues from impacting end-users or systems. Continuous monitoring provides insights into how models perform in real-world scenarios. Thus, facilitating data-driven decisions for model maintenance, updates, and retraining to ensure sustained accuracy and reliability.

17.7.2 STRATEGIES FOR MODEL RETRAINING AND UPDATES

Model retraining and updates are crucial for maintaining the relevance and accuracy of ML models over time. Strategies for retraining involve periodically revisiting models, incorporating new data, and improving their performance based on evolving trends [15]. One approach is incremental training. Where new data are continuously fed into the model to keep it updated. Another strategy involves retraining the model at scheduled intervals. By considering factors like data drift, concept drift, or changing business requirements.

Versioning models and tracking changes play a pivotal role in retraining strategies. Teams can compare performance by maintaining different versions of models and datasets. They can revert to previous versions if necessary.

Automated pipelines for retraining models streamline the process. Thus, allowing for seamless integration of new data and updated features. These pipelines manage the entire lifecycle of model updates from data collection to evaluation. They also manage retraining and deployment of updated models into production environments.

Organizations can ensure that ML models remain relevant and effective by implementing effective strategies for model retraining and updates. Organizations can thus adapt to changing conditions, new trends, and evolving user preferences over time.

17.8 AUTOMATED TESTING IN MLOPS

Automated testing is a critical component of MLOps. It ensures robustness, reliability, and accuracy of ML models throughout their lifecycle. Automated testing serves as a crucial quality assurance mechanism in

MLOps [28]. It ensures model robustness and reliability by establishing systematic and automated testing workflows. Organizations can mitigate risks, validate model performance, and maintain high-quality standards throughout the ML model lifecycle. Implementation of automated testing workflows in MLOps is challenging as it is. Investigating how automated testing safeguards model's performance is a more critical challenge.

17.8.1 ENSURING MODEL ROBUSTNESS THROUGH TESTING

Testing is indispensable to ascertain that ML models maintain accuracy, and generalize well to new data. Robust testing strategies validate models under various scenarios [38]. This enables teams to identify potential weaknesses. Different types of tests, such as unit tests, integration tests, and validation tests, play essential roles in evaluating different aspects of the model. Unit tests assess individual components, ensuring they function correctly. Integration tests verify the collaboration between different parts of the model. While validation tests assess model performance against a separate validation dataset to check generalization.

Automated testing ensures consistency and repeatability in evaluating models. These tests are executed automatically through predefined scripts and workflows. Thus, allowing for frequent and systematic validation of model behavior. This automation streamlines the testing process and enables rapid feedback to developers.

17.8.2 IMPLEMENTING AUTOMATED TESTING WORKFLOWS

Implementing automated testing workflows involves setting up pipelines and scripts. These systematically execute tests to evaluate model performance. These workflows are designed to cover a range of scenarios, including data drift, feature changes, and modifications in the model architecture. The testing workflows start with data validation to ensure that the incoming data aligns with the expected format and characteristics. Then, they proceed to feature validation, where the model's input features are checked for consistency and correctness. Subsequently, model performance tests are conducted to assess the model's performance. These tests simulate real-world scenarios to validate how the model behaves under varying conditions.

CI pipelines incorporate automated testing as an integral step. Automated tests are triggered automatically when new changes are introduced to the model. They provide immediate feedback on the impact of those changes on the model's performance. MLOps teams can identify potential issues early in the development cycle by implementing robust automated testing workflows. The motto is to ensure that models are reliable, robust, and capable of delivering accurate predictions in diverse real-world situations.

17.9 SCALABILITY AND EFFICIENCY

Scalability and efficiency are vital considerations in MLOps to accommodate growing workloads and to ensure optimal utilization of resources. Scalability and efficiency are critical aspects of MLOps that enable organizations to adapt to growing workloads and optimize resource utilization. Implementing scalable infrastructure and embracing automation and orchestration methodologies are key strategies for effectively managing the expanding demands of ML workflows [32]. The challenges of MLOps scalability and efficiency through automation and orchestration must be further researched.

17.9.1 SCALING MLOPS FOR GROWING WORKLOADS

As ML initiatives grow, the demand for computational resources, data processing capabilities, and model deployment increases. Scaling MLOps involves the ability to handle these escalating workloads effectively [26].

Horizontal scaling involves expanding resources by adding more nodes to distribute the workload. This approach is beneficial when handling increased data volume and training larger models. Vertical scaling involves increasing the capabilities of existing resources, such as upgrading the processing power and memory of a single machine. Implementing scalable infrastructure architectures facilitates flexible scaling. Architectures include cloud-based solutions and container orchestration platforms like Kubernetes [2]. These solutions allow for dynamic allocation of resources based on workload demands. They ensure efficient utilization while accommodating varying computational needs.

Automated provisioning of resources is crucial for scalability. Auto-scaling mechanisms can automatically adjust resources based on workload fluctuations. Thus, ensuring that sufficient resources are available during peak demands and reducing costs during lower usage periods.

17.9.2 EFFICIENCY GAINS THROUGH AUTOMATION AND ORCHESTRATION

Efficiency gains in MLOps are achieved through the automation of repetitive tasks and the orchestration of complex workflows [36]. Automation streamlines processes, reduces manual intervention, and minimizes human error in managing ML workflows. Model training, testing, and deployment pipelines optimized for efficiency. This can be achieved by standardizing and automating repetitive tasks. This not only saves time but also ensures consistency and repeatability in the development and deployment of models.

Orchestration tools and frameworks coordinate and manage various components of MLOps workflow [14]. They enable seamless integration of different tools, platforms, and processes involved in model development and deployment. For instance, workflow orchestrators like Apache Airflow, Kubeflow, etc., facilitates the coordination of data processing, model training, and deployment tasks.

Leveraging pre-built components and reusable modules in MLOps workflows enhances efficiency. These reusable components accelerate development by eliminating the need to recreate functionalities from scratch. Reusable components may be pre-trained models or feature extraction pipelines, etc. MLOps teams optimize resource utilization, minimize manual effort, and streamline workflows by embracing automation and orchestration. This leads to increased productivity, faster time-to-market for models, and improved overall efficiency in managing ML initiatives.

17.10 SECURITY AND COMPLIANCE IN MLOPS

Ensuring security and compliance is paramount in MLOps in order to safeguard data privacy, protect model integrity, and adhere to regulatory standards [15]. They stand as foundational principles in MLOps. Implementing robust security measures and complying with relevant regulations are essential to protect sensitive information maintain model integrity, and uphold ethical standards in ML initiatives.

17.10.1 ENSURING DATA PRIVACY AND MODEL SECURITY

Data privacy and model security are critical considerations in MLOps in order to safeguard sensitive information, to prevent unauthorized access to ML models.

Encryption techniques, such as secure data transmission (SSL/TLS) ensure that data remain protected both during transit and while stored in repositories. Implementing access controls and authentication mechanisms restrict unauthorized access to sensitive ML model components [25].

There are secure development practices such as following secure coding guidelines and regular security assessments. They help identify vulnerabilities in ML models and associated software. Conducting regular security audits and penetration testing ensures that models are robust against potential cyber threats. Moreover, adopting secure model deployment practices, such as container image signing and verifying, ensures that only trusted and authorized versions of models are deployed into production environments. Secure configuration management practices prevent misconfigurations that might expose models to security risks.

17.10.2 ADHERING TO REGULATORY COMPLIANCE

Compliance with regulatory standards, such as GDPR, HIPAA, industry-specific regulations, etc., is crucial in MLOps to ensure that ML initiatives meet legal and ethical requirements [12, 13]. Data governance frameworks that define data usage policies, data retention, and access control mechanisms help organizations comply with regulatory standards. Establishing transparent documentation and audit trails regarding data usage and model decisions is essential for demonstrating compliance.

Anonymization and pseudonymization techniques are employed to protect personally identifiable information (PII) and ensure that models are trained and deployed in a privacy-preserving manner. These techniques allow for the utilization of sensitive data while safeguarding individual privacy. Regular training and awareness programs for MLOps teams on regulatory compliance standards, ethical guidelines, and best practices is very much essential. They ensure that everyone involved is well-informed and compliant with legal requirements. Compliance with regulations not only mitigates legal risks but also enhances trust among stakeholders, customers, and users. It showcases an organization's commitment to responsible and ethical use of data and models.

17.11 MODEL VERSIONING AND ROLLBACK STRATEGIES

Model versioning and rollback strategies are crucial components that enable the effective management of ML models in MLOps. Model versioning and rollback strategies are critical components of MLOps. They provide teams with the ability to track model changes and swiftly revert to stable versions if required. They ensure reliability, traceability, and adaptability in managing ML models in dynamic production environments [37].

17.11.1 THE IMPORTANCE OF MODEL VERSIONING

Model versioning involves keeping track of different iterations, modifications, and updates made to ML models throughout their lifecycle. It enables teams to maintain a historical record of changes. Thus, facilitating traceability, reproducibility, and comparisons between different versions. Version control systems, such as Git, allow for the systematic management of model versions, code changes, configurations, and associated datasets [23]. By versioning models, data scientists and ML engineers can monitor the evolution of models. ML models can be reverted to previous versions if needed. They can also analyze the impact of changes on model performance.

Versioning ensures transparency and accountability within MLOps workflows. It helps in identifying which version of a model was deployed, what changes were made, and who made those changes. This visibility is crucial for auditing, collaboration, and troubleshooting purposes.

17.11.2 ROLLBACK STRATEGIES TO MITIGATE RISKS

Rollback strategies are contingency plans used to revert to previous model versions. Such a need can arise in case of unexpected issues, errors, and performance degradation observed in the ML model deployed in production [7]. Automated rollback mechanisms trigger the deployment of the last known stable version of the model when anomalies are detected in the current version. These mechanisms are backed by monitoring systems that continuously assess model performance and trigger a rollback if predefined thresholds or anomalies are surpassed.

A phased deployment approach minimizes the impact of potential issues by releasing new model versions to a small subset of users before a full-scale deployment. Such approaches include canary deployments, gradual rollout strategies, etc. This allows teams to monitor performance and roll back changes if issues arise. The following techniques shall be adapted to ensure a swift and controlled rollback process in case of emergencies–maintaining a comprehensive rollback plan that includes step-by-step procedures, predefined criteria for rollback initiation, communication strategy among team members, etc.

Rollback strategies play a pivotal role in reducing downtime, mitigating risks, and maintaining continuity in model performance. They provide a safety net, enabling quick recovery and minimizing the impact of unforeseen issues that may arise during model deployment or updates.

17.12 CASE STUDIES IN MLOPS EXCELLENCE

Real-world case studies serve as valuable illustrations of successful MLOps implementations [31]. They shed light on practical applications, challenges faced, and the best practices adopted [27]. Some examples of successful MLOps implementations and the lessons learned from these case studies are presented below. These case studies underscore the importance of adopting best practices in MLOps, including robust testing, automation, versioning, and collaboration. They serve as practical examples demonstrating how effective MLOps implementations can drive efficiency, reliability, and innovation in various industries.

17.12.1 REAL-WORLD EXAMPLES OF SUCCESSFUL MLOPS IMPLEMENTATIONS

Many companies across various industries have implemented MLOps to enhance their services. Some of the examples are as follows:

1. **Netflix:** Netflix utilizes MLOps to improve its content recommendation system. They continuously develop and deploy ML models to personalize recommendations for users, enhancing the user experience.
2. **Airbnb:** Airbnb employs MLOps to enhance search ranking and recommendation systems. They use ML models to provide personalized property recommendations to users based on their preferences and behaviors.
3. **Uber:** Uber uses MLOps to optimize its ride-hailing platform. They employ ML for dynamic pricing, route optimization, and ETA predictions. Thus, improving the efficiency of their services.
4. **Twitter:** Twitter employs MLOps for content moderation and user engagement. They utilize ML models to detect spam and identify relevant content. Their target is to improve user interactions on the platform.
5. **Google:** Google implements MLOps across various services, such as search, advertising, and cloud computing. They continuously refine their algorithms using ML to enhance search results, targeted ads, and other offerings.
6. **Salesforce:** Salesforce leverages MLOps for predictive analytics and customer relationship management (CRM). They use ML to provide predictive lead scoring, customer segmentation, and personalized marketing recommendations.
7. **Amazon:** Amazon utilizes MLOps for product recommendations, inventory management, and logistics optimization. Their ML-powered recommendation engines improve the shopping experience for customers.

These companies use MLOps practices to deploy, manage, and improve ML models that power various aspects of their services, enhancing efficiency, personalization, and user satisfaction.

17.12.2 LESSONS LEARNED AND BEST PRACTICES

1. **Robust Testing and Monitoring:** The case studies emphasize the significance of robust testing and continuous monitoring in ensuring the reliability and performance of deployed models. Rigorous testing frameworks and automated monitoring systems

are pivotal for maintaining the effectiveness and accuracy of models in real-world scenarios [8].

2. **Versioning and Rollback Strategies:** Implementing effective versioning mechanisms and robust rollback strategies emerged as critical elements. The ability to track changes and swiftly revert to stable model versions during unexpected issues is very much essential.
3. **Automation and Continuous Integration:** Automation played a pivotal role in enabling seamless integration and deployment of models. CI/CD pipelines facilitated efficient workflows, ensuring consistency and reliability across the model development lifecycle.
4. **Cross-functional Collaboration:** Successful MLOps implementations relied on effective collaboration among data scientists, ML engineers, operations teams, and other stakeholders. Clear communication, shared goals, and collaborative tools were instrumental in driving successful outcomes.
5. **Adherence to Security and Compliance:** Both case studies highlighted the importance of ensuring data privacy, model security, and compliance with regulatory standards. Implementing robust security measures and adhering to legal requirements is imperative for building trust and reliability in MLOps initiatives [11].

17.13 FUTURE TRENDS IN MLOPS

Exploring future trends in MLOps involves understanding emerging technologies, evolving approaches, and the dynamic landscape shaping the future of ML operations [6]. The potential trends and shifts expected to impact MLOps in the coming years are discussed below.

17.13.1 EMERGING TECHNOLOGIES AND APPROACHES

Emerging technologies, ethical considerations, and evolving approaches will shape the future (Figure 17.4). Thus, enabling organizations to harness the full potential of AI while ensuring responsible and impactful deployment in diverse real-world applications.

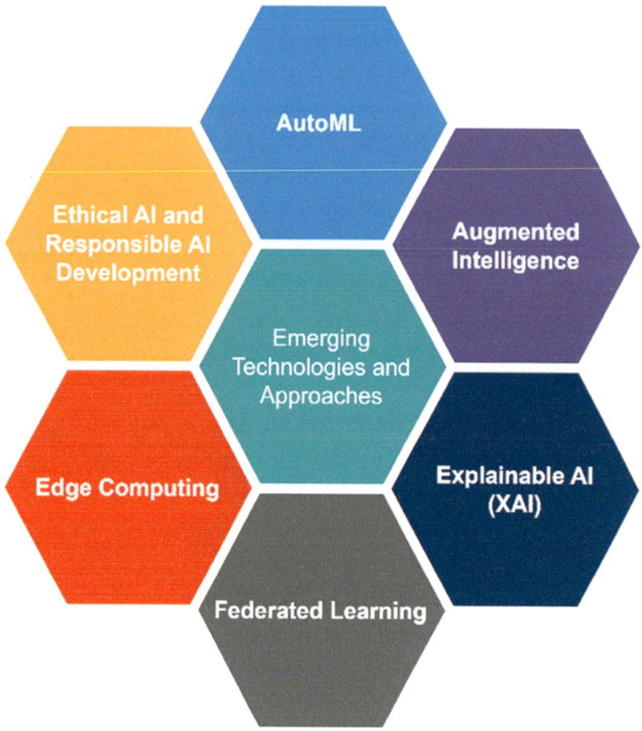

FIGURE 17.4 Emerging technologies and approaches of MLOps.

1. **AutoML and Augmented Intelligence:** Automated machine learning (AutoML) is anticipated to play a more significant role in democratizing ML by simplifying model development. Augmented intelligence works by combining human expertise with AI-driven insights. It will likely gain traction, enabling more efficient collaboration between data scientists and AI algorithms.
2. **Explainable AI (XAI):** As AI systems become more sophisticated, the need for explainable AI (XAI) grows. XAI focuses on making ML models interpretable and transparent. It allows users to understand the reasoning behind AI-generated decisions. This trend is crucial, especially in industries with high-stakes decision-making requirements, such as healthcare and finance [3].
3. **Federated Learning and Edge Computing:** In Federated Learning models are trained across decentralized devices. It will likely

gain momentum. It ensures data privacy while allowing for model improvements. Edge computing enables data processing closer to the source. It may become more prevalent, especially in scenarios requiring real-time processing and reduced latency, such as IoT and autonomous systems.

4. **Ethical AI and Responsible AI Development:** The focus on ethical considerations, fairness, and responsible AI development is expected to intensify [21]. Organizations will likely place more emphasis on building AI systems that are unbiased, ethical, and aligned with societal values.

17.13.2 THE EVOLVING LANDSCAPE OF MLOPS

The future trends in MLOps are poised to transform the landscape of MLOps (Figure 17.5) [18].

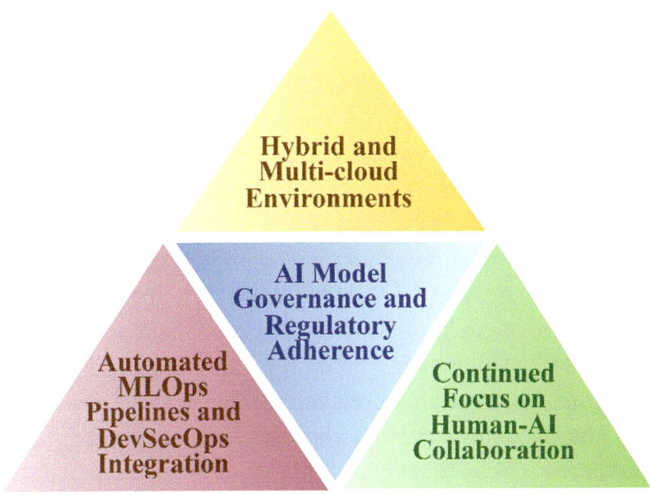

FIGURE 17.5 The evolving landscape of MLOps.

1. **Hybrid and Multi-cloud Environments:** Organizations are expected to adopt hybrid and multi-cloud strategies with the increasing complexity of AI applications. This allows them to leverage the strengths of different cloud providers. While ensuring flexibility, scalability, and cost-effectiveness in managing MLOps workflows.

2. **Automated MLOps Pipelines and DevSecOps Integration:** The evolution toward fully automated MLOps pipelines requires integrating development, security, and operations (DevSecOps). Such pipelines will streamline the end-to-end process of model development, deployment, and security. Thus, accelerating the pace of innovation.
3. **AI Model Governance and Regulatory Adherence:** As AI models are deployed at scale, AI model governance frameworks will become more crucial. Organizations will focus on ensuring compliance with regulations and ethical guidelines, implementing robust governance structures to manage the lifecycle of AI models effectively [12, 13].
4. **Continued Focus on Human-AI Collaboration:** The symbiotic relationship between humans and AI will persist [5]. Thus, emphasizing the importance of collaboration, where AI augments human capabilities rather than replacing them. This partnership will continue to drive innovation and productivity across various industries.

17.14 CONCLUSION: THE PATH TO MLOPS EXCELLENCE

MLOps, at its core, amalgamates ML with operational practices. It emphasizes collaboration, automation, and CI and deployment to streamline the development, deployment, and management of ML models. Key components, such as version control, automation, robust testing, model monitoring, and security measures lay the foundation for successful MLOps implementations. These components ensure reliability, efficiency, and compliance throughout the ML model lifecycle. CI, deployment, monitoring, and retraining are essential for maintaining model accuracy and relevance. The iterative nature of MLOps emphasizes the need for ongoing improvement, adaptation, and learning from experiences. Cross-functional collaboration between data scientists, ML engineers, operations teams, and stakeholders is critical. Establishing clear governance frameworks and ethical guidelines ensures responsible AI development and deployment.

Embarking on the journey toward MLOps excellence requires a comprehensive understanding of the key components, best practices, and evolving trends in managing MLOps. As technology evolves and new challenges arise, organizations committed to MLOps excellence must embrace continuous learning and improvement. Staying abreast of emerging technologies like

AutoML, XAI, Federated Learning, and ethical considerations allows organizations to remain agile and responsive to changing landscapes. Leveraging AI as a tool to augment human capabilities ensures more effective decision-making and problem-solving. Balancing the capabilities of AI with human expertise and ethical considerations will continue to be pivotal. Further integration of DevOps practices with MLOps, standardizing processes, and implementing governance mechanisms will streamline operations, enhance collaboration, and maintain model integrity.

KEYWORDS

- **artificial intelligence**
- **automation**
- **CI/CD pipelines**
- **learning pipeline**
- **MLOps**

REFERENCES

1. Alla, S., Adari, S. K., Alla, S., & Adari, S. K. (2021). What is MLOps? In *Beginning MLOps with MLFlow: Deploy models in AWS SageMaker, Google Cloud, and Microsoft Azure* (pp. 79–124).
2. Bahaweres, R. B., Zulfikar, A., Hermadi, I., Suroso, A. I., & Arkeman, Y. (2022). Docker and Kubernetes pipeline for DevOps software defect prediction with MLOps approach. In *2022 2nd International Seminar on Machine Learning, Optimization, and Data Science (ISMODE)* (pp. 248–253). IEEE.
3. Bodor, A., Hnida, M., & Najima, D. (2022). MLOps: Overview of current state and future directions. In *The Proceedings of the International Conference on Smart City Applications,* Cham (pp. 156–165). Springer International Publishing.
4. Calefato, F., Lanubile, F., & Quaranta, L. (2022). A preliminary investigation of MLOps practices in GitHub. In *Proceedings of the 16th ACM/IEEE International Symposium on Empirical Software Engineering and Measurement* (pp. 283–288).
5. Daniel, S. (2023). An empathetic design framework for humanity-centered AI: A preventative approach to developing more holistic, reliable, and ethical ML products.
6. Diaz-de-Arcaya, J., Torre-Bastida, A. I., Zárate, G., Miñón, R., & Almeida, A. (2023). A joint study of the challenges, opportunities, and roadmap of MLOps and AIOps: A systematic survey. *ACM Computing Surveys, 56*(4), 1–30.

7. Faubel, L., Schmid, K., & Eichelberger, H. (2023). MLOps challenges in industry 4.0. *SN Computer Science, 4*(6), 828.
8. Fursin, G. (2020). The Collective Knowledge project: making ML models more portable and reproducible with open APIs, reusable best practices and MLOps. arXiv preprint arXiv:2006.07161.
9. Garg, S., Pundir, P., Rathee, G., Gupta, P. K., Garg, S., & Ahlawat, S. (2021). On continuous integration/continuous delivery for automated deployment of machine learning models using MLOps. In *2021 IEEE Fourth International Conference on Artificial Intelligence and Knowledge Engineering (AIKE)* (pp. 25–28). IEEE.
10. Ghanta, S., Subramanian, S., Khermosh, L., Sundararaman, S., Shah, H., Goldberg, Y., Roselli, D., & Talagala, N. (2019). ML health monitor: taking the pulse of machine learning algorithms in production. In *Applications of Machine Learning* (Vol. 11139, pp. 191–202). SPIE.
11. Gift, N., & Deza, A. (2021). Practical MLOps. O'Reilly Media, Inc.
12. Granlund, T., Kopponen, A., Stirbu, V., Myllyaho, L., & Mikkonen, T. (2021). MLOps challenges in multi-organization setup: Experiences from two real-world cases. In *2021 IEEE/ACM 1st Workshop on AI Engineering-Software Engineering for AI (WAIN)* (pp. 82–88). IEEE.
13. Granlund, T., Stirbu, V., & Mikkonen, T. (2021). Towards regulatory-compliant MLOps: Oravizio's journey from a machine learning experiment to a deployed certified medical product. *SN computer Science, 2*(5), 342.
14. Hewage, N., & Meedeniya, D. (2022). Machine learning operations: A survey on MLOps tool support. arXiv preprint arXiv:2202.10169.
15. John, M. M., Olsson, H. H., & Bosch, J. (2021). Towards MLOps: A framework and maturity model. In *2021 47th Euromicro Conference on Software Engineering and Advanced Applications (SEAA)* (pp. 1–8). IEEE.
16. Kreuzberger, D., Kühl, N., & Hirschl, S. (2023). Machine Learning Operations (MLOps): Overview, definition, and architecture. *IEEE Access, 11*, 31866–31879.
17. Krishna, M. Y. S., & Gawre, S. K. (2023). MLOps for enhancing the accuracy of machine learning models using DevOps, continuous integration, and continuous deployment. *Research Reports on Computer Science, 2*(3), 97–103.
18. Lanubile, F., Martínez-Fernández, S., & Quaranta, L. (2024). Training future ML engineers: A project-based course on MLOps. *IEEE Software, 41*(2), 60–67.
19. Lima, A., Monteiro, L., & Furtado, A. P. (2022). MLOps: Practices, maturity models, roles, tools, and challenges-A systematic literature review. *ICEIS*, (1), 308–320.
20. Mäkinen, S., Skogström, H., Laaksonen, E., & Mikkonen, T. (2021). Who needs MLOps: What data scientists seek to accomplish and how can MLOps help?. In *2021 IEEE/ACM 1st Workshop on AI Engineering-Software Engineering for AI (WAIN)* (pp. 109–112). IEEE.
21. Matsui, B. M., & Goya, D. H. (2022). MLOps: A guide to its adoption in the context of responsible AI. In *Proceedings of the 1st Workshop on Software Engineering for Responsible AI* (pp. 45–49).
22. Maya, A. V., & Felipe, A. (2021). The State of MLOps.
23. Melin, P. D. (2023). Tackling Version Management and Reproducibility in MLOps.
24. Miñón, R., Díaz-de-Arcaya, J., Torre-Bastida, A. I., Zarate, G., & Moreno-Fernandez-de-Leceta, A. (2022). MLPacker: A unified software tool for packaging and deploying

atomic and distributed analytic pipelines. *In 2022 7th International Conference on Smart and Sustainable Technologies (SpliTech)* (pp. 1–6). IEEE.

25. Niemelä, P., Silverajan, B., Nurminen, M., Hukkanen, J., & Järvinen, H. M. (2022). LAOps: Learning analytics with privacy-aware MLOps. In *14th International Conference on Computer Supported Education* (Vol. 2, pp. 213–220). SciTePress.
26. Raj, E. (2021). In *Engineering MLOps: Rapidly build, test, and manage production-ready machine learning life cycles at scale*. Packt Publishing Ltd.
27. Raj, E., Buffoni, D., Westerlund, M., & Ahola, K. (2021). Edge MLOps: An automation framework for aiot applications. In *2021 IEEE International Conference on Cloud Engineering (IC2E)* (pp. 191–200). IEEE.
28. Renggli, C., Rimanic, L., Gürel, N. M., Karlaš, B., Wu, W., & Zhang, C. (2021). A data quality-driven view of MLOps. arXiv preprint arXiv:2102.07750.
29. Sothilingam, R., Pant, V., & Yu, E. (2022). Using i* to Analyze Collaboration Challenges in MLOps Project Teams.
30. Steidl, M., Felderer, M., & Ramler, R. (2023). The pipeline for the continuous development of artificial intelligence models—Current state of research and practice. *Journal of Systems and Software, 199*, 111615.
31. Subramanya, R., Sierla, S., & Vyatkin, V. (2022). From DevOps to MLOps: Overview and application to electricity market forecasting. *Applied Sciences, 12*(19), 9851.
32. Symeonidis, G., Nerantzis, E., Kazakis, A., & Papakostas, G. A. (2022). MLOps-definitions, tools and challenges. In *2022 IEEE 12th Annual Computing and Communication Workshop and Conference (CCWC)* (pp. 0453–0460). IEEE.
33. Tamburri, D. A. (2020). Sustainable MLOps: Trends and challenges. In *2020 22nd International Symposium on Symbolic and Numeric Algorithms for Scientific Computing (SYNASC)* (pp. 17–23). IEEE.
34. Testi, M., Ballabio, M., Frontoni, E., Iannello, G., Moccia, S., Soda, P., & Vessio, G. (2022). MLOps: A taxonomy and a methodology. *IEEE Access, 10*, 63606–63618.
35. Treveil, M., Omont, N., Stenac, C., Lefevre, K., Phan, D., Zentici, J., Lavoillotte, A., Miyazaki, M., & Heidmann, L. (2020). In *Introducing MLOps*. O'Reilly Media.
36. Yau, T. C. (2023). Investigate the challenges and opportunities of MLOps.
37. Zhao, Y. (2020). MLOps and data versioning in machine learning project.
38. Zwiesler, M. (2023). Implementation of MLOps. SSRN 4540074.

Index

B

C

D

E

N

O

P

T

U

V

W

Y

Z

For Product Safety Concerns and Information please contact our EU representative GPSR@taylorandfrancis.com Taylor & Francis Verlag GmbH, Kaufingerstraße 24, 80331 München, Germany